THERMOCHEMICAL KINETICS

THERMOCHEMICAL KINETICS

Methods for the Estimation of Thermochemical Data and Rate Parameters   *Second Edition*

SIDNEY W. BENSON

Professor of Chemistry
Chemistry Department
University of Southern California
Los Angeles, California

A Wiley-Interscience Publication

JOHN WILEY & SONS, New York   ·   London   ·   Sydney   ·   Toronto

Library of Congress Cataloging in Publication Data:
Benson, Sidney William, 1918–
    Thermochemical kinetics.
    "A Wiley-Interscience publication."
    Includes bibliographical references.
    1. Thermochemistry. 2. Chemical reaction, Rate of.
I. Title.
QD511.B48   1976   541'.36   76-6840
ISBN 0-471-06781-4

Printed in the United States of America

10 9 8 7 6 5 4 3 2 1

PREFACE

This book has been designed to make available to the reader techniques for the rapid and relatively quantitative estimation of thermochemical data and reaction-rate parameters for chemical reactions in the gas phase. It is assumed that the reader has an elementary background in chemistry, thermodynamics, and chemical kinetics. While some brief attention is given to elementary concepts in these fields and in statistical mechanics, the present text is not intended as a substitute for the more standard texts or advanced monographs in these areas. This will explain its rather spartan use of references. The treatment is completely molecular in its approach. It is assumed that we know the chemical identity and molecular structure of all species in our reaction system. This may place some unavoidable strain on the casual engineer or physicist reader, for which the author apologizes.

Chapter 1 reviews some elementary phenomenological relations between kinetic parameters and thermochemical data, and also the basic but confusing relations between differing choices of standard states and hence units. Chapter 2 covers in a detailed fashion the current empirical and theoretical methods for estimating the thermochemical properties, entropy (S°), enthalpy of formation (ΔH_f°), and molar heat capacity $(C_{p,T}^\circ)$ in the range 300–1500°K. The classes of species covered include molecules, free radicals, ions, and cyclic and polycyclic structures.

Chapter 3 introduces transition-state theory, which provides the basis for treating rate parameters for elementary reactions. It covers unimolecular reactions including simple fission, as well as complex reactions, isomerizations, concerted processes, and some of the basic organic name reactions. It treats only superficially the behavior of unimolecular reactions at low pressures in the "fall-off" region. Chapter 4 extends this kinetic analysis to bimolecular and termolecular reactions including metathesis, addition, and recombination. Ion–molecule reactions are considered, and there is a brief discussion of energy-transfer processes and isotope effects. Recent advances in estimating activation energies are covered as well. Chapter 5 outlines the application of these methods to the analysis of complex chain reactions (pyrolysis, cracking, hydrogenation, oxidation, and polymerization). The Appendix contains a collection of tabulated group values for different classes of organic compounds and thermochemical data for selected elements, radicals, and molecules. Every effort has been made to bring these values into agreement with current "best" results. Included also are tables for estimates of $S°$ and $C_p°$ contributions from vibration and torsional degrees of freedom as a function of temperature, frequency, and barrier.

Since the first edition of this text, there has been a considerable amount of work in the field of "thermochemical kinetics," and the present edition attempts to incorporate these results. Most noteworthy are the systematic treatments of free radicals and polycyclic structures not previously available. While some work has been done in adapting these same techniques to condensed phases, no systematic treatment yet exists. This is the reason for restricting our discussion to gas-phase systems.

I am much indebted to a number of friends, colleagues, and postdoctoral research associates who have assisted in the preparation and correction of much of the material presented in this second edition. I am especially grateful to Dr. David M. Golden, Dr. John Barker, Dr. Robert Shaw, Dr. Stephen Stein, Dr. Friedhelm Zabel, Dr. Menachem Luria, Dr. Miriam Lev-On (all of Stanford Research Institute), and to Professor H. E. O'Neal (California State University at San Diego) and Professor Alan S. Rodgers (Texas A. and M. University, College Station, Texas). Finally, my deepest appreciation to Mrs. Elaine Adkins for her good-natured and faithful typing of the many versions of the manuscript.

Sidney W. Benson

Chemistry Department
University of Southern California
Los Angeles, California
April 1976

CONTENTS

Note on Units

All units in this book, unless explicitly stated otherwise, will be kcal/mole for energy, enthalpy, free energy, and activation energy, and cal/mole-°K for entropy and heat capacity. To convert to SI units, multiply by 4.184 joules/cal.

THERMOCHEMICAL KINETICS

I

Thermodynamic Equilibria and Rates of Reaction

1.1 THERMODYNAMICS AND KINETICS

The second law of thermodynamics tells us that every closed, isolated system will approach an "equilibrium" state in which its properties are independent of time. If we know the enthalpies, entropies, and equations of state of the chemical species involved, we can predict with accuracy the chemical composition of this final equilibrium state. However, thermodynamics is unable to say anything about the time required to attain equilibrium, or about the behavior, or about the composition of the system during the period of change. These latter problems are the province of chemical kinetics, which is directly concerned with the description of chemical systems the properties of which are varying with time.

At first glance there appears to be very little in the way of a close relation between kinetics and thermodynamics. This turns out, however, not to be a correct conclusion. As we try to show in the present book, thermodynamics, or more explicitly, the thermochemical properties of substances, place a strong quantitative constraint on the kinetic parameters that are used to describe time-varying systems. The reason for this is that a "true" state of equilibrium is a dynamic state in which, on the molecular level, chemical changes are still occurring. The apparent macroscopic constancy of the composition and properties of the system arises from the fact that at equilibrium, for any given species M_j, its rate of production is precisely equal to its rate of destruction. Hence $d(M_j)/dt = 0$.

For each independent, stoichiometric chemical equation that we can

1

write for a system, all species in the equation are related to each other by the stoichiometric coefficients. If, for example, one such equation is

$$aA + bB + \cdots \underset{-1}{\overset{1}{\rightleftharpoons}} pP + qQ + \cdots , \qquad (1.1)$$

then

$$\frac{1}{p}\left[\frac{d(P)}{dt}\right] = \frac{1}{q}\left[\frac{d(Q)}{dt}\right] = \cdots = -\frac{1}{a}\left[\frac{d(A)}{dt}\right] = -\frac{1}{b}\left[\frac{d(B)}{dt}\right] \equiv R_1. \quad (1.2)$$

In this book we use the definition of R as a generalized, specific, molar rate of reaction for (1.1).

The stoichiometric relation between reactants and products means that there is one and only one independent kinetic equation per stoichiometric equation, which is needed to specify the equilibrium state. For simplicity, we can describe this by the statement $R = 0$. During the approach to equilibrium $R > 0$.

In general R will be some function of the external variables, such as volume, temperature, electric, magnetic, or gravitational fields, and also of the chemical composition of the system. Hence the statement that at equilibrium, $R_1(V, T, A, B, P, Q,$ etc.$) = 0$ implies a relation between these quantities. However, thermodynamics gives us another relation between these quantities in the form of the equilibrium constant K_1. This in turn must imply a quantitative relation between the equilibrium constant K_1 and the rate parameters that appear in the equation $R_1 = 0$. In the exceptional case that the rate law describing (1.1) takes on a very simple mass law form in terms of forward and reverse specific rate constants k_1 and k_{-1}, the relation becomes

$$K_1 = \frac{k_1}{k_{-1}}. \qquad (1.3)$$

This tells us that, of the three quantities k_1, k_{-1}, and K_1, only two are independent. A detailed knowledge of any two of them completely specifies the third, via (1.3). In general, equilibrium data are more readily available than kinetic data; so we make most frequent use of this last equation to relate forward and reverse rate constants where only one is known.

1.2 IDEAL GAS–PHASE REACTIONS

The concept of a molecule takes on a relatively unambiguous meaning only in the case of a dilute gas where individual molecules spend long periods of time isolated from each other relative to the time that they spend in a state of collision. The collision state itself needs more precise definition, and we describe it for a pair of molecules j and k as that state in which the energy of their interaction $|E_{jk}| \geq kT$.

The definition stated above is, of course, a completely arbitrary but extremely useful one, with some surprising consequences. The most obvious is that at absolute zero all molecules in a system are in a constant state of collision, and the concept of an individual molecule may either not be useful or need redefinition. Another consequence is that two ions at room temperature $(T \sim 300°K)$ in a gas phase must be looked on as being in a state of collision at a distance of 550 Å because their coulombic interaction $E_{jk} = \epsilon^2 / r_{jk} = RT = 600$ cal/mole when $r_{jk} = 550$ Å.

For the reasons given above we restrict our discussion to dilute, neutral gases in which molecular concepts take direct, simple application, and thermodynamic and kinetic parameters can be expressed in terms of simple molecular parameters. Experience shows that reactions in condensed phases between nonpolar or only slightly polar molecules do not differ greatly from gas-phase reactions. However, ionic reactions in solvents of high dielectric constant have no parallel to gas-phase reactions and are not considered.

1.3 DETAILED BALANCING

Quantum mechanics via the uncertainty principle places limits on our ability to describe a molecule precisely. In general a maximal description comprises a set of quantum numbers describing the electronic, translational, rotational, and internal vibrational states of the species. We may, if we wish, consider every such set of numbers as describing a distinct chemical entity.

In a dilute gas at equilibrium, molecular collisions will tend to exchange energy between collision partners and hence produce changes in the quantum numbers of any given molecule. Such processes may be looked on as the simplest of chemical acts.

The principle of detailed balancing asserts that at equilibrium, the specific rate of every elementary collision process is exactly equal to the specific rate of its inverse, or reverse, process. The specific rate of an elementary collision process must be proportional to the number of collisions, which in turn is proportional to the product of the concentrations of colliding species. The coefficient of proportionality we call the specific reaction-rate constant. Thus if we consider N_2 molecules in their ground $(v = 0)$ vibrational state and consider the process of collisional excitation to $v = 1$, we can describe it by the equation

$$N_2^{(0)} + M \underset{-1}{\overset{1}{\rightleftarrows}} N_2^{(1)} + M \qquad (1.4)$$

where M represents any collision partner. At equilibrium

$$R_1 = \frac{d(N_2^{(1)})}{dt} = \frac{-d(N_2^{(0)})}{dt} = 0 = k_1(M)(N_2^{(0)}) - k_{-1}(M)(N_2^{(1)}), \qquad (1.5)$$

or

$$\frac{(N_2^{(1)})}{(N_2^{(0)})} = \frac{k_1}{k_{-1}}. \qquad (1.6)$$

However, this ratio is also given by the equilibrium relation

$$\frac{(N_2^{(1)})_{eq}}{(N_2^{(0)})_{eq}} = K_1 = e^{-h\nu/kT} \qquad (1.7)$$

where ν is the vibrational frequency of N_2. Hence

$$\frac{k_1}{k_{-1}} = K_1. \qquad (1.8)$$

If the system contains many different gases M_i, each capable of transferring energy to N_2, we can write a similar set of kinetic equations for each one separately with the final result

$$\frac{k_1}{k_{-1}} = \frac{k_1'}{k_{-1}'} = \frac{k_1''}{k_{-1}''} = \cdots = K_1 \qquad (1.9)$$

where k_1' and k_{-1}' refer to molecule M_1, and so on.

If the process we are considering involves more of a chemical act, such as atom abstraction, we have a similar result. For example,

$$H + Cl_2 \underset{-1}{\overset{1}{\rightleftarrows}} HCl + Cl \qquad (1.10)$$

$$\frac{k_1}{k_{-1}} = K_1 = \frac{(HCl)_{eq}(Cl)_{eq}}{(Cl_2)_{eq}(H)_{eq}} \qquad (1.11)$$

Although the results above appear satisfactory for equilibrium systems, we may question whether they still hold for a system not at equilibrium, specifically, for systems undergoing chemical reaction.

The typical chemical rate constant, even for so elementary an act as vibrational excitation, is not a simple molecular constant but instead an average over all the equilibrium population of translational, rotational, and vibrational states of the colliding pair. It can readily be conceived that during the course of a chemical reaction, the populations of the various quantum states are different from that prevailing at true equilibrium. In such a situation the chemical rate constants may well change during the course of the reaction and thus be different from the equilibrium rate constants. What then happens to detailed balancing and to such relations as (1.8) and (1.9)?

One simple effect arises in chemical reactions that are appreciably endothermic or exothermic. An exothermic reaction liberates heat, which must escape from the reaction system by conduction and/or convection to the walls of the vessels, which themselves are in a thermostat.[1] Because conduction and convection are measurably slow processes in gases, we will find that an exothermic reaction in a gas will give rise to a temperature increase in the reacting gas, which in turn will increase the reaction-rate constant. Endothermic reactions similarly give rise to cooling effects.

The answers to difficulties raised by the foregoing are as follows. In most chemical systems, perturbations of the population of molecules among energy states are generally small and negligible. Under these conditions the specific reaction-rate constants are the same as those at equilibrium, and the relation $K_1 = (k_1/k_{-1})$ holds.[2]

In systems where the perturbations of the population are not small or negligible, the observed rate constants are different from the ones measured at equilibrium. Nevertheless the principle of detailed balancing applies to each elementary act. Thus if k_1' is the nonequilibrium, specific rate constant for elementary process 1 in a perturbed population and k_{-1}' is its inverse process,

$$\frac{k_1'}{k_{-1}'} = K_1 \tag{1.12}$$

still holds with K_1 the true equilibrium constant.

Consider strong departures from equilibrium that occur in the dissociation of diatomic or small molecules. In such systems the nonequilibrium, perturbed rate constants still must yield the equilibrium ratio for forward and inverse processes. This is especially the case in complex chemical reactions, which may take place via the production of a series of unstable free radicals or atoms. Despite the appearance of such entities and the complex rate laws they give rise to, the inverse rate constants can still be deduced via the equilibrium relationship.

---

[1] We ignore radiation, which may be important in chemiluminescent reactions and at $T > 1000°K$, such as in flames.

As an example, the rate law for the simple overall reaction of $H_2 + Br_2 \rightleftarrows 2HBr$ is given by the experimentally determined relation:

$$\frac{-d(Br_2)}{dt} = \frac{k_1(H_2)(Br_2)^{1/2}}{1 + k_2[(HBr)/(Br_2)]}. \tag{1.13}$$

At higher temperatures, where the reverse reaction is important, the rate law of the reaction of H_2 and Br_2 must be of such a form that the rate $d(Br_2)/dt$ reduces to zero at equilibrium. This requires that we write, for the reverse reaction,[3]

$$\frac{-d(HBr)}{dt} = \frac{2k_{-1}(HBr)^2/(Br_2)^{1/2}}{1 + k_2[(HBr)/(Br_2)]} \tag{1.14}$$

with

$$K_1 = \frac{k_1}{k_{-1}} = \frac{(HBr)_{eq}^2}{(H_2)_{eq}(Br_2)_{eq}}. \tag{1.15}$$

Under conditions where both forward and reverse reactions are appreciable, the correct rate law for the net reaction is:

$$\frac{-d(Br_2)}{dt} = \frac{1}{2}\frac{d(HBr)}{dt} = R_1 = \frac{k_1(H_2)(Br_2)^{1/2}}{1 + k_2[(HBr)/(Br_2)]}\left\{1 - \frac{(HBr)^2}{K_1(H_2)(Br_2)}\right\}. \tag{1.16}$$

Catalysts may be looked upon as substances that perturb the populations of chemical intermediates. Rate constants in the presence of catalysts may, of course, be markedly changed from their "normal" values. Despite this, the ratios of forward and reverse processes must still yield the true equilibrium constant. This result is comprised in the familiar rule, "A catalyst may speed up the approach to equilibrium but cannot change the equilibrium point."

1.4 THERMOCHEMICAL QUANTITIES

In dilute gases the measured equilibrium constant K_1 is related to the standard free-energy change ΔG_1° of a stoichiometric reaction by the

---

[2] Perturbations produced by temperature gradients must be treated explicitly by adding to the kinetic equations, the equations for heat conduction and convection. Similar perturbations in composition may occur and must also be treated by additional equations for diffusion of species.

[3] This is a "strong," not an absolute, requirement that rests on kinetic justification. For more details, see S. W. Benson, *Foundations of Chemical Kinetics*, McGraw-Hill, New York, 1960, Section IV.3.

relation

$$\Delta G_1^\circ = -RT \ln K_1$$
$$= -2.303 RT \log K_1$$
$$= -\theta \log K_1 \tag{1.17}$$

where 2.303 reflects the change from natural base to decadic logarithms and $\theta = 2.303 RT = 4.576 T/1000$ kcal/mole is a computationally convenient way to measure temperature.[4]

By definition,

$$\Delta G_1^\circ \equiv \Delta H_1^\circ - T \Delta S_1^\circ \tag{1.18}$$

where ΔH_1° is the standard enthalpy change in the reaction and ΔS_1° is the standard entropy change, both at the reaction temperature. Now because ΔC_p°, the molar heat capacity change in a gas reaction, is generally small, ΔH_1° and ΔS_1° may be assumed constant over any small range in temperature ($\Delta T \leq 100°K$), whereupon (1.17) yields the familiar van't Hoff relation for the variation of equilibrium constant with temperature:

$$2.303 \log K_1 = \frac{-\Delta H_1^\circ}{RT} + \frac{\Delta S_1^\circ}{R} \tag{1.19}$$

with ΔH_1° and ΔS_1° considered constant.

One of the most precise methods for obtaining ΔG_1° is by the direct measurement of K_1. If we can measure the equilibrium concentration of reactants and products, K_1 can be obtained directly, and the error in ΔG_1° is simply related to the error in the measure of K_1. If K_1 is measured to ±10%, the resultant error in ΔG_1° is ±0.1(RT), or at 500°K, ±0.10 kcal/mole. The alternative method for obtaining ΔG_1° from known values of ΔH_1° and ΔS_1° is generally subject to uncertainties at least tenfold larger, for ΔH_1° is seldom known or measurable to better than ±0.5 kcal/mole; likewise ΔS_1° is not usually known to better than ±1 cal/mole °K.

One method for obtaining ΔH_1° and ΔS_1° is from the van't Hoff relation. From (1.19) we can eliminate ΔS_1° by taking values of K_1 at two different temperatures T_1 and T_2. Then we find, on solving for ΔH_1°,

$$-\Delta H_1^\circ = R \frac{T_1 T_2}{\Delta T_{12}} \ln \frac{K_{1(1)}}{K_{1(2)}}$$
$$= R \frac{T_m^2}{\Delta T} \ln \frac{K_{1(1)}}{K_{1(2)}} \tag{1.20}$$

---

[4] At 1000°K, $\theta \sim 4.6$ kcal/mole and at 500°K, $\theta \sim 2.3$ kcal/mole. For every 4.6 kcal/mole change in ΔG_1°, K_1 changes by a factor of 10 at 1000°K and a factor of $10^2 = 100$ at 500°K.

where $T_m^2 = (T_1 T_2)$ and $\Delta T = \Delta T_{12} = T_2 - T_1$. If we can measure K_1 over a sufficiently long temperature interval so that it changes by at least a factor of e (~ 2.72), independent errors of $\pm 10\%$ in K_1 at each end of this interval will yield ΔH_1° with an uncertainty of about $\sqrt{2} \times 0.10 \times RT_m \times (T_m/\Delta T)$ where T_m is the mean temperature in the range ΔT. At $T_m = 500°K$ with $\Delta T \sim 100°K$, this becomes ± 0.70 kcal/mole. Precision of $\pm 3\%$ in K_1 will reduce this uncertainty proportionately to ± 0.2 kcal/mole.

Such precision as that achieved in the measurements above will yield average values of ΔH_1° and consequently ΔS_1° over the temperature range. It is not possible to measure changes in ΔH_1° and ΔS_1° in this range or equivalently $\langle \Delta C_p^\circ \rangle$ over the range to better than ± 7 cal/mole °K ($\pm 10\%$ in K_1) or ± 2 cal/mole °K ($\pm 3\%$ in K_1). The related uncertainties in ΔS_1° become the vector sums of the uncertainties in ΔG_1° and ΔH_1° or at $500°K$, ± 1.4 ($\pm 10\%$ in K_1) and ± 0.44 cal/mole °K ($\pm 3\%$ in K_1), respectively.

Frequently it is possible to estimate ΔS_1° to an uncertainty of better than ± 1 gibbs/mole. Under these conditions a single measurement of K_1 to within a precision of even $\pm 20\%$ (± 0.2 kcal in ΔG_1° at $500°K$) yields values of ΔH_1° with an uncertainty of only ± 0.55 kcal/mole, which is considerably better than that obtained from the van't Hoff relation. This method is referred to as the "third law method" and is of great utility because of the readiness with which ΔS_1° values can be frequently estimated. As we shall see in Chapter 2, it is generally the case that ΔC_p° can be estimated to within ± 1 over the temperature range 300–1500°K. This makes it possible to extrapolate ΔG_1°, ΔS_1°, and ΔH_1° values from any given temperature to any other temperature within this range with relatively little additional uncertainty.

1.5   STANDARD STATES

The thermochemical quantities K, $\Delta G°$, $\Delta H_f°$, $S°$, and $C_p°$ are generally tabulated for standard states of 1 atm and ideal gas. We frequently want to transform these to standard states expressed in concentration units,[5] generally 1 mole/liter (1 M). For a stoichiometric reaction

$$aA + bB + \cdots \rightleftarrows rR + qQ + \cdots \qquad (1.21)$$

the relation between equilibrium constants is[6]

$$K_p = K_c (R'T)^{\Delta n} \qquad (1.22)$$

---

[5] This has been a cause of frequent confusion. For a more detailed discussion, see D. M. Golden, *J. Chem. Ed.*, **48**, 235 (1971).

[6] R' is the ideal gas constant expressed in units of liter-atm/mole-°K.

with K_p in pressure units, K_c in concentration units, and where Δn is the mole change in the reaction as written

$$\Delta n = (r + q + \cdots) - (a + b + \cdots). \tag{1.23}$$

The relation given above comes from the ideal gas law, which relates pressure (P) and concentration (c)

$$P = cR'T. \tag{1.24}$$

For reactions in which the mole change of gases is not zero ($\Delta n \neq 0$) this transformation introduces additional corrections into the effective heat of reaction and entropy change.

By definition, the heat of reaction is given by

$$\Delta H_p^\circ = RT^2 \left(\frac{\partial \ln K_p}{\partial T} \right)_{eq}, \tag{1.25}$$

and from (1.22) this becomes

$$\Delta H_p^\circ = RT^2 \left(\frac{\partial \ln K_c}{\partial T} \right)_{eq} + \Delta nRT = \Delta H_c^\circ + \Delta nRT. \tag{1.26}$$

Thus for an increase in gas moles $\Delta n > 0$, $\Delta H_p^\circ > \Delta H_c^\circ$, whereas the converse is true for $\Delta n < 0$.

If we now represent K_p in the van't Hoff form

$$RT \ln K_p = -\Delta H_p^\circ + T \Delta S_p^\circ, \tag{1.27}$$

and substitute for K_p from (1.22) and for ΔH_p° from (1.26), and make use of the thermodynamic relation[7] $S_p^\circ = S_c^\circ + R \ln (R'T)$, then:

$$\Delta S_p^\circ = \Delta S_c^\circ + (\Delta n) R \ln (R'T)$$
$$RT \ln K_c = -\Delta H_c^\circ - \Delta nRT + T \Delta S_c^\circ \tag{1.28}$$

EXAMPLE

In the reaction $N_2O_4 \rightleftarrows 2NO_2$, $\Delta H_{300}^\circ = 13.6$ kcal/mole and $\Delta S_{300}^\circ = 41.9$ cal/mole-°K. Write the equilibrium constant K_c in van't Hoff form and calculate its value at 400°K. Neglect ΔC_p°.

From the preceding, with $\Delta n = 1$,

$$\Delta H_c^\circ = \Delta H_p^\circ - RT \qquad = 12.8 \text{ kcal/mole}$$
$$\Delta S_c^\circ = \Delta S_p^\circ - R \ln (R'T) = 35.0 \text{ gibbs/mole.}$$

Hence

$$2.303 \log K_c = -\frac{6400}{T} + 16.5; \ K_c = 1.65 \text{ moles/liter}$$

[7] In the logarithmic term R' should be expressed in liter-atm. units, that is, 0.082 liter-atm/mole-°K.

EXAMPLE

The equilibrium constant for the formation of N_2O_5 from NO_2 and O_2 via

$$\tfrac{1}{2}O_2 + 2NO_2 \rightleftarrows N_2O_5$$

is expressed in vant' Hoff form (1 mole/liter: standard state) as

$$\log K_c = \frac{2670}{T} - 9.55.$$

Calculate ΔH_p° and ΔS_p° from this expression. In this case $\Delta n = -1.5$. Hence

$$\Delta S_c^\circ = -9.55(4.575) + R\,\Delta n$$
$$= -46.7 \text{ gibbs/mole}$$
$$\Delta S_p^\circ = \Delta S_c^\circ + R\,\Delta n \ln (R'T) = -46.7 - 1.5(R \ln R'T);$$
$$\Delta H_p^\circ = \Delta H_c^\circ + \Delta n(RT) \qquad = -12.2 - 1.5RT.$$

Picking $T = 300°K$ as a mean temperature, these become

$$\Delta S_p^\circ = -56.3 \text{ gibbs/mole};$$
$$\Delta H_p^\circ = -13.1 \text{ kcal/mole}.$$

1.6 ARRHENIUS PARAMETERS

The great majority of chemical reaction rate constants lend themselves readily to expression in Arrhenius form:

$$k = Ae^{-E/RT} = A \times 10^{-E/\theta} \tag{1.29}$$

where again $\theta = 2.303RT$ kcal/mole. Experimentally it is found that A and E are constant over a small temperature range ($\sim 100°K$). Even over very large temperature ranges ($500°K$) it is found that A and E do not change very much with temperature.

By virtue of the equilibrium relation between forward and reverse rate constants we can make a very direct relation between Arrhenius parameters and thermochemical quantities. For the reaction stoichiometry

$$aA + bB + \cdots \underset{-1}{\overset{1}{\rightleftarrows}} pP + qQ + \cdots \tag{1.30}$$

$$K_1 = \frac{k_1}{k_{-1}} = \frac{A_1 e^{-E_1/RT}}{A_{-1} e^{-E_{-1}/RT}} = \left(\frac{A_1}{A_{-1}}\right) \exp\left[\frac{-(E_1 - E_{-1})}{RT}\right]. \tag{1.31}$$

Hence because $K_1 = \exp(\Delta S_1^\circ/R - \Delta H_1^\circ/RT)$, we can write

$$R \ln\left(\frac{A_1}{A_{-1}}\right) = 2.303R \log\left(\frac{A_1}{A_{-1}}\right) = \Delta S_1^\circ \tag{1.32}$$

and

$$\Delta H_1^\circ = E_1 - E_{-1}. \tag{1.33}$$

When there is no mole change in the reaction, ΔS_i° and ΔH_i° are both unaffected by changes in standard states. Because the forward and reverse rates have the same order, the Arrhenius parameters are all changed by identical amounts. When $\Delta n \neq 0$, the Arrhenius parameters for forward and reverse reactions are changed by different amounts [see (1.36) and (1.37)].

The activation energy of a chemical reaction rate is defined by the relation

$$E \equiv RT^2 \left(\frac{\partial \ln k}{\partial T} \right). \tag{1.34}$$

When E is a constant, or nearly constant over a small temperature range, the direct integration of (1.34) leads to the Arrhenius equation (1.29) with A the constant of integration. By its definition, E will have the same units as RT, kcal/mole. The A factor will have the same units as k.

For a reaction of the mth overall order, the specific rate is given by $k(C)^m$ where (C) is the concentration of active species. Hence k will have dimensions of $time^{-1} \times concentration^{(1-m)}$ because the specific rate will have dimensions of moles/liter-s. This will then also be the dimensions of A. For a first-order reaction, A has dimensions of $time^{-1}$ and is independent of standard states or units used to measure concentrations. For an mth-order reaction, however, A will depend on the units used to measure concentration, be it in moles/liter or partial pressures.

The ideal gas relationship $(P = cR'T)$ between concentration and pressure units leads to the following expression of k for an mth order-rate constant:

$$k_c = k_p (R'T)^{m-1}. \tag{1.35}$$

Hence for the Arrhenius parameters,

$$E_c \equiv RT^2 \left(\frac{\partial \ln k_c}{\partial T} \right)$$

$$= RT^2 \left(\frac{\partial \ln k_p}{\partial T} \right) + (m-1)RT$$

$$= E_p + (m-1)RT; \tag{1.36}$$

$$A_c = k_c e^{E_c/RT}$$

$$= k_p (R'T)^{m-1} \exp \left[\frac{E_p + (m-1)RT}{RT} \right]$$

$$= A_p (R'T)^{m-1} e^{(m-1)}. \tag{1.37}$$

It is more usual to measure rates in concentration units, so that, inverting the relations above, we have

$$E_p = E_c - (m-1)RT;$$

$$\ln A_p = \ln A_c - (m-1)\ln(R'T) - (m-1). \tag{1.38}$$

Note the formal resemblance between the transformations of Arrhenius parameters and the corresponding thermochemical quantities (1.26) and (1.28).

EXAMPLE

The reaction $2NO + Cl_2 \rightarrow 2NOCl$ is found to be given by a third-order rate law with a specific reaction rate constant k the Arrhenius parameters of which are given by

$$\log k = \frac{-803}{T} + 3.66.$$

The units of k and A are liter^2/mole^2-s. Calculate the parameters in units of torr^{-2}-s^{-1}. Use 400°K as the mean reaction temperature.
   We first note that $(m-1) = 2$. From the preceding

$$E_p = E_c - 2RT = 3.7 - 1.6$$
$$= 2.1 \text{ kcal/mole};$$
$$A_p = A_c(760R'T)^{-2}e^{-2} \text{ torr}^{-2}\text{-s}^{-1}$$
$$= 10^{3.66}(10^{-9.66})$$
$$= 10^{-6.00} \text{ torr}^{-2}\text{-s}^{-1}.$$

EXAMPLE

The early stages of the reaction of $H_2 + Br_2 \rightarrow 2HBr$ are described by a $\frac{3}{2}$ order rate; $R = k(H_2)(Br_2)^{1/2}$ with k, in atmosphere units, given by

$$\log k = -41.6/\theta + 11.36.$$

Calculate at 700°K the Arrhenius parameters in molar units.
   In this case $(m-1) = \frac{1}{2}$, and we have

$$A_c = A_p(e\,RT)^{+1/2} = 10^{12.46} \text{ (liter/mole)}^{1/2} \text{ s}^{-1};$$
$$E_c = E_p + \tfrac{1}{2}RT = 42.3 \text{ kcal/mole}.$$

1.7 MODIFIED ARRHENIUS EQUATION

Although the bulk of gas-phase rate constants are expressible in simple Arrhenius form, a small group of them are not. This occurs in the case of termolecular recombinations of atoms or of atoms with diatomic molecules:

$$O + O + M \rightarrow O_2 + M;$$

$$O + NO + M \rightarrow NO_2 + M.$$

In the experimental cases given above, we may find it necessary to express the rate constants in the form of a modified Arrhenius equation in which we introduce a simple temperature variation for the A factor. The equation takes the form

$$k = A'T^n e^{-E'/RT} \tag{1.39}$$

where A', E', and n are the new parameters. Such a form can be readily derived from the general definition of activation energy (1.34):

$$E \equiv RT^2 \left(\frac{\partial \ln k}{\partial T} \right).$$

If we set $E = E' + CT$ where E' and C are constants, this equation can be integrated directly to yield

$$\ln k = -\frac{E'}{RT} + \frac{C}{R} \ln T + \text{constant}. \tag{1.40}$$

Setting the constant of integration equal to $\ln A'$ and $n = C/R$, we can immediately obtain (1.39). We note that the constant C has the dimensions of a molar heat capacity, and it can be identified with an average molar heat capacity of activation, averaged over the range for which (1.39) is presumed valid. Although n tends to be small for gas-phase reactions and on the order of unity, it can become very large for ionic reactions in solution (e.g., as high as 40).

One consequence of this modified Arrhenius equation discussed above is that the A factor and activation energy are now both explicitly dependent on temperature. These must be included in any transformations between pressure and concentration units and in making relationships to thermochemical data.

When data is measured and cast in the form of the Arrhenius equation, the relation between the parameters are as follows:

$$A = A'(eT_m)^n;$$

$$E = E' + nRT_m. \tag{1.41}$$

where T_m is the experimental mean temperature.

1.8 FREE ENERGY OF ACTIVATION

The relation between equilibrium constants and forward and reverse rate constants $K_1 = (k_1/k_{-1})$ permits us to make a formal division of rate constants into parts. K_1 can be written as (1.30):

$$K_1 = \exp\left(\frac{-\Delta G_1^\circ}{RT}\right) = 10^{-\Delta G_1^\circ/\theta} \qquad (1.42)$$

where ΔG_1° is the standard Gibbs free-energy change in the stoichiometric reaction 1. We can proceed to write any rate constant k_1 in the form

$$k_1 = \nu_1 \exp\left(\frac{-\Delta G_1^*}{RT}\right) = \nu_1 10^{-G_1^*/\theta} \qquad (1.43)$$

where G_1^* can be called a free energy of activation for reaction 1, and ν_1 can be called a frequency factor for the reaction. For the reverse reaction we can write

$$k_{-1} = \nu_{-1} \exp\left(\frac{-\Delta G_1^*}{RT}\right), \qquad (1.44)$$

and if we choose to equate $\Delta G_1^\circ = (G_1^* - G_{-1}^*)$, we see that $\nu_1 = \nu_{-1}$.

If we consider G_1^* and G_{-1}^* as thermochemical quantities, we can label ν_1 and ν_{-1} as kinetic quantities. In Chapter 3 we show how transition state theory leads to such a partition of rate constants into a product of a kinetic quantity[8] and a thermodynamic quantity. Here it is sufficient to note that any kinetic quantities identified in this manner must be the same for forward and inverse processes. We may also note that rate laws, even for very complex processes, can always be written as a product of a kinetic expression (involving rate constants and concentrations) and a thermodynamic expression usually of the form

$$1 - \frac{(P)^p(Q)^q \cdots}{K_1(A)^a(B)^b \cdots} \qquad (1.45)$$

where K_1 is the equilibrium constant for the overall reaction

$$aA + bB + \cdots \overset{1}{\rightleftarrows} pP + qQ + \cdots.$$

It is this thermodynamic factor which makes the net rate go to zero at equilibrium.[9]

One consequence of this discussion above is that any parameter that does not affect the equilibrium constant but that does affect the kinetic behavior of the system must have an identical effect on both forward and reverse rate constants. This is true, for example, of catalysts, which must speed up the reverse as well as the forward rates by precisely equal factors.

---

[8] These are the quantities which involve dimensions of time.
[9] See the rate law for the $H_2 + Br_2$ reaction on p. 6.

1.9 NEGATIVE ACTIVATION ENERGIES

The general experience is that raising the temperature of a reaction leads to an increase in the specific rate constants describing the reaction. Occasionally this is not true. Thus the termolecular reaction $2NO + O_2 \rightarrow 2NO_2$ slows down as we raise the temperature, and the same is true for the reaction $O + O_2 + M \rightarrow O_3 + M$.

Now the activation energies for forward and reverse reactions are related to the enthalpy change by

$$\Delta H_1 = E_1 - E_{-1}. \tag{1.46}$$

Hence a negative activation energy for one of the steps, let us say E_{-1}, implies that $E_1 < \Delta H_1$. In the ozone dissociation, for example, ΔH_1 is the bond-dissociation energy for O_3. The significance of this is that the activation energy for O_3 decomposition is less than the bond-dissociation energy! How are we to understand this paradoxical result? Does it mean that an O_3 molecule can be decomposed by an energy less than that required to break the $O-O_2$ bond in O_3?

The answers to the questions raised above are provided by a closer look at what is meant by activation energy. As we have noted earlier, the rate constant for a chemical reaction is not a simple molecular quantity. It is instead an average of the individual reaction rate constants over the population of all the different quantum states of the system. Following a derivation given by Tolman,[10] we can write the rate constant for a reaction as

$$k = \frac{\sum p_j e^{-E_j/RT} k_j}{\sum p_j e^{-E_j/RT}} \tag{1.47}$$

where p_j is the degeneracy of the jth quantum state of the molecule with energy E_j, and k_j is the spontaneous, first-order rate constant with which molecules in the jth state with energy E_j will decompose. Then, from the definition of activation energy,

$$E \equiv RT^2 \left(\frac{\partial \ln k}{\partial T} \right)$$

$$= \frac{RT^2 \sum k_j p_j (E_j/RT^2) e^{-E_j/RT}}{\sum k_j p_j e^{-E_j/RT}} - \frac{RT^2 \sum p_j (E_j/RT^2) e^{-E_j/RT}}{\sum p_j e^{-E_j/RT}} \tag{1.48}$$

$$= \frac{\sum E_j k_j p_j e^{-E_j/RT}}{\sum k_j p_j e^{-E_j/RT}} - \frac{\sum p_j E_j e^{-E_j/RT}}{\sum p_j e^{-E_j/RT}}, \tag{1.49}$$

or

$$E = \langle E \rangle_{\text{reacting}} - \langle E \rangle_{\text{nonreacting}}. \tag{1.50}$$

[10] Richard C. Tolman, *Statistical Mechanics*, N.Y. Chem. Cat. Co., New York, 1928.

The second term on the right-hand side of (1.49) can be recognized as the definition of the average energy of the molecules in the system, whereas the first term can be identified as the average energy of the reacting molecules. Hence we find the simple result that the experimental activation energy is the difference between these two quantities.

A reacting molecule at any temperature T has a certain amount of internal thermal energy (such as vibrational). If it should react by breaking spontaneously into fragments, these fragments at their instant of formation also have a certain amount of internal thermal energy. If these two amounts of energy are equal, the activation energy for the reaction at temperature T is precisely the same as the value at absolute zero. If, on the contrary, the fragments have less energy than the nonreacting molecules, the activation energy is less than its value at absolute zero, and we have a simple explanation for our paradox. Another way of stating it is that the fragments are being formed "cold," that is, with less than the average thermal energy they have when they come to thermal equilibrium in the reaction flask. In terms of the reverse reaction of recombination, this implies that the rate of recombination is higher for fragments with less than average thermal energy. As we raise the temperature we decrease this population, and the effective rate of recombination decreases with it.

Such effects as those described above can become very pronounced at extremely high temperatures where RT is very large. Thus at 5000°K, the activation energy for NO_2 decomposition into $NO + O$ can be as much as $3RT$ or 30 kcal lower than the bond dissociation energy of 72 kcal. This could lead to an apparent activation energy for dissociation of only 42 kcal/mole.

1.10 REACTION-COORDINATE-ENERGY DIAGRAM

Relations such as that just encountered in systems with negative activation energies are best visualized in terms of what are called reaction-coordinate-energy diagrams. During the course of an elementary chemical reaction requiring activation energy, a molecule is raised from a low-energy state and a normal geometry to a high-energy state in which large changes in molecular geometry are possible. One or more of these changes in geometry can correspond to the conversion to products. This behavior is illustrated in Figure 1.1 for a hypothetical case of unimolecular isomerization of cis- to trans-butene-2.

The ordinate represents the total energy of a single molecule as a function of its "reaction coordinate," which is plotted along the abscissa. The solid line corresponds to the potential energy of the molecules at

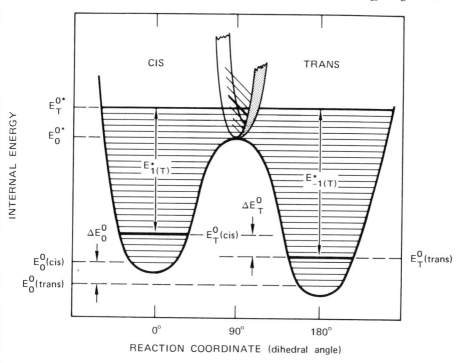

Figure 1.1   Reaction-coordinate-energy diagram for a hypothetical *cis–trans* isomerization. A number of authors have chosen to draw what they call "free-energy" diagrams rather than energy diagrams. These can be very misleading since the free energy involves entropy, which is not a property of individual molecules, but rather a property of a population. In particular, molecules from a population whose average energy is below the barrier height may decompose, whereas this is not true of molecules whose individual internal energy is below the energy barrier.

different values of the "reaction coordinate." The light shaded lines are drawn to suggest different vibrational levels of the cis and trans forms. Although the reaction coordinate in this case can be thought of simply as the angle between the planes of the two $C\diagdown^{\text{H}}_{\text{CH}_3}$ groups, it is very likely that other changes in bond angles and distances accompany the rotation of these two planes through the energy barrier at 90°. E_T° (*cis*) represents the energy of the average cis molecule at temperature T, E_T° (trans) the energy of the average *trans* molecule, and $E_T^{\circ*}$ the energy of the average reacting molecule (A^*).

The heat of reaction at absolute zero is indicated on the drawing by ΔE_0°; it is drawn between the zero vibrational levels. At any finite temperature T, the average molecules will be somewhere above these zero-vibrational levels, as is indicated by the heavy solid lines at E_T° (cis) and at E_T° (trans). Similarly the average reacting molecule at temperature $T^\circ K$ will not be precisely at the top of the energy barrier but significantly above it, as is indicated by the solid line above the barrier at $E_T^{\circ *}$. E_1^* and E_{-1}^* are, respectively, the activation energies for the cis-trans and trans-cis reaction paths at temperature T. ΔE_T° is the energy change in the reaction at temperature T.

2

Methods for the Estimation of Thermochemical Data

2.1 SOME THERMODYNAMIC RELATIONS

When a chemical reaction proceeds in the gas phase to a state of dynamic equilibrium, that state is completely described by the specification of temperature T, pressure P, and chemical composition. From P and the chemical composition we can calculate the equilibrium constant K_P and the *standard Gibbs free-energy change* ($\Delta G°$) for the process of transforming reactants to products at temperature T. Below we consider all species as ideal gases.

If the stoichiometry of the reaction above is

$$aA + bB \rightleftarrows rR + qQ, \tag{2.1}$$

$$K_P = \frac{(R)^r(Q)^q}{(A)^a(B)^b} \tag{2.2}$$

and

$$\begin{aligned}\Delta G_T^\circ &= -RT \ln K_P \\ &= r\,\Delta G_{fT}^\circ(R) + q\,\Delta G_{fT}^\circ(Q) - a\,\Delta G_{fT}^\circ(A) - b\,\Delta G_{fT}^\circ(B)\end{aligned} \tag{2.3}$$

where, in (2.2), the concentrations of each species has been expressed in units of atmospheres. Quantities such as $\Delta G_{fT}^\circ(R)$ (2.3) represent the standard Gibbs free energies of formation of 1 mole of R at the temperature T and pressure of 1 atm from the elements in their standard states at T and 1 atm pressure.

The units of K_p then are $(\text{atm})^{\Delta n}$ where

$$\Delta n = r + q - a - b \tag{2.4}$$

is the mole change in the reaction.[1] By definition

$$\Delta G_T^\circ = \Delta H_T^\circ - T \Delta S_T^\circ \qquad (2.5)$$

where ΔH_T°, the standard enthalpy change for the reaction, is related to the standard heats of formation ΔH_{fT}° by

$$\Delta H_T^\circ = r \Delta H_{fT}^\circ(R) + q \Delta H_{fT}^\circ(Q) - a \Delta H_{fT}^\circ(A) - b \Delta H_f^\circ(B), \qquad (2.6)$$

and ΔS_T°, the standard entropy change in the reaction, is related to the standard absolute entropies of the species by

$$\Delta S_T^\circ = r S_T^\circ(R) + q S_T^\circ(Q) - a S_T^\circ(A) - b S_T^\circ(B). \qquad (2.7)$$

If we know the standard entropies and the standard heats of formation at the temperature T of all reactants and products, we can calculate ΔH_T° and ΔS_T° for the reaction and from these, the equilibrium constant.

Entropies and heats of formation thus constitute our basic thermochemical data. Their variation with temperature is given by the thermodynamic relations

$$\left[\frac{\partial(\Delta H^\circ)}{\partial T} \right]_p = \Delta C_p^\circ \qquad \left[\frac{\partial(\Delta S^\circ)}{\partial T} \right]_p = \frac{\Delta C_p^\circ}{T} \qquad (2.8)$$

where ΔC_p° is the change in standard molar heat capacity for the reaction and is given by

$$\Delta C_p^\circ = r C_p^\circ(R) + q C_p^\circ(Q) - a C_p^\circ(A) - b C_p^\circ(B). \qquad (2.9)$$

In (2.9) $C_p^\circ(R)$, and so on are the standard molar heat capacities of the reactants and products and are functions of temperature, but not pressure, for ideal gases.

In order to calculate ΔH_T° or ΔS_T°, given the values of $\Delta H_{T_0}^\circ$ and $\Delta S_{T_0}^\circ$, we need to know the functional dependence of ΔC_p° on temperature over the range between T and T_0. Integration of (2.8) then gives the relations

$$\Delta H_T^\circ = \Delta H_{T_0}^\circ + \int_{T_0}^{T} (\Delta C_p^\circ) \, dT;$$

$$\Delta S_T^\circ = \Delta S_{T_0}^\circ + \int_{T_0}^{T} \left(\frac{\Delta C_p^\circ}{T} \right) dT. \qquad (2.10)$$

We see later that, although values of C_p° for individual species may be large and may change grossly over temperature intervals of 500°K, ΔC_p° for reactions tend to be very small and change very little over such

---

[1] Note that this depends on our convention for writing stoichiometric equations. In general we follow the common usage of taking the smallest set of integers.

intervals. Because of this we can, with little error, take an average value $\Delta C^\circ_{pT_m}$ over the interval $(T - T_0)$ and reduce (2.10) to

$$\Delta H^\circ_T = \Delta H^\circ_{T_0} + \Delta C^\circ_{pT_m}(T - T_0);$$

$$\Delta S^\circ_T = \Delta S^\circ_{T_0} + \Delta C^\circ_{pT_m} \ln \frac{T}{T_0} \tag{2.11}$$

$$= \Delta S^\circ_{T_0} + 2.303 \, \Delta C^\circ_{pT_m} \log \frac{T}{T_0} .$$

2.2   THE COMPENSATION EFFECT

We can improve our analysis of the effects of temperature on thermochemical quantities by using an analytical form for C°_p and ΔC°_p. Over any specified temperature range ΔT, C°_p for any molecular species usually increases monotonically with temperature. It is common to represent C°_p by a polynomial expression in the range ΔT, quadratic or cubic, in T (or ΔT). Since ΔC°_p has a much smaller variation, we can represent it with good accuracy by a linear or, at most, quadratic function of T, even over an extended range ΔT.

Let us write for the range $T_0 \le T \le T_1$; $\Delta T_1 = T_1 - T_0$

$$\Delta C^\circ_{p(T)} = a + bT + cT^2 \tag{2.12}$$

Then,

$$\Delta C^\circ_{p(T_0)} = a + bT_0 + cT_0^2 \tag{2.13}$$

and

$$\langle \Delta C^\circ_p \rangle_T = \frac{1}{\Delta T_1} \int_{T_0}^{T_1} (\Delta C^\circ_p)_T \, dT$$

$$= a + \frac{b}{2}(T_1 + T_0) + \frac{c}{3}(T_1^2 + T_0 T_1 + T_0^2) \tag{2.14}$$

We note that $\langle \Delta C^\circ_p \rangle_T$ is not very different from the arithmetic mean of $\Delta C^\circ_{p(T_0)}$ and $\Delta C^\circ_{p(T_1)}$:

$$\langle C^\circ_p \rangle_T = \tfrac{1}{2}[\Delta C^\circ_{p(T_0)} + \Delta C^\circ_{p(T_1)}] - \frac{c}{6}(\Delta T)^2 \tag{2.15}$$

For ΔH°_T and ΔS°_T we find the following [from (2.10) and (2.12)]:

$$\Delta H^\circ_T = \Delta H^\circ_{T_0} + a(T - T_0) + \frac{b}{2}(T^2 - T_0^2) + \frac{c}{3}(T^3 - T_0^3) \tag{2.16}$$

$$\Delta S^\circ_T = \Delta S^\circ_{T_0} + a \ln \left(\frac{T}{T_0}\right) + b(T - T_0) + \frac{c}{2}(T^2 - T_0^2) \tag{2.17}$$

Hence, setting $\Delta T = T - T_0$:

$$\frac{\Delta G_T^\circ}{RT} = \frac{\Delta H_{T_0}^\circ - T \Delta S_{T_0}^\circ}{RT} + \frac{a}{R}\left[\frac{\Delta T}{T} - \ln\left(\frac{T}{T_0}\right)\right] - \frac{b}{2RT}[\Delta T]^2 - \frac{c(\Delta T)^2(T+2T_0)}{6RT}$$

$$(2.18)$$

where $\Delta H_{T_0}^\circ - T \Delta S_{T_0}^\circ = \Delta G_T^{\circ\prime}$ is the uncorrected value of ΔG_T we would obtain if we had neglected ΔC_p° entirely.

Now, expanding the logarithmic term:

$$-\ln\left(\frac{T}{T_0}\right) = \ln\left(1 - \frac{\Delta T}{T}\right) \approx -\frac{\Delta T}{T} - \frac{(\Delta T)^2}{2T^2} - \cdots$$

$$(2.19)$$

we find:

$$\frac{\Delta G_T^\circ}{RT} = \frac{\Delta G_T^{\circ\prime}}{RT} - \frac{1}{2R}\left(\frac{\Delta T}{T}\right)^2\left\{a + bT + cT^2\left(1 - \frac{2}{3}\frac{\Delta T}{T}\right)\right\}$$

$$(2.20)$$

or:

$$\frac{\Delta G_T^\circ - \Delta G_T^{\circ\prime}}{RT} = -\frac{1}{2}\left(\frac{\Delta T}{T}\right)^2\left\{\frac{\Delta C_{p(T)}^\circ - \frac{2}{3}cT\,\Delta T}{R}\right\}$$

$$(2.21)$$

The term on the right-hand side of (2.21) represents a correction to $(\Delta G_T^{\circ\prime}/RT)$. Since ΔC_p° is usually small and in the range $\pm 3R$, we note that the correction term is small and much less than unity except when $(\Delta T/T) \to 1$. For small values of $\Delta T/T$ (i.e., $\leq \frac{1}{2}$) it tends to be of the order of 0.1–0.2, and hence a correction to $K_T' = \exp(-\Delta G_T^\circ/RT)$ of about 10–20%. The sign of the correction is almost entirely determined by the sign of ΔC_p°. Thus for $\Delta C_p^\circ > 0$ $(K_T/K_T') > 1$, and vice versa.

Even when $\Delta C_p^\circ(T)$ is not small and ΔH° and ΔS° both change appreciably with changes in temperature, we note that because of the quadratic dependence of the correction term on $(\Delta T/T)$, the correction to $\Delta G_T^{\circ\prime}$, and hence to K', tends to be small. The reason is inherent in the fundamental relation $\Delta G^\circ = (\Delta H^\circ - T\,\Delta S^\circ)$. Both ΔH° and ΔS° tend to change in the same way with temperature, that is, increase or decrease. In ΔG°, however, these differences tend to cancel, and so it is that in calculating ΔG_T° at one temperature T from values of $\Delta H_{T_0}^\circ$ and $\Delta S_{T_0}^\circ$ at another nearby temperature, T_0, we make relatively little error in neglecting the changes in ΔH_T° and ΔS_T° with temperature.

This is a quite general property of the free-energy function and is often referred to as the "law of compensation." It is not only true of temperature but also of pressure, solvent changes, and other external parameters of a chemical system. Any change in a chemical system that affects the ΔH of a reaction generally has a comparable effect on ΔS and hence a smaller effect on ΔG. This compensation is perhaps more easily understood in the language of statistical mechanics, where we associate decreases in enthalpy (exothermic changes) with "tighter" binding, and consequently, with less entropy ("freedom of motion").

This compensation effect explains why very simple two-parameter functions, such as the Arrhenius, Clausius–Clapeyron, and van't Hoff equations are capable of representing free energy-dependent quantities over extended temperature ranges. In the familiar semilogarithmic plots of $\log K$ (or $\log k$) against $1/T$, a systematic curvature in the experimental points can often be lost in the random error, and the data can thus still be well represented by a straight line. Note that the converse can be seriously misleading; namely, if we do observe that k (a rate constant) or K (an equilibrium constant) can be represented by a simple exponential function of temperature, we cannot conclude that ΔH and ΔS are constants. The potential error arises if we try to extrapolate K or k outside of the measured range.

2.3 SOURCES OF DATA

With roughly 100 elements in the periodic table, there are 100 atomic species, 200 possible univalent ions (positive and negative), 5050 diatomic molecules, abouit 10^6 triatomic molecules, and $\sim 10^{2n}$ n-atomic molecules. It is clearly a hopeless task to expect to have tabulated, experimental thermochemical data on all polyatomic species or even restricted subclasses of these. Fortunately, it is possible, with the techniques of statistical thermodynamics, to calculate with good precision the standard entropies and heat capacities of molecules if we know their geometrical structures and vibrational frequencies. For large numbers of simple molecules these data are available, and very accurate values of C_p° and S° over large ranges of temperature have been tabulated. This is not the case for heats of formation, and ΔH_f° values must be measured experimentally for each particular compound. However many of the more common, simpler molecules have already been measured, and, for the more complex species, a well-documented body of empirical data exists for estimating their values from the former. Also, for many of the more complex species for which structural information is not available, good empirical rules are available for estimating geometrical structure and vibrational frequencies with adequate precision to allow the calculation of S° and C_p° with good accuracy.

However even the calculations of S° aand C_p° described above are not trivial or rapid, and we make use of them only in exceptional cases. An extremely rapid and nearly as accurate method of estimation of thermochemical data is to be found in the "additivity rules," which we shall now discuss.

2.4   ADDITIVITY RULES FOR MOLECULAR PROPERTIES

Chemists and physicists have known for some time that most molecular properties of larger molecules can be considered, roughly, as being made up of additive contributions from the individual atoms or bonds in the molecule. The physical basis of such an empirical finding appears to reside in the fact that the forces between atoms in the same or different molecules are very "short range"; that is, they are appreciable only over distances of the order of 1–3 Å. Because of this, the individual atoms in a typically substituted hydrocarbon, such as isopropyl chloride, $CH_3CHClCH_3$, seem to contribute nearly constant amounts to such molecular properties as refractive index, ultraviolet and infrared absorption spectra, magnetic susceptibility, and also entropy, molar heat capacity, and even heat of formation.

Some time ago, Benson and Buss[2] showed that it was possible to make a hierarchical system of such additivity laws in which the simplest- or "zeroth-"order law would be the law of additivity of atom properties. In an atom-additivity scheme, one assigns partial values for the property in question to each atom in the molecule. The molecular property, thereby, is the sum of all the atom contributions. For an exceptional property, such as molecular weight, such a law is precise. However there is an obvious limitation on such a law. In any chemical reaction, since there is a conservation of atoms, an atom-additivity law would predict that any molecular property would be similarly conserved; that is, it is the same for reactants and products. This is certainly not the case for entropy and enthalpy in most chemical reactions, although it is roughly correct in many cases where there is no change in mole number. However even here, too many obvious exceptions exist, such as $H_2 + F_2 \rightarrow 2HF + 128$ kcal, $N_2 + O_2 \rightarrow 2NO - 43$ kcal, and $H_2 + Cl_2 \rightarrow 2HCl + 44$ kcal.

2.5   ADDITIVITY OF BOND PROPERTIES

The next higher- or first-order approximation in additivity schemes is the additivity of bond properties. Table 2.1 shows a listing of such bond contributions to C_p°, S°, and ΔH_f° of ideal gases at 25°C.

To illustrate the use of the Table 2.1, let us calculate C_p° and S° for

[2] S. W. Benson and J. H. Buss, *J. Chem. Phys.*, **29,** 546 (1958).

Table 2.1.   Partial Bond Contributions for the Estimation of C_p°, S°, and ΔH_f° of Gas-phase Species at 25°C, 1 atm[1]

Bond	C_p°	S°	ΔH_f°	Bond	C_p°	S°	ΔH_f°
C—H	1.74	12.90	−3.83	S—S	5.4	11.6	−6
C—D	2.06	13.60	−4.73	C_d—C^2	2.6	−14.3	6.7
C—C	1.98	−16.40	2.73	C_d—H	2.6	13.8	3.2
C—F	3.34	16.90	−52.5	C_d—F	4.6	18.6	−39
C—Cl	4.64	19.70	−7.4	C_d—Cl	5.7	21.2	−5.0
C—Br	5.14	22.65	2.2	C_d—Br	6.3	24.1	9.7
C—I	5.54	24.65	14.1	C_d—I	6.7	26.1	21.7
C—O	2.7	−4.0	−12.0	>CO—H^3	4.2	26.8	−13.9
O—H	2.7	24.0	−27.0	>CO—C	3.7	−0.6	−14.4
O—D	3.1	24.8	−27.9	>CO—O	2.2	9.8	−50.5
O—O	4.9	9.1	21.5	>CO—F	5.7	31.6	−77
O—Cl	5.5	32.5	9.1	>CO—Cl	7.2	35.2	−27.0
C—N	2.1	−12.8	9.3	ϕ-H^4	3.0	11.7	3.25
N—H	2.3	17.7	−2.6	ϕ-C^4	4.5	−17.4	7.25
C—S	3.4	−1.5	6.7	(NO_2)—O^4	—	43.1	−3.0
S—H	3.2	27.0	−0.8	(NO)—O^4	—	35.5	9.0

1. See Sections 2.13 and 2.14 for corrections to entropy for symmetry and electronic contributions. C_p°, and S_0° estimated from the rule of additivity of bond contributions, are good to about ±1 cal/mole-°K, but they may be poorer for heavily branched compounds. The values of ΔH_f° are usually within ±2 kcal/mole but may be poorer for heavily branched species. Peroxide values are not certain by much larger amounts. All substances are in ideal gas state.
2. C_d represents the vinyl group carbon atom. The vinyl group is here considered a tetravalent unit.
3. >CO— represents the bond to carbonyl carbon, the latter being considered a bivalent unit. This is somewhat of a "fudge" on simple bond additivity.
4. NO and NO_2 are here considered as univalent, terminal groups, but the phenyl group $\phi(C_6H_5)$ is considered a hexavalent unit.

some compounds.[3] (For convenience, we shall insert subscript T only when $T \ne 25°C$.)

EXAMPLES

1.
$$C_p^\circ(CHCl_3) = C_p^\circ(C—H) + 3C_p^\circ(C—Cl)$$
$$= 1.74 + 3 \times 4.64 = 15.66$$
$$C_p^\circ(obs) = 15.7$$
$$S^\circ(CHCl_3) = S^\circ(C—H) + 3S^\circ(C—Cl) - R \ln 3^4$$
$$= 12.90 + 3 \times 19.70 - 2.2 = 69.8$$
$$S^\circ(obs) = 70.9$$

[3] For original references to thermochemical data, see the footnote in Appendix.
[4] $R \ln 3$ is a symmetry correction, $\sigma = 3$ being the symmetry number of $CHCl_3$, arising from the threefold axis. See Section 2.14 for further discussion.

2. $$C_p^\circ(\text{2,3-dimethylpentane}) = 6C_p^\circ(\text{C—C}) + 16C_p^\circ(\text{C—H})$$
$$= 6 \times 1.98 + 16 \times 1.74 = 39.8$$
$$C_p^\circ(\text{obs}) = 39.7 \text{ (estimate only)};$$
$$S^\circ(\text{2,3-dimethylpentane}) = 6S^\circ(\text{C—C}) + 16S^\circ(\text{C—H}) - 4R \ln 3 + R \ln 2^5$$
$$= -98.40 + 206.40 - 8.8 + 1.4$$
$$= 100.6$$
$$S^\circ(\text{obs}) = 99.0.$$

3. $$\Delta H_f^\circ(\text{benzyl iodide}) = 5\,\Delta H_f^\circ(\Phi\text{—H}) + \Delta H_f^\circ(\Phi\text{—C})$$
$$+ 2\,\Delta H_f^\circ(\text{C—H}) + \Delta H_f^\circ(\text{C—I})$$
$$= 16.25 + 7.25 - 7.66 + 14.1$$
$$= +29.9$$
$$\Delta H_f(\text{obs}) = 30.4 \text{ kcal/mole.}$$

4. $$\Delta H_f^\circ(\text{2-chlorobutadiene 1,3}) = 5\,\Delta H_f^\circ(C_d\text{—H}) + \Delta H_f^\circ(C_d\text{—C}) + \Delta H_f^\circ(C_d\text{—Cl})$$
$$= 16.0 + 6.7 - 0.7 = +22.0$$
$$\Delta H_f^\circ(\text{obs}) = (\text{not known})$$

5. $$\Delta H_f^\circ(\text{ethyl acetate}) = 8\,\Delta H_f^\circ(\text{C—H}) + \Delta H_f^\circ(\text{>CO—C}) + \Delta H_f^\circ(\text{>CO—O})$$
$$+ \Delta H_f^\circ(\text{C—O}) + \Delta H_f^\circ(\text{C—C})$$
$$= -30.64 - 14.4 - 50.5 - 12.0 + 2.73$$
$$= -104.8$$
$$\Delta H_f^\circ(\text{obs}) = 103.4 \text{ kcal/mole.}$$

We find that the law of bond additivity reproduces C_p and S° values to within ± 1 cal/mole-°K on the average, but the law is poorer for very heavily branched compounds. Values of ΔH_f° are generally estimated to within ± 2 kcal/mole but again are subject to larger errors in heavily branched compounds and in compounds containing very electronegative groups such as NO_2 and F.

It is clear that bond additivity rules will give the same properties for isomeric species such as n-butene and isobutene, *cis*- and *trans*-olefins, and so on, and hence cannot be employed to distinguish differences in properties of isomers. It is generally the case that isomeric differences, which give rise to large steric effects in molecules, cannot be treated by *any* simple additivity scheme but must be treated either by exceptional rules or by individual examination.

2.6 ADDITIVITY OF GROUP PROPERTIES

The next higher- or second-order approximation to additivity behavior is to treat a molecular property as being composed of contributions due to groups. A group is defined as a polyvalent atom (ligancy ≥ 2), in a

---

[5] We substract $4R \ln 3$ for the internal symmetry of four CH_3 groups and add $R \ln 2$ for the entropy of mixing of two optical isomers.

molecule together with all of its ligands. The nomenclature we follow is to identify first the polyvalent atom and then its ligands. Thus C—(H)$_3$(C) represents a C atom connected to three H atoms and another C atom, that is, a primary methyl group. Molecules such as HOH, CH$_3$Cl, and CH$_4$ that contain only one such atom (i.e., one group), are irreducible entities and cannot be treated by group additivity. The molecules that can be treated are those with two or more polyvalent atoms. Examples of analysis of molecules into groups are as follows.

1. CH$_3$—CH$_3$: Contains two identical groups each with a carbon atom bound to a carbon atom and three H atoms. Note that all unbranched paraffin hydrocarbons contain only two groups, [C—(C)(H)$_3$] and [C—(C)$_2$(H)$_2$]. The totality of saturated paraffins is composed of four groups; the preceding two and those for tertiary and quaternary C: [C—(C)$_3$(H)] and [C—(C)$_4$].

2. CH$_3$CHOHCH$_3$: Contains four groups (i.e., four polyvalent atoms):

$$2[C—(C)(H)_3] + [C—(C)_2(O)(H)] + [O—(C)(H)].$$

With increased substitution the number of groups increases and this provides a basic limitation on the use of group properties. Thus for nonbranched, chlorinated hydrocarbons we need all the groups symbolized by [C—(C)(H)$_n$(Cl)$_{3-n}$] and [C—(C)$_2$(H)$_n$(Cl)$_{2-n}$] where $n = 0$, 1, 2, and/or 3. This is a total of seven groups or five more than are needed for the paraffins. For the branched chlorocarbons only one further group is needed [C—(C)$_3$(Cl)].

The first six tables in the Appendix list the current available values of group contributions to C_p°, S°, and ΔH_f°. Values of C_p° and S° estimated from these groups are on the average within ±0.3 cal/mole-°K of the measured values, whereas ΔH_f° estimates are within ±0.5 kcal/mole. For heavily substituted species, deviations in C_p° and S° may go as high as ±1.5 cal/mole-°K, and ΔH_f° may deviate by ±3 kcal/mole.[6]

The following examples will illustrate the application of group additivity.

EXAMPLES

1.
$$C_p^\circ(i\text{-butane}) = 3[C—(C)(H)_3] + [C—(C)_3(H)]$$
$$= 18.57 + 4.54 = 23.1$$
$$C_p^\circ(\text{obs}) = 23.1 \text{ cal/mole-°K}$$

[6] For more detailed comparisons, see the group of papers by Benson et al., in *Chem. Rev.*, **69**, 279 (1969); H. K. Eigenmann, D. M. Golden, and S. W. Benson, *J. Phys. Chem.*, **77**, 1687 (1973).

2. $$S°(\text{pentene-2}) = [C-(C_d)(H)_3] + 2[C_d-(C)(H)] + [C-(C_d)(C)(H)_2]$$
    $$+ [C-(C)(H)_3] - 2R \ln 3 \text{ (symmetry of CH}_3)$$
    $$= 30.41 + 16.0 + 9.8 + 30.41 - 4.32$$
    $$= 82.3$$
    $$S°(\text{pentene-2})(\text{obs})_{\text{trans}} = 82.0 \text{ cal/mole-°K}$$

3. $$S°(sec\text{-butyl alcohol}) = 2[C-(C)(H)_3] + [C-(C)_2(O)(H)]$$
    $$+ [C-(C)_2(H)_2] + R \ln 2 \text{ (for optical isomers)}$$
    $$- 2R \ln 3 \text{ (for CH}_3 \text{ groups)}$$
    $$= 60.8 - 11.0 + 29.1 + 9.4 + 1.4 - 4.4$$
    $$= 85.3$$
    $$S°(\text{obs}) = 85.8 \text{ cal/mole-°K}$$

4. $$\Delta H_f°(t\text{-butyl methyl ether}) = 3[C-(C)(H)_3] + [C-(C)_3(O)]$$
    $$+ [O-(C)_2] + [C-(O)(H)_3]$$
    $$= -30.2 - 6.6 - 23.7 - 10.1$$
    $$= -70.6$$
    $$\Delta H_f°(\text{obs}) = -70 \pm 1 \text{ kcal/mole.}[7]$$

2.7   THE *GAUCHE* INTERACTIONS

In Tables A.1–5 are listed two types of corrections for higher order interactions. One of these is a correction for *cis–trans* isomerization. It is not possible to include such interactions directly into the group properties because they represent the interactions of nonbonded, next-nearest neighbors. In consequence, the group properties have been tabulated for the usually more stable *trans* isomers and corrections are needed to obtain the less stable *cis* olefins.

The second correction is for *gauche* interactions of large groups, that is, anything bigger than H atoms. These again are interactions of next-nearest neighbors, and so they are not directly included in a group scheme. The method of obtaining such interactions is to write out the structure of a given molecule and then to enumerate the *gauche* configurations of every group with respect to those preceding it that have not been counted. These interactions have to do with the rotomeric configurations around the bonds. In the case of linear paraffins, the most stable conformation is one in which all heavy groups are trans to each other. This is illustrated by the three conformations shown in Figure 2.1 for *n*-butane. The two mirror-image *gauche* conformations are less stable than the trans by about 0.8 kcal mole.

Figure 2.2 shows three conformations of 1-methyl butane. The two mirror-image *gauche* conformations each have one *gauche* methyl interaction, whereas the less stable *syn* conformation has two *gauche* methyl interactions.

---

[7] Making two oxygen *gauche* corrections gives $\Delta H_f° = -69.6$ kcal/mole.

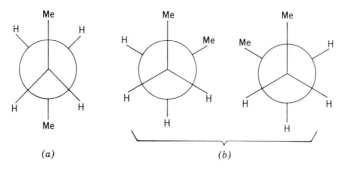

Figure 2.1   Rotomeric conformations of *n*-butane. (a) *trans;* (b) *gauche* (these are mirror images). We are looking at a projection of the molecule on a plane perpendicular to the central C—C bond. Groups shown are ligands of the C atoms comprising this bond.

A rapid method for identifying *gauche* interactions along single bonds is to draw a line skeleton formula of a molecule and count the number of hydrogen-containing groups bound to the atoms at each end of the bond. If there is only one at each end, as in *n*-butane, they can be *trans* or *gauche* to each other, as already shown (Figure 2.1). If there are two at one end and one at the other end, then there must be one *gauche* interaction (most stable form), and there may be two (least stable form), as shown in Figure 2.2 for 2-methyl butane. With three groups at one end and one at the other, there will always be two *gauche* interactions in all conformations (e.g., 2,2-dimethyl butane).

Figure 2.3 shows the skeletons of two branched hydrocarbons with numbers along the bonds indicating the minimum number of gauche interactions in the most stable form:

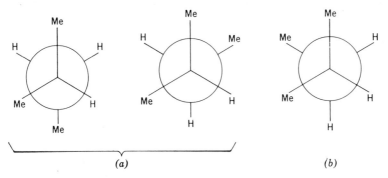

Figure 2.2   Rotomeric conformations of 2-methyl butane. (a) *gauche* (mirror images); (b) *syn.*

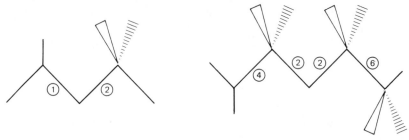

Figure 2.3 Skeleton Drawings of 2,4,4-trimethyl pentane (*a*) and 2,3,3,5,5,6,6-hepta-methyl heptane (b), showing minimum number of gauche interactions along each bond in the most stable conformation.

The evidence is fairly strong that the origin of both *cis* and *gauche* interactions lies in the repulsion of H atoms attached to too-close methyl or methylene groups. (In *gauche* *n*-butane, these would be the H atoms of the two terminal methyl groups.) If H-atom repulsions were the origin of barriers, we would not expect *gauche* interactions to involve —O—R, —N—R$_2$, or halogen atoms and this seems to be in accord with the experimental data. Thus secondary butyl alcohol has no *gauche* interactions, nor does secondary butyl amine or secondary butyl halide.

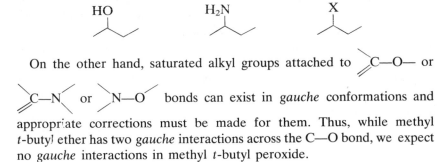

On the other hand, saturated alkyl groups attached to $\overset{\|}{\diagdown}$C—O— or $\overset{\|}{\diagdown}$C—N$\diagup$ or $\diagdown$N—O$\diagup$ bonds can exist in *gauche* conformations and appropriate corrections must be made for them. Thus, while methyl *t*-butyl ether has two *gauche* interactions across the C—O bond, we expect no *gauche* interactions in methyl *t*-butyl peroxide.

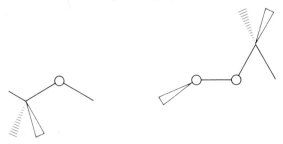

The 0.8 kcal repulsion for *gauche* interactions is assigned from a study of the conformations of alkanes and substituted alkyl cyclohexanes. It is

appropriate for H-atom repulsions on 1,4-C atoms but may not be descriptive of other types of molecules. Thus in di-*t*-butyl ether, or its homolog 2,2,4,4-tetramethyl pentane, the repulsions are between H atoms on 1,5-C atoms:

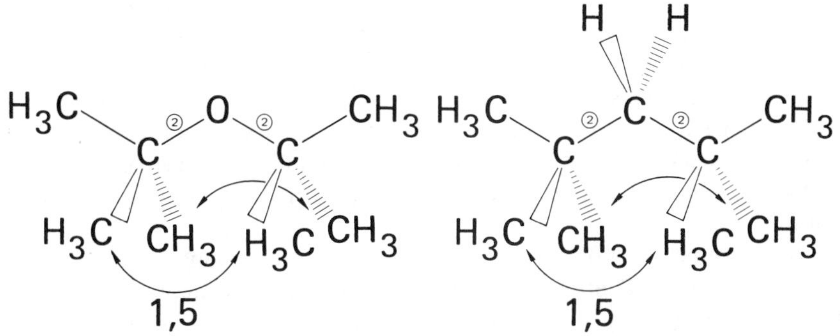

In both of these compounds, we would calculate 4 *gauche* interactions for a total of 2.0 kcal/mole (C—O *gauche*) in the ether, or 3.2 kcal/mole in the hydrocarbon. However, the proximity of the H atoms on these adjacent CH_3 groups is much closer than in *gauche* n-butane, and it is not surprising to find correspondingly larger repulsion energies. Using our normal correction for 4 *gauche* in each compound (1–4 interaction) and comparing with the observed strain energies, we arrive at an extra 1.5 kcal/mole for each of the two (1–5) interactions in the hydrocarbon and 3.5 kcal/mole each in the ether. This is probably not too surprising since the C—O—C angle in the ether is about 106° instead of the 112.5° in the hydrocarbon while the C—O distance is smaller than the C—C by 0.11 Å; both of these act to bring the pendant methyls closer together in the ether.

Although the magnitudes of even more distant interactions are explored later, it is worthy of note that the heat of isomerization of *cis*-ditertiary butyl ethylene to the trans form is −9.3 kcal/mole compared to −1.0 kcal/mole for the *cis*-dimethylethylene (that is, *cis*-butene-2) to *trans* isomerization. Interactions of such a nature cannot be anticipated from any simple scheme of corrections and in many cases can only be detected by examination of precise structural models.

2.8 NONGROUP INTERACTION, RINGS

The principle of additivity schemes rests on the assumption that whatever the additive unit, whether atom, bond, or group, its "local" properties remain unchanged in a series of homologous compounds. That is, there is no significant interaction with more distant units. Linear molecules such as the normal paraffins, ethers, and their terminally substituted derivatives [e.g., $Cl—(CH_2)_n—O(CH_2)_m—X$] are ideally suited to such an assumption, and hence it is not surprising that these fit additivity rules extremely well, namely to the precision of the experimental data. However, as soon as we introduce structural features into a molecule, which bring more distant units into proximity, we may expect departures from additivity laws. This has already been noted in the preceding section in connection with *cis* isomers for olefins and with *gauche* configurations of highly branched molecules. In order to accommodate such species into an additivity scheme we must add "corrections" for these nongroup interactions.

The extreme example of such higher order interactions is provided by ring compounds. There is no simple, natural way of incorporating ring systems even into a group additivity scheme. We can manage to do so by adding a "correction" for the structure. (The Appendix tables contain lists of such corrections for various ring structures both homo- and heterocyclic.) To evaluate a property for example, for *cis*-1-methyl,2-ethyl cyclopentane, we would add all the usual contributions for the groups present and then add corrections for the nongroup interactions:

ΔH_f°(*cis*-1-methyl, 2-ethyl cyclopentane)
$$= 2C—(C)(H)_3 + 4C—(C)_2(H)_2 + 2C—(C)_3(H)$$
$$+ C_5 \text{ ring correction} + cis \text{ correction} + gauche \text{ correction}$$
$$= -20.16 - 19.80 - 3.80 + 6.3 + 1.0 + 0.8$$
$$= -35.7$$
ΔH_f°(obs) $= -35.9$ kcal/mole.

When we are lacking a ring correction for entropy or heat capacity for a heterocyclic ring, it is a reasonable approximation to use the correction applicable for the homologous hydrocarbon ring. Thus for the $\overset{\frown}{C—C—C—C—O}$ ring correction that does not appear in the tables, we could use the cyclopentane values. This will be reasonable so long as the analog has the same number of atoms of the same size. CH_2, NH, and O will be replaceable one by the other while SiH_2, PH, and S from the second long row in the periodic table can be considered equivalent.

2.9   ESTIMATION OF THERMOCHEMICAL DATA
AT HIGHER TEMPERATURES

The additivity laws already discussed allow us to obtain ΔH_f°, S°, and C_p° at 298°K. To obtain values at higher temperatures, we need to know C_p°

as a function of T. For the hydrocarbons and simply substituted hydrocarbons, such data are available, and Tables A.1–5 list group contributions to C_p° at a series of increasing temperatures. Average values of C_p° for use in equations such as (2.11) can be quickly estimated from the tables, using linear interpolation between the listed temperatures.

EXAMPLE

Calculate $S_{840}^\circ(n\text{-}C_4H_{10}) - S_{298}^\circ(n\text{-}C_4H_{10})$.

From the Appendix tables, C_{p298}° $(n\text{-}C_4H_{10}) = 2C_p^\circ[C\!\!-\!\!(C)(H)_3] + 2C_p^\circ[C\!\!-\!\!(C)_2(H)_2]$
$$= 23.4$$

Interpolating the group values between the nearest listed temperatures of 800°K and 1000°K we estimate

$$C_{p840}^\circ = 49.4$$

As an average value over the range, we would then choose $C_{pT_m}^\circ = 36.4$ cal/mole-°K, noting that the values tend to increase faster than the average rate at the lower temperatures. From (2.11) we then find

$$S_{840}^\circ - S_{298}^\circ = (C_{pT_m}^\circ) \ln \tfrac{840}{298} = 37.7$$

For comparison the observed value is 37.0.

Alternatively, we can estimate and add increments over smaller temperature intervals:

$$S_{840}^\circ - S_{298}^\circ = \langle C_{p400}\rangle \ln \left(\tfrac{500}{300}\right) + \langle C_{p670}\rangle \ln \left(\tfrac{840}{500}\right)$$
$$= 29.6 \times 0.51 + 43.8 \times 0.51 = 37.4$$

In considering enthalpy corrections we must note the following problem. Our primary enthalpy data for compounds are standard heats of formation, which means that they refer in turn to the standard states of the elements. If we wish to compute a heat of formation at a temperature higher than 298°K, we need to obtain ΔC_{pf}° for the change in molar heat capacity in forming the compound from elements in their standard states. For a large number of compounds, tabulations are already available of ΔH_{pT}° over a range of temperatures. If we are using data from such a table, we must never use it with anything but other ΔH_{fT}° data. However if we use the present tables, in which we have listed group properties only at 298°K for ΔH_f°, it is not always necessary to compute ΔH_{fT}° because, for most purposes, we will be using these ΔH_{fT}° to obtain heats of reaction at T, and the C_{pT}° of the elements will cancel. For these latter purposes, it is sufficient to compute a value that will correspond to $\Delta H_{f298}^\circ + (H_T^\circ - H_{298}^\circ)$, the correction term being an absolute enthalpy increment for the compound. Its value is given by an equation like (2.11):

$$H_T^\circ - H_{298}^\circ = \int_{298}^{T} C_p^\circ \, dT = C_{pT_m}^\circ (T - 298). \tag{2.22}$$

To call attention to the ambiguity in such a procedure as that described above, let us label such mixed quantities "apparent heats of formation." Where actual heats of formation at T are desired, then, ΔC_{pT}° may be calculated using the values given in Table A.7, for the C_p° of the elements in their standard states.

EXAMPLE

Calculate ΔH_{f840}° (n-butane)$-\Delta H_{f298}^{\circ}$ and calculate also the enthalpy increment ($H_{840}^{\circ}-H_{298}^{\circ}$) for n-butane.

From the preceding example we note that $C_{pT_m}^{\circ}$ (n-C$_4$H$_{10}$) over this temperature interval is 36.4 gibbs/mole, so that from (2.22) above, we find

$$H_{840}^{\circ}-H_{298}^{\circ}=36.4(840-298)=19.7 \text{ kcal/mole}$$

The equation for the formation of n-butane is

$$4C(s)+5H_2(g) \rightleftarrows n\text{-C}_4H_{10}(g).$$

Hence $\Delta C_{pf}^{\circ}=C_p^{\circ}(n\text{-butane})-4C_p^{\circ}[C(s)]-5C_p^{\circ}[H_2(g)]$. Using C_p° already calculated at 298°K, and Table A.7 for the corresponding values for the elements we find that

$$\Delta C_{pf298}^{\circ}=-19.2 \quad \text{and} \quad \Delta C_{pf840}^{\circ}=-5.2$$

with $\Delta C_{pfm}^{\circ}=-12.2$ gibbs/mole being a reasonable value in this range. Hence from (2.11)

$$\Delta H_{f840}^{\circ}(n\text{-butane})-\Delta H_{f298}^{\circ}(n\text{-butane})=-12.2(840-298)$$
$$=-6.6 \text{ kcal/mole}.$$

The "observed" value from the API tables is -6.5 kcal/mole.

Note that ΔC_{pf}° can be very large and either positive or negative in value. In consequence calculations of ΔH_{fT}° may require rather sizable corrections. However it should also be noted that a large amount of the magnitude of ΔC_{pf}° is contributed by Δn, the change in moles of gases. For butane $\Delta n=-4$, and this contributes a constant -8.0 gibbs/mole to ΔC_{pf}°.

Let us now consider a typical problem, the calculation of an equilibrium constant.

EXAMPLE

Calculate K_p for the following reaction at 730°K:

$$(CH_3)_3CCH_2Cl \rightleftarrows (CH_3)_2C=CHCH_3+HCl.$$

PROCEDURE

We first calculate ΔH_f°, S°, and ΔC_p° at 298° for each of the compounds except HCl, which is listed separately, and from this compute ΔH_{298}° and ΔS_{298}°. We then compute ΔC_p° at 298°

and at 730°K and estimate $\Delta C^\circ_{pT_m}$, which we use to calculate ΔH°_{730} and ΔS°_{730} and finally ΔG_{730} and $K_{p(730)}$.

1. $\qquad \Delta H^\circ_f[(CH_3)_3CCH_2Cl] = 3 \Delta H^\circ_f[C\!-\!(C)(H)_3] + \Delta H^\circ_f[C\!-\!(C)_4]$
$\qquad\qquad\qquad\qquad\qquad + \Delta H^\circ_f[C\!-\!(C)(H)_2(Cl)] + (\text{two } gauche \text{ corrections})$
$\qquad\qquad\qquad\qquad = -30.24 + 0.50 - 15.7 + 2(0.8)$
$\qquad\qquad\qquad\qquad = -43.7 \text{ kcal/mole};$
$\qquad \Delta H^\circ_f(\text{2-methyl butene-2}) = 3 \Delta H^\circ_f[C\!-\!(C_d)(H)_3] + \Delta H^\circ_f[C_d\!-\!(C)_2]$
$\qquad\qquad\qquad\qquad\qquad + \Delta H^\circ_f[C_d\!-\!(H)(C)] + 1 \text{ } cis \text{ correction}$
$\qquad\qquad\qquad\qquad = -30.24 + 10.34 + 8.6 + 1.0 = -10.3 \text{ kcal/mole},$

and $\Delta H^\circ_f(HCl) = -22.0$ kcal/mole (from Table A.8). Thus $\Delta H^\circ_{298} = 11.4$ kcal/mole.

2. Using the same decomposition of neopentyl chloride and 2-methyl butene-2 into groups, we find for S°

$$S^\circ[(CH_3)_3CCH_2Cl] = 91.23 - 35.10 + 37.8 - 4R \ln 3 \text{ } (t\text{-butyl symmetry})$$
$$= 85.2$$
$$S^\circ(\text{olefin}) = 91.2 - 12.7 + 8.0 - 3R \ln 3 \text{ (symmetry)} + 1.2 \text{ } cis \text{ correction}$$
$$= 81.1$$

Therefore with $S^\circ(HCl) = 44.6$, we find $\Delta S^\circ_{298} = 40.5$.

3. By the same technique used above, we find $\Delta C^\circ_{p298} = 7.0 + 25.5 - 31.9 = +0.6$. At 730°K, assuming $C^\circ_p(HCl) \approx 7.0$, we find from the tables that $\Delta C^\circ_{p730} \sim 7.2 + 49.1 - 59.9 = -3.6$. Hence we can take ΔC°_{pm} as -1.5.

4. Using (2.11), we now calculate

$$\Delta H^\circ_{730} = \Delta H^\circ_{298} - 1.5 \times (0.730 - 0.298) = 10.8 \text{ kcal/mole};$$
$$\Delta S^\circ_{730} = \Delta S^\circ_{298} - 1.5 \times 2.3 \times \log\left(\tfrac{730}{298}\right) = 39.3.$$

Hence $\Delta G^\circ_{730} = \Delta H^\circ_{730} - 0.730 \Delta S^\circ_{730} = -17.9$ kcal/mole,

$$\log K_p = \frac{-\Delta G^\circ_{730}}{4.576T} = 5.35$$

and

$$K_{p(730)} = 2.2 \times 10^5 \text{ atm}.$$

2.10 SOME STATISTICAL MECHANICAL RESULTS

While additivity methods will stand us in good stead where data exist for homologous series of compounds, we will frequently be required to estimate entropies and heat capacities for unique compounds, radicals, or transition states where few or no analogs are available to guide us. In such cases we will draw on the very precise results of statistical mechanics.

We can define the molar partition function Q for an assembly of identical molecules by the equation:

$$Q = \sum_i p_i e^{-E_i/RT} \tag{2.23}$$

where p_i is the number of discrete states of the molecules which have the energy E_i in a system at thermal equilibrium T and volume V. It can then be shown that the familiar thermodynamic properties are related to Q as follows:

Helmholtz free energy: $A = -RT \ln Q$

Gibbs free energy: $\quad G = A + PV$

Internal energy: $\quad E = RT\left(\dfrac{\partial \ln Q}{\partial \ln T}\right)_V = -R\left(\dfrac{\partial \ln Q}{\partial (1/T)}\right)_V$

Enthalpy: $\quad H \equiv E + PV_{nRT} \quad n=1$

Entropy: $\quad S = \dfrac{E-A}{T}$

$$= R \ln Q + R\left(\dfrac{\partial \ln Q}{\partial \ln T}\right)_V$$

Molar heat capacity at constant volume $\quad C_V = \left(\dfrac{\partial E}{\partial T}\right)_V = R\left(\dfrac{\partial \ln Q}{\partial \ln T}\right)_V + R\left(\dfrac{\partial^2 \ln Q}{\partial (\ln T)^2}\right)_V$

$$= \dfrac{R}{T^2}\left(\dfrac{\partial^2 \ln Q}{\partial (1/T)^2}\right)_V$$

$C_P \equiv C_V + PV = C_V + R$ (for ideal gases) $\tag{2.24}$

For ideal gases the partition function Q can be factored into a product of partition functions for each of the degrees of freedom of the system. These will be $3N$ in number, where N is the number of atoms in the molecule, ion, or radical; 3 translational degrees of freedom; 0 (for atoms), 2 (for linear molecules), or 3 rotational degrees of freedom and the remainder internal vibrational degrees of freedom.

$$Q = Q_{\text{tran}} \times Q_{\text{rot}} \times Q_{\text{vibration}} \tag{2.25}$$

Where there are low-lying electron states or a nonzero spin, we can tack on an additional factor Q_{elec} for the electronic degrees of freedom.

Each of the thermodynamic functions depends on the logarithm or logarithmic derivative of Q and will thus receive additive contributions from each of the degrees of freedom. Thus:

$$S = S_{\text{tran}} + S_{\text{rot}} + S_{\text{vib}} + S_{\text{elec}} \tag{2.26}$$

The translational partition function is rigorously known; for one mole of ideal gas it is:

$$Q_{tran} = \frac{V^{N_0}}{N_0!}\left(\frac{2\pi MkT}{h^2}\right)^{3N_0/2}$$

$$E^\circ_{trans} = \tfrac{3}{2}N_0 kT = \tfrac{3}{2}RT$$

$$C^\circ_{V(tran)} = \tfrac{3}{2}R \tag{2.27}$$

$$S^\circ_{(tran)} = 37.0 + \tfrac{3}{2}R\ln\left(\frac{M}{40}\right) + \tfrac{3}{2}R\ln\left(\frac{T}{298}\right)$$

$$+ R\ln(n)$$

where N_0 = Avogadro's number 6.02×10^{23} molecules.

In the last equation for S, M is the molecular weight (amu) and n is the number of optical isomers.

For rotational degrees of freedom we have the following:

1. Linear molecule (2 degrees of freedom):

$$Q_{rot\text{-}2D} = \frac{1}{\sigma_e}\left(\frac{8\pi^2 IkT}{h^2}\right)^{N_0} \tag{2.28}$$

$$E_{rot\text{-}2D} = RT \tag{2.29}$$

$$C^\circ_{V(rot\text{-}2D)} = R \tag{2.30}$$

$$S^\circ_{rot\text{-}2D} = 6.9 + R\ln(I/\sigma_e) + R\ln(T/298) \tag{2.31}$$

2. Nonlinear molecule (3 degrees of freedom):

$$Q_{rot\text{-}3D} = \frac{\pi^{1/2}}{\sigma_e}\left(\frac{8\pi^2 I_m kT}{h^2}\right)^{(3N_0/2)} \tag{2.32}$$

$$E^\circ_{rot\text{-}3D} = \tfrac{3}{2}RT \tag{2.33}$$

$$C^\circ_{V(rot\text{-}3D)} = \tfrac{3}{2}R \tag{2.34}$$

$$S^\circ_{rot\text{-}3D} = 11.5 + \frac{R}{2}\ln\left(\frac{I_m^3}{\sigma_e}\right) + \tfrac{3}{2}R\ln\left(\frac{T}{298}\right) \tag{2.35}$$

where σ_e is the external symmetry number of the molecule (to be discussed in detail later, p. 47), I is the moment of inertia for a linear molecule about its center of mass, and I_m^3 is the product of the three principal moments of inertia for a nonlinear molecule, also about the center of gravity.

In complex molecules, one or more of the internal degrees of freedom may become a simple, one-dimensional, free rotation. In that case its properties are described by:

3. One-dimensional rotor:

$$Q_{\text{rot-1D}} = \frac{\pi^{1/2}}{\sigma_i} \left(\frac{8\pi^2 I_r kT}{h^2} \right)^{N_0/2} \tag{2.36}$$

$$E^\circ_{\text{rot-1D}} = \tfrac{1}{2}RT \tag{2.37}$$

$$C^\circ_{V(\text{rot-1D})} = \tfrac{1}{2}R \tag{2.38}$$

$$S^\circ_{\text{rot-1D}} = 4.6 + R \ln \left(\frac{I_r^{1/2}}{\sigma_i} \right) + \frac{R}{2} \ln \left(\frac{T}{298} \right) \tag{2.39}$$

where σ_i is the symmetry of the internal rotation (p. 43), and I_r is the reduced moment of inertia for the internal rotation (see p. 44).

Finally, for each vibrational degree of freedom we have:

$$Q_{\text{vib}} = (1 - e^{-h\nu/kT})^{-N_0} \tag{2.40}$$

with the appropriate thermodynamic contributions to be evaluated from equations (2.24). ν is the frequency of the vibration and, in general, the contributions of the vibrational degrees of freedom to S°, C°_p, and E° tend to be small. The only exception to this occurs when $h\nu \ll kT$, so that the exponent can be expanded in equation (2.40), and we find:

$$Q_{\text{vib}} \xrightarrow[\substack{\text{classical} \\ \text{limit}}]{} \left(\frac{kT}{h\nu} \right)^{N_0} \tag{2.41}$$

$$E^\circ_{\text{vib}} \longrightarrow RT \tag{2.42}$$

$$C^\circ_{\text{vib}} \longrightarrow R \tag{2.43}$$

$$S^\circ_{\text{vib}} \longrightarrow R \ln \left(\frac{kT}{h\nu} \right) + R \tag{2.44}$$

In the opposite extreme, when $kT \ll h\nu$, then $Q_{\text{vib}} \to 1$ and all of the thermodynamic functions for vibration approach zero.

2.11 STRUCTURAL METHODS OF ESTIMATING C°_p

Lacking values for bonds or groups to estimate C°_p, we can make very good estimates by assigning vibrational frequencies to the molecule and thereby obtain the vibrational contribution to C°_p. This is usually the only difficult part of the calculation. From the methods of statistical mechanics we know that C°_p of an ideal gas can be broken down as follows:

$$C^\circ_p = C^\circ_p(\text{tran}) + C^\circ_p(\text{rot}) + C^\circ_p(\text{vib})$$
$$+ C^\circ_p(\text{elec}) + R. \tag{2.45}$$

Where quantum effects are not important $C^\circ_p(\text{translation}) = \tfrac{3}{2}R$, and $C^\circ_p(\text{rotation}) = R$ for linear molecules or $\tfrac{3}{2}R$ for nonlinear molecules.

C_p°(electronic) is usually zero except for a handful of odd-electron species or species such as O_2, and even then it is small. C_p° then becomes

1. Linear molecules:

$$C_p^\circ = \tfrac{7}{2}R + C_p^\circ(\text{vib}) + C_p^\circ(\text{elec}).\qquad(2.46)$$

2. Nonlinear molecules:

$$C_p^\circ = \tfrac{8}{2}R + C_p^\circ(\text{vib}) + C_p^\circ(\text{elec}).\qquad(2.47)$$

For small molecules containing H and D atoms and only one other heavy atom we can usually neglect C_p°(vibration) at 298°K, as well as C_p°(electronic), and we find from (2.46) and (2.47) that species such as HCl and DBr, have C_{p298}° of about 7 cal/mole-°K. whereas for H_2O, NH_3, NO_3, CH_4, and so on, 8 cal/mole-°K is usually within 10% of the observed C_{p298}° (see Tables A.9, A.10, and A.11).

For larger molecules the vibrational frequencies can be assigned on a very simple empirical basis. First let us note that in a molecule containing N atoms, there are $(3N-5)$ fundamental frequencies for linear molecules and $(3N-6)$ for nonlinear ones. These frequencies can be further subdivided into three classes: (a) stretching frequencies, (b) bending (or deformation) frequencies, and (c) hindered internal rotations. Because N atoms in a molecule must be connected by at least $(N-1)$ bonds, there will be $(N-1)$ stretching modes, each corresponding to the stretching of a characteristic bond and the remaining $(2N-4)$ or $(2N-5)$ frequencies belong to the class of deformations plus hindered rotations. In simple rings N atoms will be connected by N bonds, so that they have one more stretching mode than a noncyclic system.

Hindered (or free) internal rotations or torsions can only occur around single bonds, and their number is given by the number of single bonds connecting polyvalent atoms. Thus a saturated, substituted paraffin $C_nH_{2n+2-m}X_m$, where X is a univalent atom, will have $(n-1)$ single bonds connecting the n carbon atoms and $(n-1)$ internal rotations. If rings exist, of course, as in cyclohexane and benzene, internal rotations are not possible, so ring compounds must be treated as special cases.

One further distinction is worth making, namely the separation of frequencies into those involving H and D atoms and all of the rest. The reason for this is that H and D atom frequencies are generally very high because of the low mass of H (or D) and at low temperatures contribute very little to C_p°. This separation can be effected by treating the molecule as if it were composed only of heavy atoms. If it has, for example, P heavy atoms, it then has $(3P-6)$ or $(3P-5)$ related frequencies of which $(P-1)$ are stretches and $(2P-5)$ or $(2P-4)$ are deformations.

Table 2.2 shows a decomposition of vibrational frequencies for a few

Table 2.2. Classification of Vibrational Frequencies of Some Selected Molecules

Molecules	Number of Atoms N	Total Number of Frequencies	Numbers of Stretching Modes	Number of Bending Modes
C_2H_6	8	18	$7\begin{cases}6\ \text{C—H}\\1\ \text{C—C}\end{cases}$	$11\begin{cases}1\ \text{int. rot. (all H)}\\10\ \text{def.}\quad\text{(all H)}\end{cases}$
Isopropyl chloride	11	27	$10\begin{cases}7\ \text{C—H}\\1\ \text{C—Cl}\\2\ \text{C—C}\end{cases}$	$17\begin{cases}2\ \text{int. rot. (all H)}\\15\ \text{def.}\begin{cases}12\text{-H}\\3\ \text{heavy}\end{cases}\end{cases}$
Methyl cyclohexane	21	57	$21\begin{cases}14\ \text{C—H}\\7\ \text{C—C}\end{cases}$	$36\begin{cases}1\ \text{int. rot. (all H)}\\35\ \text{def.}\begin{cases}27\text{-H}\\8\text{-heavy}\end{cases}\end{cases}$

(Structural diagram for isopropyl chloride)

```
 H   H  H
 |   |  |
H—C—C—C—H
 |   |  |
 H   Cl H
```

(Structural diagram for methyl cyclohexane, C_7H_{14})

Heptene-2

CH₃—CH=CH—CH₂—CH₂—CH₂ (with CH₃ branch)

$\mathrm{CH_3\!-\!CH\!=\!CH\!-\!CH_2\!-\!CH_2\!-\!CH_2}$, branch $\mathrm{CH_3}$

(C_7H_{14})

| 21 | 57 | $20\begin{cases}14\ C\!-\!H\\ 6\ C\!-\!C\end{cases}$ | $37\begin{cases}5\ \text{int. rot.}\begin{cases}2\text{-H}\\3\text{-heavy}\end{cases}\\ 32\ \text{def.}\begin{cases}26\text{-H}\\6\text{-heavy}\end{cases}\end{cases}$ |

Methyl methacrylate

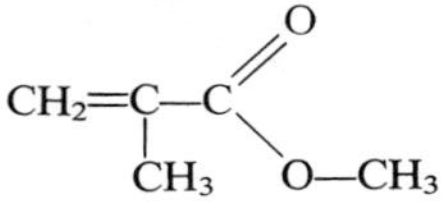

$\mathrm{CH_2\!=\!C}$ with $\mathrm{CH_3}$ and $\mathrm{C(\!=\!O)\!-\!O\!-\!CH_3}$

($C_5H_8O_2$)

| 15 | 39 | $14\begin{cases}8\ C\!-\!H\\ 3\ C\!-\!C\\ 3\ C\!-\!O\end{cases}$ | $25\begin{cases}4\ \text{int. rot.}\begin{cases}2\text{-H}\\2\text{-heavy}\end{cases}\\ 21\ \text{def.}\begin{cases}14\text{-H}\\7\text{-heavy}\end{cases}\end{cases}$ |

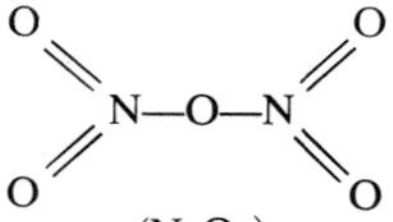

(N_2O_5) (structure: O=N–O–N=O with two O)

| 7 | 15 | $6\{6\ N\!-\!O\}$ | $9\begin{cases}2\ \text{int. rot. (all heavy)}\\ 7\ \text{def. (all heavy)}\end{cases}$ |

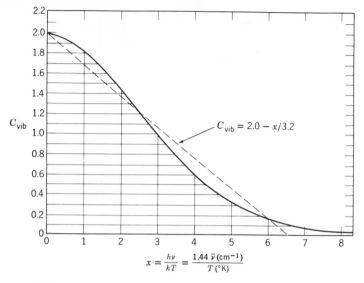

$$x = \frac{h\nu}{kT} = \frac{1.44 \; \tilde{\nu} \, (\text{cm}^{-1})}{T \, (^\circ\text{K})}$$

Figure 2.4   Variation of C_{vib} with frequency and temperature.

selected molecules. In Table A.13 (Appendix) are shown some characteristic bending and stretching frequencies, which are probably accurate to ±10%. Table A.15 in the Appendix lists the contributions of a typical harmonic oscillator to C_p° at different temperature. Figure 2.4 shows the same data in graphic form. The values are tabulated in terms of the dimensionless ratio $x = (ch\tilde{\nu}/kT) = (1.44\tilde{\nu}/T)$ with $\tilde{\nu}$ in cm^{-1}, T in °K, k the Boltzmann constant, h Planck's constant, and c the velocity of light.

2.12   VIBRATIONAL CONTRIBUTIONS TO C_p°

An internal harmonic mode of oscillation of a molecule with frequency $\tilde{\nu}$ (cm^{-1}) contributes to vibrational heat capacity, C_{vib} with

$$\frac{C_{\text{vib}}}{R} = \frac{x^2 e^x}{(e^x - 1)^2} \tag{2.48}$$

where $x = (h\tilde{\nu}c/kT)$. To orient ourselves let us note that (C_{vib}/R) goes monotonically from zero at $x \gg 1$ to 1 at $x = 0$. Some typical values taken from Figure 2.4 are 0.06 for $x = 6$, 0.30 for $x = 4$, 0.71 for $x = 2$, and 0.90 for $x = 1.0$. At 300°K the quantity (kT/hc) is approximately 210 cm^{-1} so that $x \sim 15$ for H stretching modes ($\tilde{\nu} \sim 3000$ cm^{-1}), and $x \sim 6$ for H

deformations ($\tilde{\nu} \sim 1300 \, \text{cm}^{-1}$). We see that both types of vibration will contribute very little to C_p^o at 300°K. An error of $\pm 10\%$ in $\tilde{\nu}$ will contribute an error of about ± 0.04 gibbs/mole to C_p^o when x is in the range $1 \leq x \leq 4$, although outside these limits C_p^o is much less sensitive to x.

At 300°K only frequencies in the neighborhood of 200–1000 cm^{-1} are in a region where appreciable errors in C_p^o arise from even a $\pm 20\%$ error in $\tilde{\nu}$. At 600°K only $\tilde{\nu}$ values of 400–2000 cm^{-1} have this sensitivity. Above 1500°K almost all except H stretching frequencies are nearly fully excited and contribute close to their maximum of R to C_p^o. Note in Figure 2.4 that the linear relation for $C_{vib}(x)$ incurs a maximum error of 0.1 cal/mole-°K over the entire range.

Hindered internal rotations have more complex contributions C_{ir} to C_p^o. They vary from zero at very low temperatures where V_0, the height of the energy barrier to rotation, is very large compared to RT and reach a maximum contribution, which can approach or slightly exceed R, for very heavy rotors with large moments of inertia. This occurs at intermediate, or even low, temperatures. C_{ir} then decreases to $R/2$ for a free internal rotation at very high T. Molecules with practically free internal rotation are nitromethane (CH_3NO_2) and toluene (C_6H_5—CH_3). Table A.16 gives the contribution to C_p^o per rotor as a function of (V_0/RT) and Q_f, the partition function for the free rotor (2.36):

$$Q_f = \frac{\pi^{1/2}}{\sigma_i} \left(\frac{8\pi^2 I_r' kT}{h^2} \right)^{1/2}$$

$$= \frac{2.8}{\sigma_i} \times 10^{19} (I_r' T)^{1/2} = \frac{3.6}{\sigma} \left[I_r \left(\frac{T}{100} \right) \right]^{1/2} \tag{2.49}$$

where I_r' is the reduced moment of inertia in g-cm^2, σ_i is its internal symmetry number, and I_r has units of amu-Å^2.

For symmetrical coaxial rotors, as in ethane, I_r can be calculated directly from the formula

$$\frac{1}{I_r} = \frac{1}{I_1} + \frac{1}{I_2} \tag{2.50}$$

where I_1 and I_2 are the moments of inertia of each top about its symmetry axis. In C_2H_6 where $I_1 = I_2$

$$I_r' = \frac{I_1}{2} = \tfrac{3}{2} m_H r_{C-H}^2 \sin^2 (180 - \theta) = 2.66 \times 10^{-40} \, \text{g cm}^2/\text{molecule}$$

where $\theta = 109°$ is the H—C—C angle in C_2H_6 and $r_{C-H} = 1.09$ Å.

For light tops attached to a heavy molecule, I_r can be taken as the moment of the top directly. For the unsymmetrical cases with two tops, the moment of inertia of each top may be calculated about an axis parallel to the bond joining them and passing through its own center of gravity.[8] The reduced moment I_r is then found by inserting these moments into (2.50). Where one of the tops has a much greater moment of inertia than the other, the reduced moment of interia may then be taken as equal to that of the lighter top about the actual bond. Thus in treating the molecule CI_3OCl composed of the CI_3 top (symmetrical, heavy) and the unsymmetrical, much less massive, O—Cl top, we take the CO axis as the spin axis and calculate the moment of inertia of the OCl top about the CO axis.

Current evidence suggests that the barriers to rotation about single bonds arise from the repulsive interactions of atoms attached at the ends of these bonds. In the case of C_2H_6, CH_3NH_2, and CH_3OH, it would be the repulsions of the H atoms attached to the C—C, C—N, and C—O bonds, respectively. In the case of neopentane $(CH_3)_4C$, there are no H atoms on the central C atom, and the barrier presumably arises from the repulsion of the H atoms on the terminal methyl groups.

Because repulsive interactions are exceptionally sensitive to distance, varying either exponentially or as some high algebraic power (e.g., $1/r^{12}$), there is no simple method for estimating barriers. Table A.19 lists some typical values and can be used to estimate barriers in related classes of compounds for molecules not listed. Care must be taken in approximating the near-neighbor bonded and nonbonded interactions.

In trying to visualize the internal motions of isolated molecules in space, remember that the conservation laws are strong constraints. Thus the center of gravity must remain fixed in space through all motions and angular momentum must be conserved. In consequence, light atoms in any concerted motion (vibration or torsion) will move faster than heavier atoms and will also have larger amplitudes of motion. To take an extreme example, in the internal rotation around the B—B bond in the molecule HBI—HBI, the heavy I atoms will execute small oscillating motions between their closest (*cis*) and farthest (*trans*) positions, while the BH groups will appear to concertedly rotate and twist about the I $\cdots$ I axis.

EXAMPLE

Calculate the contribution to C_p° of the hindered rotation in CH_3CHCl_2 at 300°K and 500°K. Because the $CHCl_2$ end is so heavy, we can treat this molecule as having a CH_3

---

[8] D. R. Herschbach, H. S. Johnston, K. S. Pitzer, and R. E. Powell, *J. Chem. Phys.*, **25**, 736 (1956).

group with $I_r' = 5.3 \times 10^{-40}$ g cm^2, $I_r = 3.0$ amu-Å^2. We then find from (2.49) that $Q_{500} = 4.7$; $Q_{300} = 3.6$, and with $V_0 \sim 3.5$ kcal, we see from Table A.16 that C_{p300}°(int. rot.) = 2.1 and C_{p500}°(int. rot.) = 2.1 cal/mole-°K.

EXAMPLE

Calculate $C_p^\circ(500°K)$ of CH$_3$CHCl$_2$, using average frequencies. C$_2$H$_4$Cl$_2$ has eight atoms, including four heavy atoms. It will have $3 \times 8 - 6 = 18$ frequencies, as follows:

	Average Frequency (cm^{-1})	x	C_p°(vib.) (in gibbs/mole)
7 stretches: 4C—H	3100	8.9	0.1
1C—C	1000	2.9	1.0
2C—Cl	650	1.87	3.0
11 Deformations			
1 int. rot.: ($V = 3.5$ kcal)		—	2.1
2C—C—Cl bends	400	1.15	3.6
1Cl—C—Cl bends	280	0.81	1.9
3H—C—H bends	1450	4.2	1.5
2H—C—C bends	1150	3.3	1.7
2H—C—Cl bends	1150	3.3	1.7
			17.3

The average frequencies are assigned from Table A.16 and x calculated at 500°K. The C_p°(vib) contributions are then taken from Table A.15. The total value of C_{p500}°(vib) thus calculated is 17.3. Adding 7.9 for other contributions, we find C_{p500}°(C$_2$H$_4$Cl$_2$) = 25.2 compared to 24.8 cal/mole-°K from group additivity values (also from other measurements).

The bookkeeping system we have been employing for hindered internal rotations is to use the free rotor as a starting point and then make corrections for the restriction due to the barrier. This is the simplest and most accurate method for most purposes. Frequently, however, it is useful, when doing computer programming, to simulate the hindered internal rotation by a torsion frequency. For not too high temperatures, such that $T < (V_0/R)$, the errors in such a procedure are not too great so long as there are not too many such rotors in the molecule. The frequency of such an oscillator is then given by:

$$\nu = \frac{n}{2\pi}\left(\frac{V_0}{2I_r}\right)^{1/2} \tag{2.51}$$

where n is the foldedness of the barrier, that is, the number of potential maxima achieved in one complete rotation of the rotor,[9] and V_0 is the barrier height. For the CH_3 rotor in ethane at 300°K, $V_0 \approx 3$ kcal, $n = 3$, and we find $\bar{\nu} \sim 320$ cm^{-1}. If we consult Table A.15, we note that this would have $C_p^\circ = 1.65$ gibbs/mole, whereas Table A.16 gives $C_p^\circ = 2.1$ gibbs/mole for this rotor. The discrepancy diminishes with increasing temperature. At 600°K, for example, they both give the same result. At very high temperatures, however, the frequency model leads to $C_p = 2$, whereas the free rotor $C_p \to 1$.

2.13 ELECTRONIC CONTRIBUTIONS TO C_p°

Radicals or molecules like NO, NO_2, or ClO_2 have odd numbers of electrons and in consequence may have low-lying electronic states. A few radicals, such as CH_2 and CF_2, are believed to have excited electronic states only a few kcal/mole above their ground states. Our information about such low-lying electronic states is very limited, but fortunately not too many species have such states.

Low-lying states can contribute significantly to the entropy of the species, but they contribute very little to C_p°. This can be illustrated by choosing a species with only one electronic state at an energy E_e above the ground state and a statistical weight (degeneracy) relative to the ground state g_e. Then we can show that it contributes to C_p° the amount $C_p^\circ(\text{elec})$ given by [see (2.48)]:

$$\frac{C_p^\circ(\text{elec})}{R} = \frac{g_e x^2 e^{-x}}{(1 + g_e e^{-x})^2} \tag{2.52}$$

where $x = (E_e/RT)$. $C_p^\circ(\text{elec})/R$ has a maximum at a value of x between 2 (for $g_e \ll 1$) and $\sim \ln g_e$ (for $g_e \gg 1$). At $g_e = 1$ the maximum occurs at $x = 2.3$ with $(C_p^\circ/R)_{\max} \sim 0.41$. When g_e is small, the maximum is proportionately smaller, and (C_p°/R) never becomes appreciable. However for

---

[9] If we expand the potential barrier in a Fourier series, then $V(\theta) = A + \sum_n (a_n \sin n \cdot \theta + b_n \cos n(\theta))$. It is customary to take the most stable form as our zero of energy so that at $\theta = 0$, $V(0) = 0$. If the torsion is symmetric about $\theta = 0$, then the sin terms all vanish and we can rewrite:

$$V(\theta) = \sum_n \frac{V_n}{2}(1 - \cos n\theta)$$

For the rotation of a CH_3 group, the first term to have a nonzero V_n is the third term $(V_3/2)(1 - \cos 3\theta)$, and because of the symmetry, the only terms to appear will be those for which n is a multiple of 3. For the rotation of a CH_2, as in an olefin, the first term is the V_2 term and only even terms appear (2θ, 4θ, 6θ, etc.).

When the maxima do not have the same value as in CH_2ClCH_2Cl, it is not a large error to average the various maxima and minima and approximate the barrier by a cos 3θ term.

$g_e \gg 1$, $(C_p^\circ(\text{elec})/R)_{max}$ is approximately $2.3[\log g_e]/4$ or less than 0.6 if $g_e = 10$, which would be unusually high.

Because $C_p^\circ(\text{elec})$ is zero at both very low and very high temperatures, we see that it never gets much beyond about 1 gibbs/mole at temperatures of the order of $(E_e/2R)$, and thus we will usually make what amounts to a small error in neglecting it.

2.14 SYMMETRY NUMBERS AND ISOMERS

From a statistical point of view, the molar entropy of a compound is given by $R \ln W$ where W is the number of distinguishable configurations that the compound can have. These configurations must be compatible with certain imposed constraints, such as the chosen standard molar volume and temperature, and must be suitably weighted by the Boltzmann energy distribution $e^{-E_i/RT}$, where E_i is the energy of the ith configuration.

The so-called "translational entropy" of an ideal gas represents the contribution to $R \ln W$ of the various configurations achieved by different assignments of 1 mole (N_0) of particles within the standard volume of 24.6 liters at 25°C.

Gibbs was the first to point out that any permutation of the identical molecules would lead to an indistinguishable configuration of the gas. Because, for N_0 molecules in a box, there are $N_0!$ permutations, he corrected this simple calculation of W, defined above, by dividing it by $N_0!$. This is the origin of the famous "entropy of mixing" paradox. There is no such correction for distinguishable molecules.

The rotational entropy of an ideal gas arises from the contributions to $R \ln W$ of the various configurations achieved by changing the orientation of the molecular axis of a polyatomic molecule in space. For molecules made up of nonidentical atoms, such as H—O—Cl, or N—N—O, this poses no problem. However, if the molecule contains identical atoms, we must correct for those indistinguishable configurations, which merely correspond to having permuted the positions of identical atoms. In principle this correction could be as large as $(n_a!n_b! \cdots)$, where n_a, n_b, and so on are the number of indistinguishable atoms of type a, b, and so on in the molecule. In practice this is hardly ever the case because the rigidity of the molecule rules out certain permutations.

For homonuclear diatomic molecules such as H_2 and O_2 we have two identical atoms that can be permuted by a simple 180° rotation. Hence we get the full correction of 2!. In a molecule such as O_3, only the two end O atoms are identical, so the correction is again $2! = 2$. However, in the hypothetical cyclic molecule O—O—O, where all O atoms are identical, we would have the full correction of $3! = 6$. In CH_4 with four identical H atoms, we might expect a maximum correction of $4! = 24$. However not all permutations of the four atoms can be arrived at by simple rotations. In particular, half of the permutations that would correspond to an inversion of the molecule through the central C atom are not possible and so the correction is $4!/2 = 12$. A flabby CH_4 molecule that could invert rapidly would require the full correction.

The correction to the rotational entropy due to these features of indistinguishable atoms is spoken of as a symmetry correction and is made by subtracting $R \ln \sigma$ from the rotational entropy. σ is the symmetry number and is defined as the total number of *independent* permutations of identical atoms (or groups) in a molecule that can be arrived at by simple rigid rotations of the entire molecule. Some typical symmetry numbers are: (a) $H_2O(2)$; (b) $SO_2(2)$; (c) SO_3 planar (6); (d) NH_3 nonplanar (3); (e) $CH_4(12)$; (f) $SF_6(24)$; (g) benzene, $C_6H_6(12)$; (h) cyclopropane, $C_3H_6(6)$; (i) cyclobutane, nonplanar (4); (j) $C_2H_4(4)$; (k) $H_2O_2(2)$; (l) cyclobutadiene, planar (8).

A simple method of estimating σ is to multiply the symmetries of each of the independent symmetry axes. Thus CH_4 $\sigma = 12$ can be looked upon as having four independent threefold symmetry axes. Each one goes through a different C—H axis and involves rotation of the remaining CH_3 group about this axis. Hence $\sigma = 4 \times 3 = 12$. SF_6 has six fourfold axes. Hence $\sigma = 6 \times 4 = 24$. The trianglular bi-pyramid, PF_5, with two apical and three equatorial F atoms, has one threefold and one twofold axis.[10] Hence $\sigma = 3 \times 2 = 6$. Benzene has six twofold axes. Hence $\sigma = 6 \times 2 = 12$.

For molecules with hindered internal rotation, such as ethane, there is a contribution to the external rotation of six due a twofold axis and a threefold symmetry axis. In addition, the internal rotation of the CH_3 group has a threefold axis. We shall combine these and represent the overall $\sigma = \sigma_{ext} \times \sigma_{int} = 18$. On the same basis all the normal alkanes have $\sigma = 18$, due to an external twofold axis, and $\sigma = 9$ for the two terminal CH_3 groups. For neopentane, C_5H_{12}, $\sigma_{ext} = 12$, as does in CH_4, whereas $\sigma_{int} = 3^4$ for the four methyl groups. Hence $\sigma = 12 \times 3^4 = 972$, an extremely large correction.

For readers who are familiar with point groups and their symmetry designations, Table 2.3 summarizes the symmetry numbers of molecules belonging to the different groups.

A different type of correction accompanies isomeric species. Let us suppose that we have two isomeric forms of a molecule that have the same ΔH_f° but are physically distinguishable. Thus they can be optical isomers, rotational conformers, and so on. If we have an equilibrium

---

[10] Alternatively, we can look upon it as having three independent twofold axes.

Table 2.3.   Symmetry Numbers for Various Point Groups[a]

Point Group	σ	Point Group	σ	Point Group	σ
C_1, C_i, C_s	1	$D_2, D_{2d}, D_{2h} \equiv V$	4	$C_{\infty v}$	1
C_2, C_{2v}, C_{2h}	2	D_3, D_{3d}, D_{3h}	6	$D_{\infty h}$	2
C_3, C_{3v}, C_{2h}	3	D_4, D_{4h}, D_{4h}	8	T, T_d	12
C_4, C_{4v}, C_{4h}	4	D_6, D_{6d}, D_{6h}	12	O_h	24
C_6, C_{6v}, C_{6h}	6	S_6	3		

[a] Taken from G. Herzberg, *Molecular Spectra and Molecular Structure*, Van Nostrand, Princeton, N.J., 1945. C_j indicates a j-fold axis of symmetry. D_j denotes a molecule of the class C_j with a j-fold axis and J twofold axes at right angles to the C_j axis and at equal angles to each other. T denotes tetrahedral symmetry, and O, octahedral symmetry.

mixture of such species, we must add a term to $S°$ of $R \ln 2$ gibbs/mole to represent the entropy of mixing of these species. If there are n such species, *all of the same energy content*, the correction is $R \ln n$ given by the formula for 1 mole of mixture:

$$\Delta S_{\text{mixing}} = -R \sum_i n_i \ln (n_i)$$

where n_i is the equilibrium mole fraction of the ith form. When they all have the same energy so that $n_i = 1/n$, this becomes the familiar $R \ln n$.

When we calculate entropies for such species as *sec*-butyl iodide, we shall add $R \ln 2$ to the estimate from the group tables to correct for such optical isomerism. Similarly, species such as ROOH and H_2O_2, in which the O—H and O—R bonds are nearly at right angles, exist in right- and left-hand forms and have a higher entropy by $(R \ln 2)$. This is also true of 1,3-disubstituted allenes.

EXAMPLE

Estimate the entropy of dimethyl peroxide from groups

$$S° = 2S°[C—(O)(H)_3] + 2S°[O—(C)(O)] - R \ln \sigma + R \ln 2$$
$$= 60.82 + 18.8 - 5.8 + 1.4 = 75.2$$

The symmetry number corrections that we have been making correspond to a high temperature, or alternatively, to the classical limits in statistical mechanics. As we approach $0°K$, all partition functions approach unity, including the functions of translation, rotation, and internal rotation. Under these conditions we would have to use the actual quantum-mechanical partition functions rather than our "high" temperature approximations with their symmetry corrections. Fortunately, even

$200°K$ is sufficiently high for our approximation to hold, and so long as we stay above those temperatures, we commit no serious error. For translation and rotation (except for H_2 and D_2) even $20°K$ is a "high" temperature.

One last point regarding our use of symmetry numbers requires further discussion and this had to do with the effect of vibrational motion on symmetry. We assign the NH_3 molecule a value of $\sigma = 3$. However NH_3 undergoes an internal inversion at a rate of about $10^{10}\,s^{-1}$ and one may ask if this doesn't justify assigning $\sigma = 6$, since the inversion makes possible all six permutations of the three identical H atoms. Our answer depends on comparing the frequency of this motion with that of the rotational motion to which our σ is a correction. At $300°K$ the rotational motion of NH_3 about its symmetry axis (energy of $\frac{1}{2}RT$) is about 10^{12} cps. This is so much faster than the inversion that we are justified in treating NH_3 as pyramidal. If, however, the inversion had a frequency close to the rotational frequency, then we would find it justifiable to treat the molecule as being dynamically flat with $\sigma = 6$. The result of so rapid an inversion, however, would be for there to be a very low-frequency vibration corresponding to the "umbrella" motion with a consequently high contribution to the vibrational entropy. This is the situation that exists in the case of the $\dot{C}H_3$ radical which we treat as dynamically flat with $\sigma = 6$. Yet it has an unusually low "umbrella" frequency of about $560\,cm^{-1}$. This can be compared to the much higher "umbrella" frequency of NH_3 of about $900\,cm^{-1}$.

A similar situation exists for cyclopentane which is nonplanar and has $\sigma = 2$ and two optical isomers. However, the puckering motion of the ring is so fast that we can treat the molecule as dynamically flat ($\sigma = 10$, no optical isomers). The abnormally low-frequency puckering mode of about $70\,cm^{-1}$ makes a contribution to the vibrational entropy which just about compensates for the lowering contributed by $\sigma = 10$. This puckering is called a "pseudo-rotation."

2.15   ESTIMATIONS OF C_{pT}° FROM MODEL COMPOUNDS

We have noted that it is possible to estimate C_{pT}° from structural considerations. However the calculation is time consuming, and more often we can do just as well by making comparisons with known compounds or compounds for which group values are available. The same is true for ΔH_{f298}° and for S_{298}°. Let us consider some examples.

Suppose we wish to estimate S_{298}° of a compound such as $CH_3CH{=}NH$ for which no data exist. We start by taking the simplest molecule having a similar mass and structure with known properties. In this case, this would be $CH_3CH{=}CH_2$ for which $S_{298}^{\circ} = 63.8$ gibbs/mole.

We now consider the corrections we need to apply. These can be classified as follows (see p. 37): (a) translational (depends on total mass), (b) rotational (depends on moments of inertia), (c) vibrational (depends on frequencies), (d) symmetry including optical centers, and (e) internal rotations.

$S^\circ_{(\text{trans})}$ varies as $\frac{3}{2}R \ln M$ where M is the molecular weight. If we change M to M', $S^\circ_{(\text{trans})}$ changes by an amount $\frac{3}{2}R \ln(M'/M)$. If $M'/M = 1.10$, this becomes an increase of $\sim 0.15R = 0.3$ gibbs/mole. In the example above $(M'/M) = 43/42$ and $\Delta S^\circ_{(\text{trans})} \cong 0.08$ gibbs/mole, which is negligible.

$S^\circ_{(\text{rot})}$ varies as $\frac{1}{2}R \ln(I_A I_B I_C)$, where I_A, I_B, and I_C are the three principal moments of inertia of the molecule. $\Delta S^\circ_{(\text{rot})}$ becomes $\frac{1}{2}R \ln(I'_A I'_B I'_C/I_A I_B I_C)$. Each I varies as $\Sigma m_i r_i^2$ where m is the mass of the heavier atoms and r_i the Cartesian projection (xy, yz, zx) of their distances from the center of gravity. Unless we have grossly changed m_i or r_i, in going from our model compound, $\Delta S^\circ_{(\text{rot})}$ will be very small. Note that a doubling of one principal moment of inertia increases $S^\circ_{(\text{rot})}$ by $\frac{1}{2}R \ln 2$ or only 0.7 gibbs/mole. In the case chosen, the change in $S_{(\text{rot})}$ is negligible.

In changing from a $=CH_2$ group to $=NH$, we have lost one atom and hence three internal vibrations. These are a C—H stretch and two C—C—H bends. From Table A.13 we estimate these at 3100 and 1100 cm^{-1}, respectively. At 298°, $x = (ch\tilde{\nu}/kT)$ becomes 14.5 and 5.3, and from Table A.17, we estimate a loss of 0.06 gibbs/mole from the loss of the H atom.

In the case above, there is no change in symmetry or in internal rotations, so that the estimated value of S°_{298} (CH$_3$—CH=NH) is 63.8 gibbs/mole with an estimated uncertainty of not more than ±0.5.

By the method described above it is not surprising that S°_{298} (CH$_3$—CO—CH$_3$) is 70.5, and that for structurally similar isobutene it is 70.6.

We can make a further comparison of S°_{298} (CH$_3$CHO) = 63.2 and S°_{298} (CH$_3$CH=CH$_2$) = 63.8 gibbs/mole. From comparing S°_{298} (CH$_3$CH$_2$OH) = 67.3 with S°_{298} (CH$_3$CH$_2$CH$_3$) = 64.6, we note a symmetry difference of $R \ln(18/3) = 3.6$, favoring the alcohol and giving a difference of only 0.9 gibbs/mole in the symmetry corrected (intrinsic) entropies.[11] Once again, there are no other major corrections. Small corrections for the change in hindered rotation barriers and number of hydrogen atoms approximately cancel.

In Table 2.4 we list values of C_p° and S° for structurally similar molecules over the range 300°K to 1000°K. They are arranged in various groupings to point out the effect of adding H atoms to a heavy skeleton or

---

[11] $S^\circ_{(\text{int})} = S^\circ_{(\text{obs})} + R \ln(\sigma/n)$.

Table 2.4   Some Examples of Structural Similarities in C_p^o and S^o

		$C_p^o (C_{vib})^c$			$S^{o\,d}$		
Molecule (σ total)		300°K	600°K	1000°K	300°K	600°K	1000°K
A. N≡N	(2)	7.0 (0.1)	7.2 (0.3)	7.8 (0.9)	45.8	50.7	54.5
C≡O	(1)	7.0 (0.1)	7.3 (0.4)	7.9 (1.0)	47.3	52.2	56.0
H—C≡N	(1)	8.6 (1.7)	10.6 (3.7)	12.3 (5.3)	48.3	54.9	60.7
H—C≡C—H	(2)	10.6 (3.7)	13.9 (7.0)	16.3 (9.4)	48.0	56.6	64.3
B. O═C═O	(2)	8.9 (2.0)	11.3 (4.4)	13.0 (6.1)	51.1	58.1	64.3
O═N≡N	(1)	9.3 (2.4)	11.6 (4.7)	13.1 (6.2)	52.6	59.8	66.1
C. H$_2$C═O	(2)	8.5 (0.6)	11.5 (3.6)	14.8 (6.9)	52.3	59.1	65.8
H$_2$C═CH$_2$	(4)	10.3 (2.4)	16.9 (9.0)	22.4 (14.5)	52.4	61.7	71.8
H$_3$C—C≡CH	(3)	14.6 (6.7)	21.8 (13.9)	27.7 (19.8)	59.4	71.9	84.5
HO—OH	(2)^a	10.3 (2.4)	13.3 (5.4)	15.0 (7.1)	55.7	63.9	71.2
D. O—N═O	(2)^b	8.9 (1.0)	11.0 (3.1)	12.5 (4.6)	57.4	64.2	70.2
O$_3$	(2)	9.4 (1.5)	11.9 (4.0)	13.2 (5.3)	57.1	64.5	71.0
F—O—F	(2)	10.4 (2.5)	12.5 (4.6)	13.3 (5.4)	59.2	67.2	73.8
F—CH$_2$—F	(2)	10.3 (2.4)	15.7 (7.8)	20.0 (12.1)	59.0	67.9	77.0
E. O—Cl—O	(2)^b	10.9 (3.0)	12.7 (4.8)	13.4 (5.5)	61.5	71.9	78.7
O═S—O	(2)	9.5 (1.6)	11.7 (3.8)	13.0 (5.1)	59.4	66.7	73.0
CH$_3$—S—CH$_3$	(18)	17.8 (9.9)	27.0 (19.1)	35.2 (27.3)	63.3	78.7	94.6

^{a} H$_2$O$_2$ exists in a skew form with optically active isomers. Hence S^o includes a term $R \ln 2$ due to entropy of mixing.

^{b} S^o includes a term $R \ln 2$ due to electron spin.

^{c} Values in parentheses are $C_p^o - C_{p(trans)}^o - C_{p(rot)}^o - R$.

^{d} These are absolute entropies. To obtain intrinsic entropies, add $R \ln (\sigma/n_i g_e)$ where g_e is the electronic degeneracy.

of adding internal rotations. Because translational and rotational contributions to C_p^o above 300°K are constant, they have been subtracted and the values in parenthesis give the residues due to internal degrees of freedom.

In Group A we note that C_p^o (298°K) increases by about 2 eu per H atom added for the linear molecules, which is about the increase obtained on extending the linear molecule by one heavy atom (Group B). In contrast to this we note C_p^o only gains 1 eu per H atom in going from CH$_2$O to C$_2$H$_4$. Comparing C_{p298}^o of CO$_2$ and CH$_3$CCH, we note an increase in the internal contributions of 1.2 eu per H atom. We find about the same increment in comparing ClO$_2$ with (CH$_3$)$_2$S.

In comparing entropies, we should correct S^o for symmetry, electronic spin, and optical isomers. We do this by adding to the listed S^o, $R \ln \sigma$, subtracting $R \ln g_e$ and subtracting $R \ln n_i$. The symmetry number σ includes, for convenience, internal symmetry arising from hindered rotors, as in $(CH_3)_2S$. The electronic degeneracy g_e is usually equal to $2j+1$, where j is the total electronic angular momentum. For polyatomic molecules, it is generally taken as $2S+1$, where S is the total spin. The quantity n_i is the total number of energetically equivalent optical isomers. We call these corrected values the intrinsic entropies or S_{int}°.

For CH_2O at $298°K$, $S_{int}^{\circ} = 53.7$ and for C_2H_4, $S_{int}^{\circ} = 55.2$. For H_2O_2, $S_{int}^{\circ} = 55.7$, and for NO_2, $S_{int}^{\circ} = 57.4$. For O_3, $S_{int}^{\circ} = 58.5$, and for CH_2F_2, $S_{int}^{\circ} = 60.4$. Comparing $S_{int}^{\circ}((CH_3)_2S)$ of 74.2 with $S_{int}^{\circ}(ClO_2) = 63.7$, we see that the internal rotations and vibrations associated with the CH_3 groups contribute about 5.3 per CH_3. Most of this arises directly from the internal rotations. From Table A.18 we can obtain 5.8 per CH_3 group at $300°K$ (for zero barrier).

2.16 FREE RADICALS

Except for a few stable free radicals such as NO, NO_2, ClO_2, and NF_2 there are very few direct thermochemical data on free radicals. The best values for heats of formation are almost invariably deduced from kinetic measurements of bond-dissociation energies, whereas entropies and heat capacities are only calculable from statistical formulas. Because of this, there are large gaps in our knowledge of the thermochemical data on free radicals. Heats of formation usually are uncertain to about ± 1 kcal/mole in the best cases and can be considerably worse. However there is every indication that the laws of group additivity apply to radicals, so that we can deduce ΔH_f° for all of the paraffin radicals from our knowledge of ΔH_f° for $CH_3\dot{C}H_2$; $(CH_3)_2\dot{C}H$ and $(CH_3)_3\dot{C}.$[12]

The estimation of the entropies of radicals can be deduced by group additivities from the properties of the simple radicals, but these latter must be calculated from assumed structures and frequency assignments. Although there is good ground for believing that substituted methyl radicals are nearly planar, and consequently have a higher symmetry number than a nonplanar radical, this is still a gross uncertainty in our

---

[12] This is not strictly true because there is no stable compound corresponding to the hypothetical formula $\dot{C}H_2$—$\dot{C}H_2$ if, by analogy with the substituted X—CH_2—CH_2—X, we treat the odd electron as though it were the same as a substituents X. A true group treatment would require values for all of the groups $\dot{C}$—$(C)(H)_2$, $\dot{C}$—$(C)_2(H)$, $\dot{C}$—$(C)_3$, C—$(\dot{C})(H)_3$, C—$(\dot{C})(C)(H)_2$, C—$(\dot{C})(C)_2(H)$, and C—$(\dot{C})(C)_3$. However the assumption that groups such as C—$(\dot{C})(C)_2(H)$ have the same partial heat of formation as the corresponding nonradical, paraffin groups C—$(C)_3(H)$ seems to be valid within the limits of present data, and so we shall employ it.

Table 2.5

Species	CH_3CH_3	CH_3NH_2	CH_3OH	$CH_2{=}CH_2$	$CH_3\dot{C}H_2$	CH_3F
σ, spin	18, 0	3, 0	3, 0	4, 0	6, $\frac{1}{2}$	3, 0
S°_{298}	54.9	58.1	57.3	52.5	58.1	53.3
S°_{298} (intrinsic)	60.6	60.3	59.5	55.3	60.3	55.5

estimates. In similar fashion, we usually assume only spin degeneracy and assign an entropy of $R \ln 2$ to odd electron radicals, such as $\dot{C}H_3$, $CH_3\dot{C}O$, and $\dot{C}H_2{-}CH{=}CH_2$. The existence of low-lying electronic states is always a source of uncertainty in our estimates of S° and C°_p. Still larger uncertainties are related to the assignment of barriers to internal rotation in complex radicals such as t-butyl, $(CH_3)_3\dot{C}$, and allyl, $\dot{C}H_2{-}CH{=}CH_2$.

With the warnings we give above in mind, we can assign values to C°_p and S° for radicals by analogy with related compounds. Thus we will assume that $\dot{C}H_3$ radical has the same value of C°_{p298} as CH_4 and NH_3, both of which are 8.5. The expected uncertainty here can only arise from low-lying electronic states and is clearly small. The value of S°_{298} ($\dot{C}H_3$), using spin $= \frac{1}{2}$, $g_e = 2$ and $\sigma = 6$, can be assigned as 46.0 by analogy with S°_{298} (NH_3) $= 46.0$, $\sigma_{NH_3} = 3$. Further comparisons can be made with S°_{298} (CH_4) $= 44.5$, $\sigma_{CH_4} = 12$; S°_{298} (H_2O) $= 45.1$, $\sigma_{H_2O} = 2$. A lower limit of 45.7 can be fixed by correcting the translational entropy of NH_3 from $M = 17$ to $M = 15$.

For the more complex $CH_3\dot{C}H_2$ radical ($\sigma = 6$), the comparison we give in Table 2.5 can be made. It is unlikely that the assigned S° intrinsic ($\dot{C}_2H_5$) is in error by more than ± 1.0. (Note: We have assumed the same barrier to rotation as in CH_3NH_2.)

Conjugated radicals offer more of a problem because they have stiffer structures. Thus the conjugated allyl radical $\dot{C}H_2{-}CH{=}CH_2$ is estimated to have a stabilization energy due to conjugation of about 11 kcal/mole. This would represent an extra barrier to rotation of the $\dot{C}H_2$ group, which can be compared to a barrier of about 2 kcal/mole for the CH_3 group in the model compound $CH_3{-}CH{=}CH_2$. Again the analogies with similar compounds given in Table 2.6 provide a quick basis for assigning a value of about 3 to this loss in free CH_3 rotation.

Comparing the intrinsic entropies of C_3H_6 and C_3H_8, we note a difference of 4.2, whereas CH_3OCH_3 is only 0.9 less than C_3H_8. Correcting for the mass difference, we can assign about 0.2 (Table A.17) to the vibrational modes due to two H atoms, and the remaining 4.0 to the internal rotation. Applying this to the allyl radical, S° (allyl) $= 63.0$, after we subtract an additional 0.1 for the loss of one H atom and add 1.0 for

Table 2.6

Species	$CH_3CH_2CH_3$	CH_3NHCH_3	CH_3OCH_3	CH_3CH_2F
σ, spin	18, 0	9, 0	18, 0	3, 0
S^o_{298}	64.6	65.3	63.7	63.3
S^o_{298} (intrinsic)	70.3	69.7	69.4	65.5

Species	$CH_3\!\!-\!\!CH\!\!=\!\!CH_2$	$\dot{C}H_2\!\!-\!\!CH\!\!=\!\!CH_2$	$CH_2\!\!=\!\!C\!\!=\!\!CH_2$	CH_3CH_2F
σ, spin	3, 0	2, $\frac{1}{2}$	4, 0	3, 0
S^o_{298}	63.9	63.0 (63.2)^a	58.3	63.3
S^o_{298} (intrinsic)	66.1	63.0 (63.2)^a	61.1	65.5

^{a} Calculated from Table 2.14 (radical additivities).

an 11 kcal barrier to internal rotation of one conjugated $\dot{C}H_2$ group (Table A.20).

Where structural analogs are not readily available, values of S^o and C^o_p for radicals $R\cdot$ can be estimated from their molecular hydrogenated parents RH. We do this by making corrections for the various degrees of freedom lost or altered significantly when we remove the H atom.

The ethyl radical provides a useful example. Starting with the parent C_2H_6, for which S^o and C^o_p are available over a broad range, we can arrive at the corrections at 300°K given in Table 2.7. These give $S^o_{p300}(C_2H_5\cdot) = S^o_{300}(C_6H_6) + 3.5 = 58.4$ and $C^o_{p300}(C_2H_5\cdot) = 12.5$.

Vibration assignments made above are taken from Table A.13, and the corresponding S^o and C^o_p from Tables A.17 and A.15, respectively. The

Table 2.7.   Estimates of $S^o_{300}(\dot{C}_2H_5)$ and C_{p300} by differences from C_2H_6

	ΔS^o	ΔC^o_p
Translation	Negligible (-0.1)	None
Rotation^a	Negligible (-0.27)	None
Vibration		
One C—H stretch (3100 cm^{-1})	0	0
Two H—C—H bends (1450 cm^{-1})	0	-0.2
Hindered rotation: reduced moment	-0.25	
Barrier change from	$+0.5$	0
3 kcal $\rightarrow$ 2 kcal		
Symmetry and spin	$+3.6$	None
Totals	$+3.5$	-0.2

^{a} If the C—C$\cdot$ bond is 1.45 Å rather than 1.54 Å, there is an additional correction of about -0.3 due to lower moment.

barrier corrections to $S°$ and $C_p°$ are taken from Tables A.20 and A.16, respectively.

Most of the $C_p°$ and $S°$ values for the radicals listed in Table A.12 have been arrived at in the manner described above, as have the data on biradicals we use in Chapter 3. Note that $S°_{300}(C_2H_5)$, derived in this manner, is somewhat higher than the value assigned on the basis of structural analogy. A value halfway in between is probably the best.

2.17 VIBRATIONAL CONTRIBUTIONS TO ENTROPY

Vibrational contributions to $S°$ are very easily estimated from the methods of statistical mechanics, and they are given by the following formula for a simple harmonic oscillator with frequency $\bar{\nu}$ (cm^{-1}) (p. 36):

$$\frac{S°}{R} = \ln Q_v + \frac{\partial \ln Q_v}{\partial \ln T}$$

where Q_v is the vibrational partition function (omitting zero-point energy);

$$Q_v = (1 - e^{-x})^{-1}; \qquad x = \frac{ch\bar{\nu}}{kT}.$$

Hence, on substitution,

$$\frac{S°}{R} = -\ln(1 - e^{-x}) + \frac{x}{e^x - 1}. \qquad (2.53)$$

Values of $(S°/R)$ are shown plotted in Figure 2.5 as a function of x, and, for easier use, $S°$ is tabulated in Table A.17 for different frequencies and temperatures. In the limit that $x \leq 1$, these equations reduce to

$$Q_v = \frac{1}{x}\left(1 + \frac{x}{2} + \frac{x^2}{12} + \right) \approx \frac{1}{2} + \frac{1}{x}; \qquad (2.54)$$

$$\frac{S°}{R} = -\ln\left[x\left(1 - \frac{x}{2}\right)\right] + \left(1 + \frac{x}{2} + \frac{x^2}{6} + \right)^{-1}$$

$$= 1 - \ln x + \frac{x}{2} + \frac{5x^2}{24} + \frac{x^3}{24}$$

$$\approx 1 - \ln x + \frac{x}{2}. \qquad (2.55)$$

Even when $x = 1$, the error in the approximate formulas given [(2.54) and (2.55)] amount to an error of less than 5% in Q_v and 0.4 in $S°$.

Tables A.17 and A.13 together make it possible to estimate vibrational

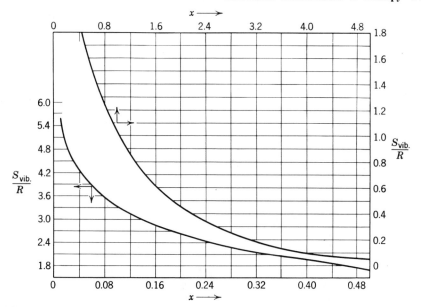

Figure 2.5 Vibrational entropy of harmonic oscillators as a function of reduced frequency.

entropies for molecules or radicals over the temperature range 300–1500°K with great facility because interpolations can easily be made from them.

EXAMPLE

Calculate the absolute vibrational entropy of CH_3Br at 750°K. With five atoms, we have $15 - 6 = 9$ internal frequencies, of which four are stretches and five are deformations. We assign them as listed in Table 2.8 from Table A.13. S_v° can be compared with the more precise estimate from known frequencies of 5.3 gibbs/mole.

Table 2.8

	Frequency	$S_v^\circ(750°K)^a$
Three C—H stretches	3100	0.2
One C—Br stretch	560	2.0
Three H—C—H bend	1450	1.5
Two H—C—Br bends[b]	1150	1.5
Total S_v°		5.2

[a] From Table A.17.
[b] These are actually CH_3 rocking modes.

When internal rotations are involved, we use (2.49) to calculate Q, the partition function for a free rotor and then, using the rules suggested in Section 2.12, assign a barrier to rotation. From these assignments we use Table A.18 to find the free rotor entropy and then Table A.19 to assign corrections.

EXAMPLE

Estimate the entropy contribution due to the hindered rotation in C_2H_6 at 800°K. From Table A.18, we see that the entropy of a free CH_3 rotating against a heavy group is 6.8 gibbs/mole at 800°K. However in C_2H_6 the reduced moment of inertia I_r is just one half this value, and because $S_f^\circ \propto \frac{1}{2}R \ln I_r$, this introduces a correction of $-\frac{1}{2}R \ln 2 = -0.7$ gibbs/mole for a value of 6.1 gibbs/mole. The barrier $V = 3.0$ kcal, and $V/RT = 1.8$. From Table A.18, Q_f, the partition function at 800°K, becomes

$$Q_f(800) = \frac{1}{\sigma} \frac{Q_f(600°K)}{(2)^{1/2}} \times \left(\frac{800}{600}\right)^{1/2} = \frac{1}{3} \times \frac{15.6}{1.41} \times 1.16 = 4.2.$$

The factor of $2^{1/2}$ corrects for the reduced moment. From Table A.20, we now find the correction of $S_f^\circ = -0.3$, so that the final value is 5.8. Note that, although we have used the symmetry to make the correction to S_f°, the value of 5.8 does not contain it because of our usual method of making internal and external symmetry corrections at the same time. If we were to make it now, the value would become 3.6. For comparison the frequency method $(320\,\text{cm}^{-1})$ gives 3.2 (Table A.17).

EXAMPLE

Estimate the entropy of hindered rotation in 1,2-dibromoethane at 500°K. This is an unsymmetrical top with two different rotomeric forms, *gauche* and *trans*, each of which would have different moments of inertia and different barriers.

The stable form is the *trans* form shown in Figure 2.6. The crosses indicate approximately the center of gravity of each CH_2Br group, and the dashed line through them, the instantaneous torsion axis. The C—H bonds symmetrically located at angles of ~55° to the plane are shown in projection on the plane. By simple trigonometry, we estimate the C—Br bond forms an angle of ~20° to the spin axis SS (using 110° for the "tetrahedral" angle C—C—Br). We also estimate the center of gravity of each group is about 0.29 Å from the Br and approximately along the C—Br bond. Then:

$$I_{CH_2Br} \approx 81(r_{Br-s})^2 + 12(r_{C-s})^2 + 2(r_{H-s})^2$$
$$\approx 49.0(\text{amu-Å}^2)$$

and by symmetry, $I_r = \frac{1}{2}I_{CH_2Br} = 24.5$. Using (2.49) with $\sigma_i = 3^*$, we find $Q_f(500°K) = 13.3$, and from Table A.19 we estimate a barrier of 3.5 kcal/mole (EtBr). From (2.49) $(\sigma_i = 3)^*$, we estimate $S_{f(\text{int})}^\circ(500°K) = 6.1$. Finally, from Table A.20 we estimate a correction of about 0.9 for the barrier of 3.5 kcal/mole for a net contribution of $S^\circ(500°K) = 5.2$.

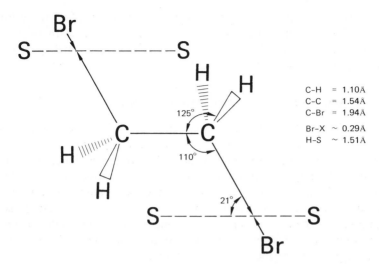

Br
S
S
H
H
125°
C
C
H
110°
H
21°
S
S
Br
C-H = 1.10Å
C-C = 1.54Å
C-Br = 1.94Å
Br-X ~ 0.29Å
H-S ~ 1.51Å

Figure 2.6*a* Projection of *trans* conformation of CH_2Br—CH_2Br (1,2-dibromoethane) on the plane of heavy atoms. Rotation axis indicated by dashed line *SS*.

Br
H
Br
C
C
H
H
H

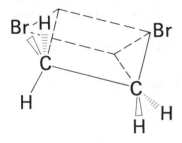

Figure 2.6*b* The *gauche* conformation of $BrCH_2$–CH_2Br showing nonplanarity of heavy atoms.

Br
X
H
H
H
H
X
Br

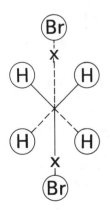

Figure 2.6*c* Newman projection of *trans* form. *X* indicates center of gravity of CH_2Br.

This use of $\sigma_i = 3$ amounts to equating the entropies of all three forms, the one *trans* form and the two *gauche* forms. Because there are two mirror-image *gauche* forms, we would calculate for their entropy the above 5.2 for the torsion and add $R \ln 2 = 1.4$ for optical isomers.

2.18 STRAIN ENERGY AND RESONANCE ENERGY

The term "strain energy" is frequently used in speaking of highly branched or cyclic compounds. Such a term can only have meaning relative to some reference state, which is arbitrarily assigned zero strain. Because unbranched hydrocarbons and their derivatives fit the additivity laws for ΔH_f° so well, they are taken as the "unstrained" standards. On this basis we can assign a strain energy E_s to a compound as the difference between its observed ΔH_f° and its value estimated from the group-additivity relations using the group values which have been, of course, derived from the unstrained standards.

Thus E_s for the alicyclic rings $(CH_2)_m$ is given by

$$E_s[(CH_2)_m] = \Delta H_f^\circ[(CH_2)_m] - m\,\Delta H_f^\circ[C-(C)_2(H)_2]. \qquad (2.56)$$

The principle described above permits us to calculate strains in cyclic olefins, aromatics, heterocyclic compounds, and polycyclic compounds. Table 2.9 illustrates these calculations.

It will be noted that the strain energy in benzene relative to butene-1 is

Table 2.9. Strain Energies in Some Compounds[a]

Compound	ΔH_f° (obs) (kcal/mole)	ΔH_f° (calc) (kcal/mole)	E_s (kcal/mole)
Cyclopropane	12.7	−14.9	27.6
Ethylene oxide	−12.5	−40.1	27.6
Cyclopentane	−18.5	−24.8	6.3
Cyclopentene	7.9	2.5	5.4
Cyclopentadiene-1,3	32.0	26.0[b] (29.6)	6.0 (2.4)
Cyclohexane	−29.5	−29.7	0.2
Bicycloheptane (2,2,1)	−12.4	−27.7	15.3
Benzene	19.8	42.3[b] (52.8)	−22.5 (−33.0)
Furan	−8.4	−2.2	−6.2

[a] Note that the nongroup interaction corrections in Table A.1 correct for these strains.

[b] These values are derived from group values based on butadiene. However, butadiene has conjugation energy of 3.6 kcal based on butene-1. The values in parentheses correct for this conjugation energy and are based on trans olefins.

negative (!), that is, -33.0 kcal/mole. This is very close to the stabilization or resonance energy of 36.0 kcal/mole of benzene computed from the heat of hydrogenation and referenced to cyclohexene. If we correct for the resonance energy, we calculate a strain energy of 3.0 kcal/mole in benzene. However, if we examine this separation in some detail, we shall find that there is no way of truly identifying strain and resonance energies separately in benzene.

We may speak of "resonance energy" or "stabilization energy" or more generally "interaction energy" as the deviation of the observed ΔH_f° of a compound from that obtained from additivity rules based on groups whose own interaction energy is arbitrarily defined as zero. Because there is no absolute energy, all such concepts, such as that of potential energy, must always invoke some standard and quite arbitrary zero.

Once again the remarkable fit of the unbranched paraffins to the group-additivity rule provides a basis for measuring interaction energy. A simple example is provided by butadiene, $CH_2\!\!=\!\!CH\!\!-\!\!CH\!\!=\!\!CH_2$, which is made up of four groups: $2[C_d\!\!-\!\!(H)_2] + 2[C_d\!\!-\!\!(C_d)(H)]$. The $C_d\!\!-\!\!(C_d)(H)$ group we find from the tables has a ΔH_f° of $+7.0$ kcal/mole. This can be compared with related group $C_d\!\!-\!\!(C)(H)$, which has $\Delta H_f^\circ = +8.8$ kcal/mole. Twice the difference between those of $2 \times 1.8 = 3.6$ kcal represents the difference in interaction of a double bond with an adjacent saturated group and with another double bond. This is frequently called a resonance energy of two conjugated double bonds. For purely empirical purposes, it is better referred to as a double-bond interaction energy.

If we compare our definitions of "strain" and "interaction" energies, we note that there is no way of separating the two. In effect the difference between an observed and a calculated ΔH_f° gives only a single result, and it is a combination of all the effects we can identify. Only if we have some independent basis for separating strain and resonance can such identification be made.

For benzene, if we consider the hypothetical molecule made of alternate single and double bonds, we can calculate a net stabilization of the real benzene of 33.0 kcal. This value would be based on unconjugated *trans* olefins, and if we corrected by 3×1 kcal/mole to allow for three *cis* olefins in the hypothetical benzene, we arrive at the frequently quoted 36 kcal. However we have no way of really setting the strain equal to zero because we have no independent reference point for the strain energy.

It is only because we have no reason to expect "resonance" interaction in cyclopentane that we assign all of the difference in ΔH_f° to strain, although again it is only because we do not expect any strain in our hypothetical alternately bonded benzene that we assign all of the difference in ΔH_f° to resonance.

Ring compounds introduce a further difficulty when we reference their "strain" energies to groups derived from open-chain compounds. If we hydrogenolyze a ring compound by breaking open one of its ring bonds and form an open-chain molecule, then in principle, we remove all of its strain energy. If we compare this heat of hydrogenolysis of a ring compound with the analogous open chain compound, the difference in heats of hydrogenolysis should equal the strain energy.

Thus if we compare the hydrogenolysis of a paraffin ring $(CH_2)_n$ with an open-chain compound, C_nH_{2n+2}:

$$\text{cyclo-}(CH_2)_n(g) + H_2(g) \overset{1}{\rightleftharpoons} CH_3(CH_2)_{n-2}CH_3(g)$$

$$CH_3(CH_2)_{n-2}CH_3(g) + H_2(g) \overset{2}{\rightleftharpoons} 2CH_3(CH_2)_{\frac{n}{2}-2}CH_3(g)$$

What we have been calling $\Delta H(\text{strain})$ is given by:

$$\Delta H_{\text{strain}} = \Delta H_2 - \Delta H_1$$

It essentially involves the conversion in each case of two $C-(C)_2(H)_2$ groups into two $C-(C)(H)_3$ groups, but in reaction (1) we lose the ring contribution to ΔH as well. However we notice another difference,[13] namely, that the two reactions involve differences in mole changes. $\Delta n_1 = -1$ and $\Delta n_2 = 0$; because of this, ΔH_1 differs from ΔH_2 by $(\Delta n_1 - \Delta n_2)$ $PV = (\Delta n_1 - \Delta n_2)RT$ (for ideal gases) $= -RT$. Hence we should subtract (at 300°K) 0.6 kcal/mole from our strain energies for saturated single rings if we are trying to make theoretical interpretations of our strain energies. For polycyclic structures we should subtract RT for each ring present. Note that we would avoid this problem if we used internal energies E instead of enthalpies H in our discussions.[14] The general utility of the latter, however, more than compensates for any such ambiguity.

There is a further difficulty associated with interpreting strain energies for ring compounds, and this has to do with thermal energy and zero-point energy. The $\Delta H_f^{\circ}(300°K)$ of a gas molecule includes contributions from thermal energy in going from 0°K to 300°K, and at 0°K it includes a contribution from the zero-point kinetic energy of the atoms. CH_2 groups in ring compounds have less thermal energy at 300°K than in open-chain compounds because of a lower net heat capacity. In compensation, however, they have a higher zero-point energy. As we shall see later, the

---

[13] B. Nelander and S. Sunner, *J. Chem. Phys.*, **44**, 2476 (1966).
[14] By definition, $H \equiv E + PV$ and for ideal gases, $H = E + RT$.

closure of an open-chain compound into a ring structure generally involves the conversion of a low-frequency torsion (hindered rotation) into a higher frequency out-of-plane bending motion. The net change in zero-point energy is about $150 \, \text{cm}^{-1}$ (i.e., $\sim 0.4 \, \text{kcal/mole}$), with a compensating change in thermal energy at 300°K, due to differences in C_p° of about $0.25 \, \text{kcal/mole}$. Depending on the stiffness of the ring, this might introduce about 0.1–$0.2 \, \text{kcal/mole}$ of net "strain" per CH_2 group in a ring structure. This is usually obscured by other effects.

Ring structures illustrate further the *"gauche"* effect we have already observed in branched alkanes. If the *gauche* interactions were not a repulsion of too close H atoms but were instead due to general *gauche-ness* of large groups, we might then expect six *gauche* interactions ($4.8 \, \text{kcal/mole}$ strain) in cyclohexane from adjacent CH_2 in the ring. However cyclohexane has a negligible strain energy of $0.2 \, \text{kcal/mole}$ (Table 2.10) and actually, $-0.4 \, \text{kcal/mole}$ if we correct by RT as noted earlier.

2.19 PI-BOND ENERGIES

There is one additional energy useful to define in discussing radicals. This is the "pi-bond energy," which is associated with multiple bonding. If we consider the very simplest example of C_2H_6, we observe that in the process of removing two H atoms to form C_2H_4, the bond-dissociation energies[15] are not equal. The first bond-dissociation energy is $98 \, \text{kcal/mole}$, corresponding to the heat of reaction at 298°K for

$$CH_3CH_3 \rightarrow CH_3\dot{C}H_2 + \dot{H} - 98 \, \text{kcal.}$$

The second bond dissociation is $38.5 \, \text{kcal/mole}$, corresponding to ΔH_f° for

$$CH_3\dot{C}H_2 \rightarrow CH_2{=}CH_2 + \dot{H} - 38.5 \, \text{kcal.}$$

If bond additivity were obeyed, we would expect that these two bond-dissociation energies would be equal. We explain the fact that they are not equal by noting that, in removing the second H atom, we have also formed a double bond. It is useful to define E_π°, the strength of this multiple or "pi bond," as the difference in bond-dissociation energies of the two H atoms.

$$E_\pi^\circ = DH^\circ(C_2H_5{-}H) - DH^\circ(C_2H_4{-}H) = 59.5 \, \text{kcal.} \qquad (2.57)$$

---

[15] We are using the conventional definition of bond-dissociation energy DH°, namely the standard enthalpy change for the reaction in which the designated bond is broken.

For unsymmetrical compounds, such as propylene, we can define the pi-bond energy as the difference in bond strengths of the same C—H bond, when the other C—H bond is intact or broken. Diagrammatically,

$$CH_3\dot{C}HCCH_3 + H$$

$$CH_3—CH_2—CH_3 \qquad\qquad CH_3CH{=}CH_2 + 2H.$$

$$CH_3CH_2—\dot{C}H_2 + H$$

By Hess's law, $DH_1 + DH_2 = DH'_1 + DH'_2$, so that $E_\pi(\text{propylene}) = DH_1 - DH'_2 = DH'_1 - DH_2$.

E_π° corresponds to the energy required to "break" the multiple bond in a double- or triple-bonded compound to form the corresponding biradical. There is no reason to expect that these biradicals correspond to metastable or observable species or to known spectroscopic states. There are no potential energy minima associated with such biradicals, which would be the criterion for stability. They are in this sense, better labeled, "hypothetical" biradicals in the same sense that a "noninteracting" butadiene whose properties we can derive from butene-1 is a hypothetical, not a real species. Symbolically, we can write pi-bond energies for the following processes:

$$E_\pi^\circ(C_2H_4) + CH_2{=}CH_2 \rightleftarrows \dot{C}H_2—\dot{C}H_2;$$
$$E_\pi^\circ(HCN) + HC{\equiv}N \rightleftarrows H\dot{C}{=}\dot{N}.$$

For olefins we expect that the pi-bond energies will correspond to the minimum activation energies required to convert cis to trans species, and in the few cases where data are available this has turned out to be true.

It should be noted that the energy required to break $CH_2{=}CH_2$ into $2\ddot{C}H_2$ radicals is precisely the sum of $E_\pi^\circ(C_2H_4) + DH^\circ(\dot{C}H_2—\dot{C}H_2)$. However, there is no independent method whereby $DH^\circ(\dot{C}H_2—\dot{C}H_2)$ can be measured, and in fact when $\Delta H_f^\circ(\ddot{C}H_2)$ is available, this relation can be used to calculate $DH^\circ(\dot{C}H_2—\dot{C}H_2)$. Using $\Delta H_f^\circ(\ddot{C}H_2) = 86$ and $\Delta H_f^\circ(C_2H_4) = 12.5$ kcal/mole, we note that $DH^\circ(CH_2{=}CH_2) = 160$ kcal/mole, so that, with $E_\pi^\circ(C_2H_4) = 59.5$, we deduce $DH^\circ(\dot{C}H_2—\dot{C}H_2) = 110$ kcal/mole. Although this seems rather high compared to $DH^\circ(CH_3—CH_3) = 88$ kcal/mole, it is in line with C—C single-bond dissociation energies connecting C atoms in sp^2 hybridization states. Thus in butadiene-1,3, $DH^\circ(C_2H_3—C_2H_3) = 111$ kcal, in biphenyl $DH^\circ(\Phi—\Phi) =$

116 kcal/mole, and in the radical $\dot{C}H_2CH_3$, $DH°(\dot{C}H_2—CH_3) =$ 94 kcal/mole.

It should be further noted that we may expect the most stable conformation $\dot{C}H_2—\dot{C}H_2$ biradical to be flat by analogy with the planar structures observed for such compounds as butadiene, acrolein biacetyl, and glyoxal, which also have (sp^2) C atoms joined by a single bond. We can estimate from the stabilization energy of butadiene that it requires 3.6 kcal to bend the two $\dot{C}H_2$ groups to the perpendicular conformation.

The fluorocarbons form an interesting contrast to the hydrocarbons. Although the relevant $\Delta H_f°$ are known with much less accuracy, we can estimate from the data in the Appendix that $DH°(C_2F_5—F) =$ 126 kcal/mole and $DH°(\dot{C}_2F_4—F) = 76$ kcal/mole making $E_\pi°(C_2F_4) =$ 50 kcal/mole. One consequence of this is that $DH°(\dot{C}F_2—\dot{C}F_2) =$ 20 kcal/mole, an extraordinarily small value when compared with 100 kcal/mole for $DH°(\dot{C}H_2—\dot{C}H_2)$ and 96 kcal/mole for $DH°(CF_3—CF_3)$. The chief cause of this small dissociation energy in C_2F_4 and the hypothetical $\dot{C}F_2—\dot{C}F_2$ is the enhanced stabilization energy of the $\ddot{C}F_2$ biradical. From the differences in bond strengths of $DH°(\dot{C}F_3—F) =$ 128 kcal/mole and $DH°(\dot{C}F_2—F) = 88$ kcal/mole, the extra stability of $\ddot{C}F_2$ is estimated at about 40 kcal/mole. A similar stabilization energy in the NF_2 radical results in a very low N—N bond-dissociation energy in N_2F_4 of about 22 kcal/mole, compared to $DH°(NH_2—NH_2) =$ 70 kcal/mole.

2.20 ENTROPIES OF RING COMPOUNDS

Although there are a reasonable number of enthalpy data on ring compounds, entropy data are, on the whole, sparse. Thus we shall be forced, in many instances, to estimate entropies and molar heat capacities for ring compounds from either frequency assignments or model compounds. This is particularly true for the case of polycyclic ring systems where almost no $C_p°$ or $S°$ data exist.

As a starting point, let us compare some simple ring systems with the corresponding open chain compounds. In Table 2.10, we compare the intrinsic entropies $S_{int}°$ of a number of ring compounds and open-chain compounds. The last column in the table lists the ratio of the entropy difference to the number of hindered rotations in the open chain that have been "frozen" in ring formation. This entropy decrement is about 4.7 ± 0.3 for all but the C_4, C_5, and C_8 rings. It is smaller for these latter, in agreement with the observation that these rings have an unusually low frequency-puckering mode called a pseudorotation, which is responsible for an excess entropy in these systems.

Table 2.10. Comparisons of the Entropies of Some Ring Compounds with Open Chain Compounds

Ring (σ)		$S_{int}{}^a$	Open-Chain Compound (σ)		$S_{int}{}^a$	$\Delta S_{int\,rot}{}^b$	Residual[c] Ring Entropy
C_3H_6	(4)	55.3	C_2H_6	(18)	60.7	5.4	
$CH_3CH{=}CH_2$	(3)	66.1	C_3H_8	(18)	70.3	4.2	
Trans-Butene-2	(18)	76.7	n-C_4H_{10}	(18)	79.7	3.0	
C_3H_6	(6)	60.4	C_3H_8	(18)	70.3	4.9	0
$\overline{CH_2CH_2O}$	(2)	59.5	C_2H_5OH	(3)	69.7	5.1	
			CH_3OCH_3	(18)	69.5	5.0	
$\overline{CH_2CH_2NH}$	(1)		$C_2H_5NH_2$	(3)	(69.9		
			$(CH_3)_2NH$	(9)	69.8		
$\overline{CH_2CH_2S}$	(2)	62.5	$(CH_3)_2S$	(18)	74.1	5.8	
			C_2H_5SH	(18)	73.0	5.3	
C_4H_8	(8)	67.6	n-C_4H_{10}	(18)	79.7	4.0	2.7
C_5H_{10}	(10)	74.6	n-C_5H_{12}	(18)	89.1	3.6	5.2
C_6H_{12}	(6)	74.9	n-C_6H_{14}	(18)	98.5	4.7	1.0
C_7H_{14}	(1)	81.8	n-C_7H_{16}	(18)	107.9	4.4	3.0
$C_8H_{16}{}^d$	(8)	91.9	n-C_8H_{18}	(18)	117.3	3.6	9.1

[a] $S_{int}^{\circ} = S^{\circ} + R \ln \sigma$.
[b] $\Delta S_{int\,rot} = [S_{int\,(open\,chain)}^{\circ} - S_{int\,(ring)}^{\circ}]/(n-1)$, where n is the number of atoms in the ring.
[c] Defined with reference to cyclopropane as $(n-1)[4.9 - \Delta S_{int\,rot}^{\circ}]$.
[d] Crown form.

As a reference point, the "two-membered rings" (i.e., simple olefins) have been included. We note that they show a rapid change in entropy loss on going from ethylene to *trans*-butene-2. This arises from the sharp drop in torsion frequency going from C_2H_4 to *trans*-C_4H_8—2, which itself begins to approach the frequency equivalent to a hindered rotation. Thus in tetramethylethylene the reduced mass of the torsion has become so large that the double bond torsion has a frequency of only about $200\,\text{cm}^{-1}$ and a corresponding entropy at 300°K of 2.2.

Table A.20 in the Appendix lists the torsion frequencies for some substituted olefins.

We can now use these entropy decrements in ring formation to estimate ring entropies, starting with open chain compounds.

EXAMPLE

Estimate the entropy of spiropentane at 300°K.

Let us start with the model compound, ethyl cyclopropane, the entropy of which we can estimate from groups as

$$S_{int} \left(\triangle \diagdown \right) = 3S°[C—(C)_2(H)_2] + S°[C—(C)_3)(H)]$$

$$+ S°[C—(C)(H)_3] + S°(\triangle) \text{ correction}$$

$$= 28.3 - 12.1 + 30.4 + 32.1 = 78.7.$$

Using the entropy decrement in closing to a three-membered ring, of $-2 \times 4.9 = -9.8$, this becomes 68.9, and correcting for symmetry ($\sigma = 4$), we find $S° = 66.1$ compared to an observed value of 67.5. Alternatively, starting with $\boxed{\diagdown\diagup}$, we obtain 64.1.

EXAMPLE

Estimate the entropy of (2,1,2)bicycloheptene-1  at 300°K. We do

this by two independent paths, one starting with ethylcyclopentene, the other with methyl-cyclohexene.

Starting with $\pentagon\diagup$, we calculate $S°_{int}$ from groups

$$S°_{int} = S°[C—(C)(H)_3] + 2S°[C—(C)_2(H)_2] + S°[C—(C_d)(C)_2(H)]$$
$$+ S°[C—(C_d)(C)(H)_2] + 2S°[C_d—(C)(H)] + \text{ring correction}$$
$$= 30.4 + 18.8 - 11.7 + 9.8 + 16.0 + 25.8 = 89.1$$

After closing the new ring, we lose two rotations characteristic of a "tight" (i.e., nonpucker-ing) C_5 ring. We use the value of -2×4.9 for this ring decrement to give a value for the bicyclic ring of 79.3.

Starting with $\hexagon$—CH_3, we calculate from groups the value of $S°_{int} = 84.2$.

Note that this has precisely the same groups as the isomeric C_5 ring and differs only in the ring correction. After closing the second ring by forming the bridge CH_2, we lose one rotation, or -4.9 to yield $S°_{int}$ of 79.3, in good agreement with the preceding value.

Some idea of the extra entropy of the C_4 and C_5 rings can be obtained by comparing the intrinsic entropy changes observed by inserting a double bond. This should "tighten" these rings and eliminate the low frequency "pucker."

Table 2.11 compares the intrinsic entropies of some saturated and unsaturated rings. We see that, whereas there is a very small loss in entropy in dehydrogenating the C_3 and C_6 rings, there is an appreciable loss in doing the same for the C_4 and C_5 rings. In the change from cyclohexene to benzene, we introduce two more double bonds, and it is quite likely that most of the overall loss of 5.0 gibbs/mole comes from the last step in forming benzene.

Table 2.11. Intrinsic Entropies of Some Saturated and Unsaturated Rings

Ring	(σ)	S°_{int}	Ring-(H_2)	(σ)	S°_{int}	ΔS°_{int}
C_3H_6	(6)	60.4	C_3H_4	(2)	59.8	-0.6
C_4H_8	(8)	67.6	C_4H_6	(2)	64.4	-3.2
C_5H_{10}	(10)	74.6	C_5H_8	(2)	70.6	-4.0
C_5H_8	(2)	70.6	C_5H_6	(2)	67.0	-3.6
C_6H_{12}	(6)	74.9	C_6H_{10}	(2)$(n=2)$	74.3	-0.6
C_6H_{10}	(2)	74.3	[benzene]^a	(12)	69.3	-5.0

^{a} Here we have lost $2H_2$.

2.21 POLYCYCLIC RING ESTIMATES FROM OPEN CHAINS

These simple observations on the entropies of hydrocarbon rings can be extended[16] to construct a set of explicit rules, permitting the estimation of both entropies and heat capacities of cyclic compounds.

In the comparison of an open-chain alkane with its cyclic homologue, we can think of the cyclization as involving the loss of two kinds of rotors: (a) the two end methyl groups and (b) the heavier internal groups. Table 2.13 shows the ΔS°_{300} and ΔC°_{pT}, which can be assigned to each of these structural changes in estimating S°_{300} and C°_{pT} of a cyclic compound starting from an open chain. Table 2.12 also includes the corrections to be made for the introduction of a double bond in a ring. Different rings have

Table 2.12. Entropy and Heat-capacity Changes in Cyclization^a

Groups	ΔS°_{300}	ΔC°_p $(T^\circ K)$				
		300	500	800	1000	1500
One-end CH_3	-5.2	-3.3	-4.5	-6.1	-7.1	-8.5
Internal group	-4.9	-0.8	-0.6	0.0	0.1	0.7
Two-end CH_3^{b}	-9.3	-4.6	-5.0	-6.0	-6.6	-7.9
Introduction of double bond in ring^c	-0.4	-1.4	-3.8	-6.3	-7.5	-9.3

^{a} These changes are standardized for the very "stiff" rings, cyclopropane and cyclohexane. All other rings will need ring corrections.
^{b} A distinction is made between cyclizing a chain across its ends involving two methyl groups, as opposed to joining an end CH_3 group to one of the atoms in the chain.
^{c} These corrections arise from the loss of the two H atoms. For ΔC_{pT} they would thus reach a maximum of $-6R = -12.0$ at $T > 2000^\circ K$.

[16] H. E. O'Neal and S. W. Benson, *J. Chem. Eng. Data*, **15**, 266 (1970).

different intrinsic entropies and heat capacities arising from their different degrees of stiffness. These values are based on the stiff rings, cyclopropane and cyclohexane, which then have zero correction. For all other rings, we will make appropriate corrections as shown in Table 2.13. While this system is more complex than the simple system just considered in the last section, it is also more accurate.

EXAMPLE

Estimate S_{300}° and C_{p300}° for methyl cyclobutane.
  If we start with n-pentane, we find, using groups, that:

$$S_{int}^{\circ} = 89.1 - \Delta S^{\circ}(\text{end Me}) - 2 \Delta S^{\circ}(\text{internal group}) + \Delta S^{\circ}(\text{tightness C}_4)$$
$$= 89.1 - 5.2 - 9.8 + 1.9 = 76.0$$

since $\sigma = 3$, $S^{\circ} = 73.8$, and

$$C_{p300}^{\circ} = 28.9 - 3.3 - 1.6 + 0.2 = 24.2$$

Starting with 2-methyl butane, we find using groups:

$$S_{int}^{\circ} = 88.5 - \Delta S^{\circ}(\text{2 end Me}) - \Delta S^{\circ}(\text{internal group}) + \Delta S^{\circ}(\text{tightness C}_4)$$
$$= 88.5 - 9.3 - 4.9 + 1.9 = 76.2$$
$$C_{p300}^{\circ} = 28.6 - 4.6 - 0.8 + 0.2 = 23.4$$

These are both in good agreement with each other and with the values obtained directly from groups for the cyclic compound:

$$S_{int}^{\circ} = 76.4 \qquad C_{p300}^{\circ} = 22.6$$

EXAMPLE

Estimate S_{300}° and C_{p800}° for cyclo-octatetraene.
  If we start with n-octane, we find from groups:

$$S_{int}^{\circ} = 117.3 - \Delta S^{\circ}(\text{2 end Me}) - 5 \Delta S^{\circ}(\text{internal group}) + \Delta S^{\circ}(\text{tightness C}_8)$$
$$- 4 \Delta S^{\circ}(\text{double-bond creation}) - \Delta S^{\circ}(\text{tightness of 4 double bonds})$$
$$= 117.3 - 9.3 - 24.5 + 8.4 - 1.6 - [3.2 + 4.2 + 1.0 + 3.4]$$
$$= 78.5 \text{ (observed } S_{int}^{\circ} = \textbf{78.4})$$

since $\sigma = 4$ (tub form), $S_{300}^{\circ} = 75.7$, and

$$C_{p800}^{\circ} = 92.4 - 6.0 - 5(0.0) - 0.4 + 4(-6.3) + [0.2 + 0.1 + 0.0 - 0.2]$$
$$= 60.9 \text{ (observed } C_{p800}^{\circ} = 62.2)$$

  Note that in the procedures just described, we make all of the corrections successively in forming the ring from the open chain, and then add in any double bonds and accompanying tightness corrections successively and cumulatively.

  These same methods can now be applied to the formation of polycyclic rings starting with either open chains or monocyclic rings. In doing this we follow the same procedure as just employed but in addition we need to add corrections for changes in tightness due to the connection between

Table 2.13. Tightness Ring Corrections for ΔS°_{300} and $\Delta C^\circ_p(T^\circ K)$ on Cyclization Changes

Ring System	Double Bonds	ΔS°_{300}	ΔC°_p $(T^\circ K)$ 300	500	800	1000	1500
C_3	0	0	0	0	0	0	0
	1	0	0	0	0	0	0
C_4	0	1.9	0.2	0.1	0	0	0
	1	−1.9	−0.2	−0.1	0	0	0
	2	−1.4	−0.7	−0.3	−0.1	−0.1	0
C_5	0	3.9	−0.8	−0.9	−1.0	−1.0	−1.0
	1	−3.9	0.8	0.9	1.0	1.0	1.0
	2	−2.8	−1.4	−0.6	−0.2	−0.2	0
C_6	0	0	0	0	0	0	0
	1	0	−1.0	−1.0	−1.0	−1.0	−1.0
	2	$0(−1.4)^c$	$0(−0.7)^c$	$0(−0.3)$	$0(−0.1)$	0 (0)	0 (0)
	3	−2.8	−1.4	−0.6	−0.3	−0.2	−0.1
C_7	0	2.5	0.3	0.1	−0.2	−0.5	−0.7
	1	−2.5	−0.3	−0.1	0.2	0.5	0.7
	2	$0(−1.4)^c$	$0(−0.7)^c$	$0(−0.3)$	$0(−0.1)$	0 (0)	0 (0)
	3	−1.4	−0.7	−0.3	−0.1	−0.1	0
C_8	0	8.4	0.8	0.5	−0.4	−0.8	−1.4
	1	−3.2	−0.3	−0.1	0.2	0.4	0.7
	2^d						
	(1, 3)	−4.2	−0.4	−0.3	0.1	0.3	0.6
	(1, 4)	−1.0	−0.1	−0.1	0.1	0.1	0.1
	(1, 5)	0	0	0	0	0	0
	3	−3.8	−1.3	−0.6	−0.2	0	0
		$(−1.0)^d$	(−0.1)	(−0.1)	(0)	(0)	(0)
	4	−3.4	−1.5	−0.7	−0.2	0	0
Spirane^b		1.4	0.7	0.3	0.1	0.1	0

^{a} These corrections are standardized on the tightest rings, cyclopropane and cyclohexane, which thus have zero corrections. Double bond corrections are cummulative, that is, for two double bonds, we add the corrections for the first, second, and so on.
^{b} These corrections apply only when both rings are C_4 or smaller.
^{c} Values in parentheses are for conjugated double bonds.
^{d} Index numbers show position of double bonds in ring.

the old and new rings. We will treat three categories of ring connection, a shared-corner correction, a shared-edge correction, and a methylene bridge. The correction for shared corners applies only to C_3—C_3, C_3—C_4, and C_4—C_4 ring fusions, and are shown in Table 2.13 (spirane). The correction for a shared edge is treated as a fractional double-bond tightening for the ring in question with the fraction T_f given by the formula:

$$\text{Fractional tightening for } C_n \text{ ring sharing an edge} = T_f = \frac{8-n}{5} \quad \text{for} \quad 4 \le n \le 8$$

For $n \ge 9$ it is assumed that no tightening exists. For $n = 4$, $T_f = 0.8$, and for $n = 3$, no correction is needed, the ring already being at maximum tightness.

For CH_2 bridges across a C_n ring, we assume that the C_n ring is tightened by the equivalent of two conjugated double bonds. Once we have decided the equivalent tightening, the necessary corrections are taken from Table 2.13.

EXAMPLE

Estimate S°_{300} and C°_{p300} for 2,2,1-bicycloheptane (norbornane).

Starting with 1,3-dimethyl cyclopentane, we find from groups:

$$S^\circ_{\text{int}} = S^\circ_{\text{int}}(1,3\text{-dimethyl}-C_5) - \Delta S^\circ(2 \text{ end Me}) + \Delta S^\circ(1,3\text{-double bonds in } C_5)$$
$$= 92.2 - 9.3 + (-3.9 - 2.8) = 76.2$$

since $\sigma = 2$, $S^\circ_{300} = 74.8$, and

$$C^\circ_{p300} = 31.5 - 4.6 + (0.8 - 1.4) = 26.3.$$

Starting with methyl cyclohexane we find from groups:

$$S^\circ_{\text{int}} = S^\circ_{\text{int}}(\text{methyl } C_6) - \Delta S^\circ(1 \text{ end Me}) + \Delta S^\circ(1,3 \text{ double bonds in } C_6)$$
$$= 84.2 - 5.2 - 2.4 = 76.6 \text{ eu}; \quad S^\circ_{300} = 75.2$$
$$C^\circ_{p300} = 32.4 - 3.3 - 1.7 = 27.4.$$

Both results are in good agreement with the observed values:

$$S^\circ_{300} = 74.2 \quad \text{and} \quad C^\circ_{p300} = 26.1.^{17}$$

2.22 GROUPS FOR RADICALS

Thermodynamically, it is most useful to define the bond-dissociation energy (BDE) of a compound AB in which groups A and B are connected by a bond as the standard enthalpy change in the reaction:

$$AB \overset{1}{\rightleftharpoons} A^{\cdot} + B^{\cdot}$$

Groups A and B may be atoms or free radicals. If we can measure ΔH°_1, then we have measured the bond-dissociation energy, or preferably, bond-dissociation enthalpy $[DH^\circ(A-B)]$. Data on BDE then permit us to obtain heats of formation of free radicals, such as $A^{\cdot}$ if $\Delta H^\circ_f(AB)$ and $\Delta H^\circ_f(B^{\cdot})$ are known:

$$\Delta H^\circ_1 = \Delta H^\circ_f(A^{\cdot}) + \Delta H^\circ_f(B^{\cdot}) - \Delta H^\circ_f(AB)$$

If $A^{\cdot}$ is an alkyl radical, such as $\dot{C}H_2CH_3$ (ethyl), then we can treat the atom bearing the odd electron as a new element and decompose the properties of the radical into groups. To do this rigorously would require almost more data than is currently available, and so one makes use of an empirical assumption.[18] The assumption is that the law of bond additivity holds for alkyl radicals or for radicals to which are attached saturated alkyl substituents. In the case of the hydrocarbon, it is equivalent to saying that primary C—H bond strengths are the same, all secondary C—H bond strengths are the same, and all tertiary C—H bond strengths are the same. This seems to work within the accuracy of the data on free radicals, which is of the order of ± 1.0 kcal/mole in the best cases. It is, however, a restricted empirical finding and we must be cautious in applying such "group" values, particularly for ΔH°_f. Bond additivity works very much better for entropies and C°_p, and so we may feel more secure in using these quasigroup values for those properties. Table 2.14 shows some group values for organic radicals derived in this way. The entropy includes the electronic degeneracy of the odd electron, and so it is not necessary to add it to the final result.

---

[17] Continuing from the above, we find for tetrahydrodicyclopentadiene(exo) a value of $S^\circ_{298} = 89.4$ in excellent agreement with the measured value of 88.8 [See R. H. Boyd et al., *J. Phys. Chem.*, **75**, 1264 (1971)].

[18] H. E. O'Neal and S. W. Benson, "Thermochemistry of Free Radicals," Ch. 17 in J. K. Kochi, (ed.), *Free Radicals*, Wiley, New York, 1973.

Table 2.14. Free-radical Group Additivities[a]

Radical	ΔH_f°	S°	C_p°						
			300	400	500	600	800	1000	1500
[·C—(C)(H)$_2$]	35.82	30.7	5.99	7.24	8.29	9.13	10.44	11.47	13.14
[·C—(C)$_2$(H)]	37.45	10.74	5.16	6.11	6.82	7.37	8.26	8.84	9.71
[·C—(C)$_3$]	38.00	−10.77	4.06	4.92	5.42	5,75	6.27	6.35	6.53
[C—(C·)(H)$_3$]	−10.08	30.41	6.19	7.84	9.40	10.79	13.02	14.77	17.58
[C—(C·)(C)(H)$_2$]	−4.95	9.42	5.50	6.95	8.25	9.35	11.07	12.34	14.25
[C—(C·)(C)$_2$(H)]	−1.90	−12.07	4.54	6.00	7.17	8.05	9.31	10.05	11.17
[C—(C·)(C)$_3$]	1.50	−35.10	4.37	6.13	7.36	8.12	8.77	8.76	8.12
[C—(O·)(C)(H)$_2$]	6.1	36.4	7.9	9.8	10.8	12.8	15.0	16.4	—
[C—(O·)(C)$_2$(H)]	7.8	14.7	7.7	9.5	10.6	12.1	13.7	14.5	—
[C—(O·)(C)$_3$]	8.6	−7.5	7.2	9.1	9.8	11.1	12.1	12.3	—
[C—(S·)(C)(H)$_2$]	32.4	39.0	9.0	10.6	12.4	13.6	15.8	17.4	—
[C—(S·)(C)$_2$(H)]	35.5	17.8	8.5	10.0	11.6	12.3	13.8	14.6	—
[C—(S·)(C)$_3$]	37.5	−5.3	8.2	9.8	11.3	11.8	12.2	12.3	—
[·C—(H)$_2$C$_d$)]	23.2	27.65	5.39	7.14	8.49	9.43	11.04	12.17	14.04
[·C—(H)(C)(C$_d$]	25.5	7.02	4.58	6.12	7.19	8.00	9.11	9.78	10.72
[·C—(C)$_2$(C$_d$)]	24.8	−15.00	4.00	4.73	5.64	6.09	6.82	7.04	7.54
[C$_d$—(C·)(H)]	8.59	7.97	4.16	5.03	5.81	6.50	7.65	8.45	9.62
[C$_d$—(C·)(C)]	10.34	−12.30	4.10	4.71	5.09	5.36	5.90	6.18	6.40
[·C—(C$_B$)(H)$_2$]	23.0	26.85	6.49	7.84	9.10	9.98	11.34	12.42	14.14
[·C—(C$_B$)(C)(H)]	24.7	6.36	5.30	6.87	7.85	8.52	9.38	9.84	10.12
[·C—(C$_B$)(C)$_2$]	25.5	−15.46	4.72	5.48	6.20	6.65	7.09	7.10	6.94
[C$_B$—C·]	5.51	−7.69	2.67	3.14	3.68	4.15	4.96	5.44	5.98
[C—(·CO)(H)$_3$]	−5.4	66.6	12.74	14.63	16.47	18.17	21.14	23.27	—
[C—(·CO)(C)(H)$_2$]	−0.3	45.8	12.7	14.5	15.8	16.8	19.2	20.7	—
[C—(·CO)(C)$_2$(H)]	2.6	(23.7)	(11.5)	(12.8)	(14.3)	(15.5)	(17.4)	(18.5)	—
[·N—(H)(C)]	(55.3)	30.23	5.38	5.67	5.89	6.09	6.60	6.97	7.74
[·N—(C)$_2$]	(58.4)	10.24	3.72	4.13	4.38	4.53	4.86	4.95	4.91
[C—(·N)(C)(H)$_2$]	−6.6	9.8	5.25	6.90	8.28	9.39	11.09	12.34	—
[C—(·N)(C)$_2$(H)]	−5.2	−11.7	4.67	6.32	7.64	8.39	9.56	10.23	—
[C—(·N)(C)$_3$]	(−3.2)	34.1	4.35	6.16	7.31	7.91	8.49	8.50	—
[·C—(H)$_2$(CN)]	(58.2)	58.5	10.66	12.82	14.48	15.89	18.08	19.80	—
[·C—(H)(C)(CN)]	(56.8)	40.0	9.1	11.4	13.1	14.4	16.3	17.4	—
[·C—(C)$_2$(CN)]	(56.1)	19.6	8.8	10.4	11.3	12.3	13.7	14.5	—
[·N—(H)(C$_B$)]	38.0	27.3	4.6	5.4	6.0	6.4	7.2	7.7	8.6
[·N—(C)(C$_B$)]	42.7	(6.5)	(3.9)	(4.2)	(4.7)	(5.0)	(5.6)	(5.8)	(5.9)
[C$_B$—N·]	−0.5	−9.69	3.95	5.21	5.94	6.32	6.53	6.56	—
[C—(CO$_2$·)(H)$_3$]	−47.5	71.4	14.4	17.8	20.4	23.1	27.1	29.6	—
[C—(CO$_2$·)(H)$_2$(C)]	−41.9	49.8	15.5	18.5	20.3	22.3	27.5	27.2	—
[C—(CO$_2$·)(H)(C)$_2$]	−39.0	−12.1	(4.5)	(6.0)	(7.2)	(8.0)	(9.3)	(10.1)	(11.2)
[C—(N$_A$)(H)$_3$]	−10.08	30.41	6.19	7.84	9.40	10.79	13.02	14.77	17.58
[C—(N$_A$)(C)(H)$_2$]	−5.5	9.42	5.50	6.95	8.25	9.35	11.07	12.34	14.25
[C—(N$_A$)(C)$_2$(H)]	−3.3	−12.07	4.54	6.00	7.17	8.05	9.31	10.05	11.17
[C—(N$_A$)(C)$_3$]	−1.9	−35.10	4.37	6.13	7.36	8.12	8.77	8.76	8.12
[N$_A$—C]	32.5	8.0	4.0	4.4	4.7	4.8	5.1	5.3	5.2
[N$_A$—(N$_A$·)(C)]	74.2	36.1	7.8	8.2	8.4	8.6	8.9	9.0	9.0

[a] Values in parentheses are best guesses.

Table 2.14 (*contd.*)

Mass Corrections in Conjugated Systems

(a) If the masses on each side of a resonance stabilized bond in a radical have the same number of C atoms (i.e., are roughly equal), add 0.7 to the entropy; for example

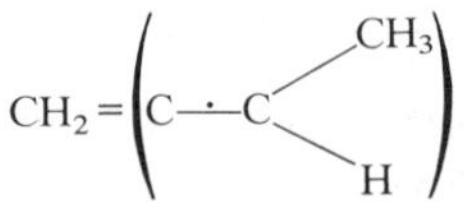

$$CH_2 = \left(C \cdots C \begin{array}{c} CH_3 \\ \\ H \end{array} \right) \quad m_1 = CH_2, \; m_2 = (CH_3 + H).$$

(b) If the masses (as above) differ by one C atom, add 0.3 to $S°$, for example, ($CH_2 = C \cdots CH_2$) $m_1 = CH_2$, $m_2 = H_2$.

Internal Rotation Barrier Corrections (for "Standard" Correction, see text):

	$S°$	$C_p°$ at $T°K$ (as listed below)						
		300	400	500	600	800	1000	1500
$V_0(R^·) = \frac{2}{3}V_0(RH)$								
$(CH_3 \dashv\!\!\!- R)\sigma = 3$	+0.5	0	−0.1	−0.2	−0.3	−0.3	−0.3	−0.2
$(\infty \dashv\!\!\!- R)\sigma = 3$	+0.5	0.1	0.1	0	−0.2	−0.3	−0.4	−0.4

Perhaps the largest uncertainty arising in the estimation of the thermochemistry of free radicals has to with estimates of their symmetry and in assigning barriers to rotation. In the case of $\dot{C}H_3$ radicals, current evidence suggests that the radicals are planar or near-planar, and we have adopted this for all alkyl radicals. By way of contrast, $\dot{C}F_3$ and $\dot{C}Cl_3$ radicals, on the basis of EPR spectra, have been judged pyramidal. Hence $\sigma = 6$ for $\dot{C}H_3$, while $\sigma = 3$ for $\dot{C}F_3$ and $\dot{C}Cl_3$. Doubly substituted (with electronegative elements) alkyl radicals will be treated as nonplanar (e.g., $\dot{C}HF_2$), and singly substituted (e.g., $\dot{C}H_2Cl$) as planar. The evidence for this is not strong, and the assignment thus has some uncertainty attached to it.

Barriers to rotation are even more difficult to assign since there are little or no experimental data available for assignments. We are thus reduced to analogies, and the best ones are expected to be those based on neighboring isoelectronic structures. In the simplest case of the $\dot{C}H_3$

radical, we expect the structure to be between the pyramidal NH_3 molecule and the flat BH_3. The experimental evidence on CH_3 suggests it is flat, but the data are not accurate enough to rule out a cone angle of, for example, $80°$ (H—C—H angle $= 117°$) which would put the C atom only 0.19 Å above the plane of the H atoms. The zero point amplitudes of motion of the H atoms are sufficiently large as to make this a dynamically plane structure (i.e., very rapid inversion frequency). It could then be argued that halogen atom substitution could accentuate the pyramidal shape on the grounds that there is a repulsion is expected between the lone electron on carbon and the lone pairs on the halogens. The nonpolar, carbon-substituted alkyl radicals might then exist in a dynamically near-planar structure like $\dot{C}H_3$.

Such a view would be compatible with current thermochemical data, which suggests that alkyl-substituted radicals are dynamically nearly flat and can be assigned a symmetry number on that basis, but have a barrier that would be more typical of the nearest olefin or nitrogen analogue. Thus for the CH_3—$\dot{C}H_2$ radical we might assign a $\sigma = 6$ and a barrier such as that in propylene ($V_0 = 2.0$ kcal/mole) or in CH_3NH_2 ($V_0 = 1.9$ kcal/mole). This would be contrary to the expectation that for rotors which have a barrier with sixfold or higher symmetry, the barriers are found to be zero (e.g., $B(CH_3)_3$, ϕCH_3, CH_3NO_2, CH_3BF_2, etc.).

A few example calculations should serve to illustrate the use of the group-additivity table.

EXAMPLE

t-Butoxy Radical $(CH_3)_3CO\cdot$

	Contributions		
Groups	$\Delta H_f^\circ(300)$	$S^\circ(300)$	$C_p^\circ(300)$
$3[C\text{—}(C)(H)_3]$	−30.24	91.2	18.57
$[C\text{—}(C)_3(O\cdot)]$	8.6	−7.5	7.2
Symmetry ($\sigma = 3^4$)		−8.7	
Totals	−22.6	75.0	25.8

Rotational barriers of the methyl groups in the radical are probably the same as those in t-butyl alcohol. Therefore no additional barrier corrections need to be made.

EXAMPLE

2-Butyl Radical CH_3—$\dot{C}H$—CH_2CH_3

	Contributions		
Groups	$\Delta H_f^\circ(300)$	$S^\circ(300)$	$C_p^\circ(300)$
$[C—(C)(H)_3]$	−10.08	30.41	6.19
$[C—(C^\cdot)(H)_3]$	−10.08	30.41	6.19
$[C—(H)_2(C)(C^\cdot)]$	−4.96	9.42	5.50
$[^\cdot C—(H)(C)_2]$	37.45	10.74	5.16
Barrier $(CH_3 —\!\!\!\{\ R)_{V(RH)\rightarrow\ V(R\cdot)}$		.50	0.0
Corrections $(CH_3\dot{C}H\!\!-\!\!\!\{CH_2CH_3)_{V(RH)\rightarrow\frac{2}{3}V(R^\cdot)}$		.50	0.1
Symmetry $(\sigma = 9)$		−4.40	
Totals	2.3	77.6	23.1

There are two groups bound to the radical center; therefore, there are two barrier corrections. The corrections made above assume that the barriers to rotation of the two groups are $\frac{2}{3}$ their height in the alkane (*n*-butane). In this event, barrier corrections are essentially independent of the type of rotor and the height of the barrier and can be taken directly from the additivity table.

EXAMPLE

1-pentyl-3,5-diradical $\cdot\diagdown\diagup\diagdown\cdot$ (*singlet*)

This is a conjugated biradical, and we will asume the species is a singlet as formed in the vinylcyclopropane ring-opening reaction.

	Contributions		
Groups	$\Delta H^\circ(300)$	$S^\circ(300)$	$C_p^\circ(300)$
$[C_d—(H)_2]$	6.26	27.61	5.10
$[C_d—(C^\cdot)(H)]$	8.59	7.97	4.16
$[C^\cdot—(C_d)(C)(H)]$	25.5	$(7.02 − 1.4)^a$	4.58
$[C—(C^\cdot)_2(H)_2]^b$	−4.95	9.42	5.50
$[C^\cdot—(C)(H)_2]$	35.82	$(30.7 − 1.4)^a$	5.99
Symmetry $(\sigma = 2)$		−1.4	
Totals	71.2	71.5	25.3

[a] The singlet diradical has an electronic degeneracy of unity; therefore, the radical entropy groups must be *corrected* for the doublet degeneracy built into the tabulated values.

[b] Groups bonded to the radical center are the same as the molecular groups, for instance, $[C—(C^\cdot)_2(H)_2]\equiv[C—(C)_2(H)_2]$.

EXAMPLE

t-Butyl radical (CH₃)₃C·

Assume that *t*-butyl is flat, $\sigma = 6 \times 3^3$, $V_0 = \frac{2}{3} V_0(RH)$ for all methyl rotors in the radical.

	Contributions		
Groups	$\Delta H_f^\circ(300)$	$S^\circ(300)$	$C_p^\circ(300)$
$3[C-(C^\cdot)(H)_3]$	−30.24	91.2	18.57
$[C^\cdot-(C)_3]$	38.00	−10.77	4.06
$3(CH_3-\!\!\{-\infty)$		3(0.6)	3(0)
$V(RH) = 4.7 \rightarrow V(R^\cdot) = 3.0$			
Symmetry		−10.1	
Totals	8.24	72.1	22.6

Table 2.15 shows the method applied to the oxacyclopentyl-1 radical $\overline{O-\dot{C}H-CH_2-CH_2-CH_2}$; where one group and the ring correction are not listed, we must adopt values from comparable structures.

Table 2.15.   Thermochemistry of the Oxa-cyclopentyl-1 Radical
$\overline{O-\dot{C}H-CH_2-CH_2-CH_2}$

	Contributions		C_{pT}°		
Groups	$\Delta H_{f,300}^\circ$	ΔS_{300}°	300°K	500°K	800°K
$C-(C)_2(H)_2$	−4.95	9.4	5.50	8.25	11.07
$C-(C)(\dot{C})(H)_2$	−4.95	9.4	5.50	8.25	11.07
$C-(O)(C)(H)_2$	−8.1	9.8	5.0	8.3	11.1
$O-(C)(\dot{C})^a$	−23.2	8.7	3.4	3.7	4.4
$\dot{C}-(O)(C)(H)^b$	31.4	8.9	4.2	6.1	7.9
Ring correction^c	5.3	26.5	−6.3	−4.7	−2.9
Symmetry $(\sigma = 1)$	—	—	—	—	—
Totals	−4.5	72.7	17.3	29.9	42.6

^{a} These values are set equal to O—(C)₂ group values. This is an arbitrary convention.
^{b} Group values are assigned as follows: (a) ΔH_f° is set to give the experimental value of ΔH_f° for the radical and (b) ΔS° and ΔC_{pT} contributions are taken from the N—(C)₂(H) group, which has some valency and about the same masses and bond lengths.
^{c} ΔH_f° contribution (i.e., strain) is taken as the average of the saturated $\overline{O-CH_2)_4}$ ring (5.9 kcal) and $\overline{O-CH_2CH_2CH=CH}$ (4.7 kcal). S° and C_{pT}° contributions are taken as average of cyclopentane and cyclopentene.

3

Arrhenius Parameters for Gas Phase Unimolecular Reactions

3.1 ELEMENTARY PROCESSES IN GAS REACTIONS

We adopt the point of view in this book that in a dilute gas there are only two types of simple kinetic processes. The first of these is a process undergone by an energetically activated chemical species when it is completely isolated from other gas phase species; this is called a unimolecular process. It may correspond to an internal rearrangement of atoms, the breaking of a bond, the rotation of a group, as in cis–trans isomerization, or very simply the internal redistribution of energy.

The second process is bimolecular and requires the collision of two chemical species. The combination of these two chemical species will be referred to as a collision complex. By virtue of our first definition, it follows that the subsequent chemical fate of the collision complex corresponds to a unimolecular process. Thus bimolecular chemical reactions are composite acts consisting of the formation of a collision complex followed by a unimolecular kinetic process.

Although we may consider a termolecular process as the joint, "simultaneous" collision of three species to form a termolecular collision complex, it is simpler to picture the formation of a termolecular collision complex as proceeding kinetically in two bimolecular stages. The mechanism of formation of a termolecular complex ABC can be written as

$$A + B \leftrightarrows [AB] + C$$
$$A + C \leftrightarrows [AC] + B \leftrightarrows [ABC] \to \text{products} \qquad (3.1)$$
$$B + C \leftrightarrows [BC] + A$$

where brackets signify collision complexes.

3.2   TRANSITION-STATE THEORY OF RATE PROCESSES

The simplest chemical rate process is a unimolecular reaction. However in order for a unimolecular reaction to occur, the reacting molecule must accumulate sufficient internal energy to break the necessary bonds or undergo the internal rearrangement. This activation energy is generally accumulated in a sequence of atypical or "lucky" collisions with average molecules in the gas or with the vessel walls. The typical collision, of course, corresponds to one in which the energy-rich molecule loses part of its energy to its energy poor collision partner.

If we observe an energy-rich, polyatomic molecule in a gas, we may expect to see, in the (relatively) long intervals between collisions, that the energy will redistribute itself over the various parts of the molecule as the various atoms execute their not quite harmonic vibrations. At STP there is about $10^{-9.3}$ s between gas collisions, whereas a $600 \, \text{cm}^{-1}$ vibration will execute $600 \times 3 \times 10^{10} = 1.8 \times 10^{13}$ oscillations/s or about 10 000 oscillations between successive collisions. As a result, we may expect to see the energized molecule take a great many different configurations. If any one of them corresponds to the localization of enough of the energy to break a bond, to dissociate a fragment, or to rotate a group past a barrier, chemical reaction occurs.

The most useful quantitative model for the reaction described above is provided by the "transition-state theory" originally developed by Henry Eyring and his co-workers.[1] It proposes that molecules having the required energy and the conformation corresponding to the internal energy barrier for chemical reaction be considered in "virtual" equilibrium with "normal," unexcited species. The rate of the chemical reaction is then given by the product of the concentration of these transition-state molecules by the rate constant for their passage over the energy barrier. For the simple case of a unimolecular isomerization, $A \rightarrow B$ we can represent the scheme as

$$A \leftrightharpoons A^*$$
$$A^* \leftrightharpoons A^{\ddagger} \leftrightharpoons B \tag{3.2}$$

where A^* represents a molecule of A with sufficient internal energy to isomerize, and $A^{\ddagger}$ represents the same A^* molecule in the geometrical conformation corresponding to the top of the barrier or transition state. This is illustrated in the energy-reaction coordinate plot of Figure 3.1.

Note that because, in principle, all reactions are reversible, we must have a totally symmetrical scheme for B and the corresponding transition

[1] S. Glasstone, K. J. Laidler, and H. Eyring, *The Theory of Rate Processes*, McGraw-Hill, New York (1941).

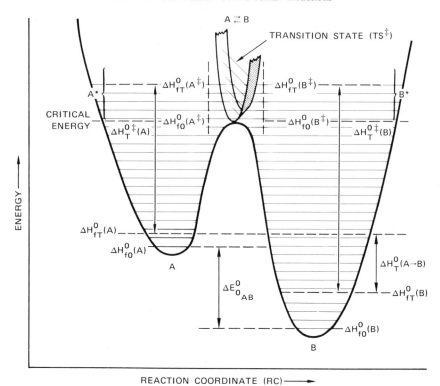

Figure 3.1   Energy-reaction coordinate diagram for an exothermic, unmolecular isomerization. The reaction coordinate is the internal coordinate or combination of internal coordinates that measures uniquely the progress of the reaction. The solid curve gives the potential energy of the molecule corresponding to zero kinetic energy in the reaction coordinate for the appropriate value on the abscissa of the reaction coordinate. For *cis–trans* isomerization, RC may be the dihedral angle.

state $B^{\ddagger}$ must be identical to $A^{\ddagger}$ in every respect except the phase of motion of its atoms. The phase of motion of the atoms of $B^{\ddagger}$ will correspond to change from a B state toward an A state and conversely for the phases of $A^{\ddagger}$. In the simplest case of a cis–trans isomerization, for example $CHD{=}CHD$, we have

The transition state is identified as that state in which the planes of the end methylene groups are at $90°$ to each other. At this point, one direction of rotation corresponds to cis $\rightarrow$ trans, and the opposite rotation corresponds to trans $\rightarrow$ cis. Note that this configuration has two optical isomers.

In energy space the activation barrier corresponds to a "relative" maximum in the potential energy of the molecule. If we express V, the potential energy of the molecule, as a function of $q_1 \cdots q_{3N-6}$, the $3N-6$ internal "normal" coordinates of the system, then,

$$\left(\frac{\partial V}{\partial q_i}\right)_{\ddagger} = 0 \text{ at the barrier for all } q_i \ (i = 1, \ldots, 3N-6).$$

The barrier has a further property, namely that of being a saddle point, that is, a passage in energy space from reactants to products. This means that if we station ourselves at the barrier, the potential energy, rises like a mountain pass about us no matter which direction we take, except one. This one unique direction, which we can thus define as the reaction coordinate $q^{\ddagger}$, is the only internal, normal coordinate of the system for which the potential energy is a maximum. For all other internal coordinates the potential energy is a minimum at the barrier; that is, any motion away from the barrier along the other coordinates leads to an increase in potential energy of the molecule. It is for this reason that the barrier, or as we shall call it, the transition state, is referred to as a saddle point.

If $k_A^{\ddagger}$ represents the specific rate constant for the passage over the barrier, the rate of isomerization is given by (neglecting the back reaction)

$$\frac{d(B)}{dt} = -\frac{d(A)}{dt} = k_A^{\ddagger}(A^{\ddagger}). \tag{3.3}$$

For equilibrium of $A^{\ddagger}$ and A with equilibrium constant $K_A^{\ddagger}$, $(A^{\ddagger}) = K_A^{\ddagger}(A)$; hence

$$-\frac{d(A)}{dt} = k_A^{\ddagger} K_A^{\ddagger}(A), \tag{3.4}$$

so that the first-order, isomerization rate constant k_A is given by

$$k_A \equiv -\frac{d(A)}{(A)\,dt} = k_A^{\ddagger} K_A^{\ddagger}. \tag{3.5}$$

If we could look at the equilibrium system $A \rightleftarrows B$, we would see equal numbers of molecules crossing the barrier in each direction. We could consider that motion across the barrier corresponded to a normal coordinate of the composite transition-state species consisting of equal numbers of $A^\ddagger$ and $B^\ddagger$. The frequency $\nu^\ddagger$ associated with a round trip $A^\ddagger$ to $B^\ddagger$ and back to $A^\ddagger$ can be assigned to this normal coordinate, and we note that $k_A^\ddagger = k_B^\ddagger = 2\nu^\ddagger$. Further, the total concentration of transition-state species is $(A^\ddagger) + (B^\ddagger)$, and because $(A^\ddagger)_{\text{eq}} = (B^\ddagger)_{\text{eq}}$, we could have originally written $(A^\ddagger) = \frac{1}{2}K^\ddagger(A)$ where $K^\ddagger = K_A^\ddagger + K_B^\ddagger$ is the true "equilibrium constant" for all transition-state species $A^\ddagger$ and $B^\ddagger$ at the top of the barrier. Hence we can rewrite (3.5)

$$k_A = \nu^\ddagger K^\ddagger. \tag{3.6}$$

The important assumption made in transition-state theory is that, at compositions far from equilibrium, for example, when back reaction $B \rightarrow A$ is negligible, and $(B^\ddagger) \ll (A^\ddagger)$, the concentration of $A^\ddagger$ is the same as it would be if $(B^\ddagger) = (A^\ddagger)$. This further implies that every $(B^\ddagger)$ that becomes (A^*) is principally deactivated to (A) and contributes negligibly to the net rate of production of $(A^\ddagger)$. This seems quite reasonable. It is also subject to experimental verification. If this condition were not satisfied, then an increase in concentration of inert gases added to the system could deactivate a newly formed B^* as it was about to return through the barrier to A^*. This would cause a net increase in the rate of reaction. The observation that the reaction rate is independent of the concentration of foreign, inert gases is thus a test of this assumption of equilibrium.

If we now factor out of $K^\ddagger$ the internal vibration coordinate that corresponds to passage across the barrier, we have, from statistical mechanics,

$$K^\ddagger = K^{\ddagger\prime}(1 - e^{-h\nu^\ddagger/kT})^{-1}, \tag{3.7}$$

and when $h\nu^\ddagger \ll kT,^2$ we can expand the exponential, so that

$$K^\ddagger \approx K^{\ddagger\prime}\left(\frac{kT}{h\nu^\ddagger}\right). \tag{3.8}$$

Substituting into (3.6) we obtain the transition state result

$$k_A = \left(\frac{kT}{h}\right)K^{\ddagger\prime}. \tag{3.9}$$

---

[2] The condition $h\nu^\ddagger < kT$ is equivalent to saying that $(\partial^2 V/\partial q^{\ddagger 2})$, the curvature at the saddle point in the direction of $q^\ddagger$, is not as sharp as for most normal molecular vibrations, or equivalently that the barrier is relatively flat. This seems very reasonable and for reaction coordinates involving motions of atoms heavier than D, it is hard to see how even its failure can cause a very serious error. For H or D atom motion, however, it may involve a more serious chance of error. In the absence of any real evidence bearing on this point, we shall ignore this prospect.

To avoid clumsy notation, we shall omit the prime and write $K^{\ddagger}$, it being understood that the internal coordinate corresponding to motion over the barrier has been factored out.

An additional factor, $\kappa \leq 1$, the transmission coefficient, is usually included in (3.9) and (3.3). It makes allowance for the possibility that a fraction $(1-\kappa)$ of transition-state complexes $A^{\ddagger}$ may be reflected back toward A^{*} before completing their passage to B^{*}. Very little is known about κ. It is a very complex function of the shape of the potential energy surface at the top of the barrier. It is the general practice to set $\kappa = 1$, and for simplicity, we have chosen to omit it.

The only classical way in which a reactant species could pass across the barrier and then be reflected back again would be if there were some reflecting wall, very close to the barrier and normal to the reaction coordinate. While this is possible in principle, it is difficult to imagine a molecular configuration with such tortuous properties, and it is doubtful if it is a serious likelihood in real systems. However, even such a possibility would be sensitive to the concentration of foreign gases, as noted above.

Although we have derived (3.9) for an isomerization, there is no difference in principle for any other type of unimolecular process and therefore it becomes a quite general result. Because bimolecular or termolecular events can be looked upon as involving preequilibrium formation of "collision" complexes, these can be absorbed into $K^{\ddagger}$, so that (3.9) becomes applicable to processes of any complexity that proceed through a single rate determining step, the transition-state barrier.

Thus for the bimolecular reaction

$$A + B \leftrightarrows [AB]^{\ddagger} \rightarrow \text{products}$$

the bimolecular, specific rate constant k_B is given by

$$k_B = \left(\frac{kT}{h}\right) K^{\ddagger}_{AB}$$

where $K^{\ddagger}_{AB}$ is the equilibrium constant for formation of $(AB^{\ddagger})$ from A and B.

The importance of the transition-state result is that all "equilibrium"-rate constants can be written as a product of a universal frequency factor (kT/h), which has the value $6.3 \times 10^{12}\,\text{s}^{-1}$ at $T = 300°K$ and a thermodynamic factor $K^{\ddagger}$, which depends only on the structure and energy of the transition state. The rate is thus independent of the details of the collision process.

Because we can write

$$-RT \ln K^{\ddagger} = \Delta G^{\circ \ddagger} = \Delta H^{\circ \ddagger} - T\,\Delta S^{\circ \ddagger}. \tag{3.10}$$

We may expect $K^{\ddagger}$ to be susceptible to analysis and correlation by the same additivity laws used for our ordinary thermochemical quantities.

3.3 UNIMOLECULAR REACTIONS—CATEGORIES

We classify unimolecular reactions into the following categories: (a) simple fission, which involves the breaking of a single bond; (b) complex fission, which involves the breaking of two or more bonds; and (c) isomerization, which involves only internal rearrangement of atoms; these may be as simple as a rotation of groups or more usually, a complex reorganization. Some examples from each of the above categories are:

Simple fission:

$$C_2H_5I \rightarrow C_2H_5 + I$$
$$CH_3{-}CH_3 \rightarrow 2CH_3$$

Complex fission:

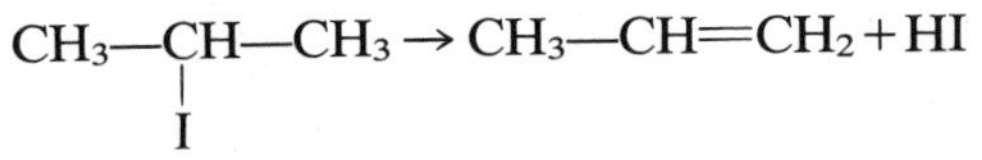

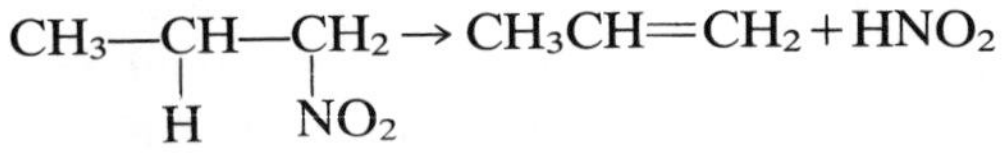

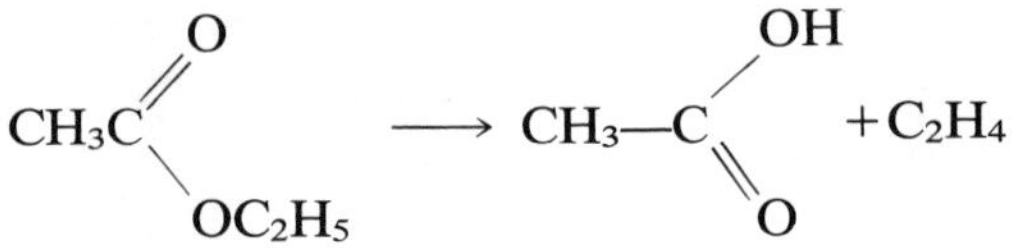

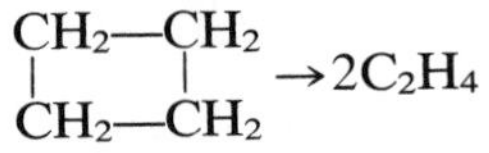

$$CH_3{-}\underset{\underset{I}{|}}{CH}{-}CH_3 \rightarrow CH_3{-}CH{=}CH_2 + HI$$

$$CH_3{-}\underset{\underset{H}{|}}{CH}{-}\underset{\underset{NO_2}{|}}{CH_2} \rightarrow CH_3CH{=}CH_2 + HNO_2$$

$$CH_3C\underset{OC_2H_5}{\overset{O}{\diagup}} \longrightarrow CH_3{-}C\underset{O}{\overset{OH}{\diagup}} + C_2H_4$$

$$\underset{CH_2{-}CH_2}{\overset{CH_2{-}CH_2}{|\quad\quad|}} \rightarrow 2C_2H_4$$

Isomerization:

$$\underset{CH_3}{\overset{H}{\diagdown}}C{=}C\underset{CH_3}{\overset{H}{\diagup}} \quad (cis) \longrightarrow \underset{H}{\overset{CH_3}{\diagdown}}C{=}C\underset{CH_3}{\overset{H}{\diagup}} \quad (trans)$$

$$\underset{CH_2{-}\!\!-\!\!-CH_2}{\overset{CH_2}{\diagup\diagdown}} \rightarrow CH_3{-}CH{=}CH_2$$

$$\underset{CH}{\overset{CH_2}{\diagdown}}\underset{\underset{CH_3}{\overset{|}{CH}}}{\overset{CH_2}{\diagup}}\underset{CH_2}{\overset{CH}{\diagdown}} \longrightarrow \underset{CH}{\overset{CH_2}{\diagdown}}\underset{CH_2}{\overset{CH_2}{\diagup}}\underset{\underset{CH_3}{\overset{|}{CH}}}{\overset{CH}{\diagdown}}$$

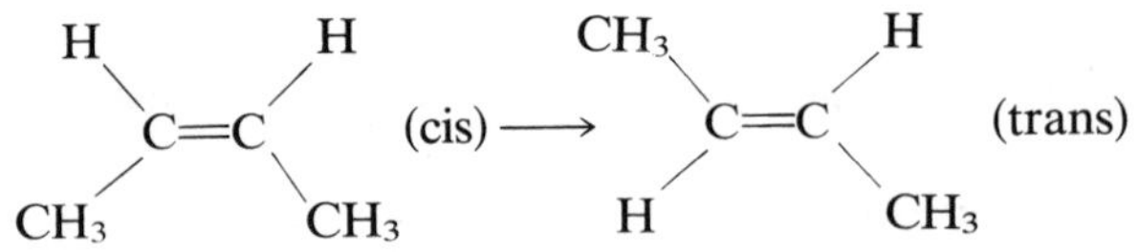

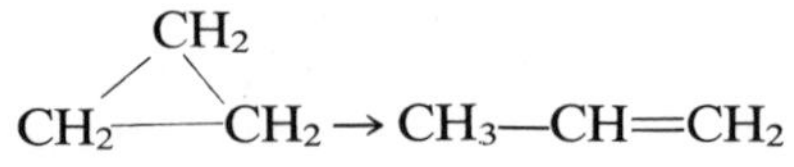

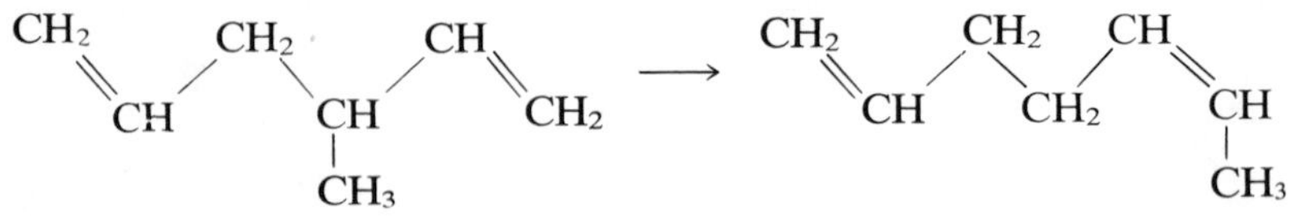

The simple fissions appear quite generally to be resolvable into a single elementary bond-breaking act. In contrast, the more complex fissions involving more than two bonds can be comprised of many consecutive elementary steps. Alternatively complex fission reactions may involve the simultaneous (concerted) breaking and/or making of two or more bonds.[3] Similarly, simple cis–trans isomerizations involving more than one bond usually may be considered to involve a sequence of resolvable, elementary acts. We consider each of these categories separately.

3.4 SIMPLE FISSION OF ATOMS FROM MOLECULES

For very simple molecules containing fewer than five atoms, the process of simple fission is likely to turn out to be controlled by the rate of accumulation of internal energy by collision and hence become a bimolecular process. This is true for the dissociation of diatomics such as H_2 or I_2 and for the triatomics such as O_3, NO_2, N_2O, and Cl_2O. Tetratomics such as F_2O_2 and NO_2Cl usually decompose bimolecularly but may become intermediate between unimolecular and bimolecular kinetics.

Most molecules with five to eight atoms will be of intermediate molecularity. More complex molecules are generally in the first-order, unimolecular region with a rate controlled by bond fission. However, at sufficiently high temperatures the decompositions of *all* molecules become collision controlled and hence bimolecular.

For the simple bond-fission processes, the transition-state result gives us a facile starting point. Consider the unimolecular, simple fission of AB. We can represent the process by

$$AB \rightleftarrows AB^{\ddagger} \to A + B.$$

Then from (3.9) the first-order rate constant for decomposition of AB is given by

$$k_{AB} = \left(\frac{kT}{h}\right) K^{\ddagger}_{AB},$$

or in thermodynamic language

$$k_{AB} = \frac{kT}{h} \exp\left(\Delta S^{\ddagger}_{AB}/R\right) \exp\left(-\Delta H^{\ddagger}_{AB}/RT\right). \tag{3.11}$$

---

[3] In the unimolecular fission of symmetrical trioxane, cyclo-$(CH_2O)_3 \to 3CH_2O$, three single bonds are broken concertedly with the formation of three pi bonds.

If we use the Arrhenius equation $Ae^{-E/RT}$ for k_{AB} as a point of comparison, we can identify

$$A = \left(\frac{ekT_m}{h}\right) \exp\left(\Delta S^{\ddagger}_{AB}/R\right); \qquad E = \Delta H^{\ddagger}_{AB} + RT_m \qquad (3.12)$$

where T_m is the mean temperature at which the experiments have been carried out (see Sec. 1.6).

We note that because (ekT_m/h) will be in the range of $10^{13.55}$ s^{-1} for most experiments (600°K), A will differ from this usual frequency factor only to the extent that $\Delta S^{\ddagger}_{AB}$ differs from zero. When, for example, $AB^{\ddagger}$ is "bigger" and "looser" than AB we may expect $\Delta S^{\ddagger}_{AB} > 0$ and in consequence an abnormally large A factor. When, in contrast, $AB^{\ddagger}$ is more compact and "stiffer" than AB, we may expect $\Delta S^{\ddagger}_{AB} < 0$ and an abnormally small A factor. These possibilities are categorized as "loose" or "tight" transition-state complexes, respectively.

We can expect in simple fission reactions, which involve detachment of single atoms from a large molecule, that $\Delta S^{\ddagger}_{AB}$ will be small except for a statistical factor. Thus in the reaction $C_2H_6 \rightarrow \dot{C}_2H_5 + H$ any of six equivalent H atoms may become detached, and this leads to a symmetry contribution to $\Delta S^{\ddagger}_{AB}$ of $R \ln 6$.[4] This is also called the reaction path degeneracy. Optical isomerism in the transition state can also increase the number of reaction paths.

This is about as far as simple reasoning will take us. To proceed further we need to know in some detail the structure of the transition state. This in turn means that we need to know how $CH_3CH_2 \cdots H$ looks at every stage in the process of breaking the C—H bond and creating the $\dot{C}_2H_5$(ethyl) radical plus a separated H atom. A theoretical solution to this problem is out of the question at this time. The alternative to which we can have recourse is to use what information does exist for the breaking of simple bonds, and it comes from the wealth of information, mostly spectroscopic, on diatomic molecules. For one of these, H_2, we even have an accurate solution of the Schroedinger equation.

3.5   POTENTIAL ENERGY OF SIMPLE BONDS

Current experimental data suggest that the potential of interaction between atoms in a diatomic molecule can be represented by a simple Lennard–Jones (V_{LJ}) or Morse (V_M) or Buckingham exponential (V_{BE})

---

[4] Note that $K^{\ddagger}$ contains this factor $(\sigma_{AB}/\sigma^{\ddagger}_{AB})$ directly. For more discussion, see S. W. Benson, *J. Am. Chem. Soc.*, **80**, 5151 (1958) and S. W. Benson and W. B. DeMore, *Ann. Rev. Phys. Chem.*, **16**, 433 (1965).

function which depends only on the internuclear distance r. If we express the internuclear distance in terms of the dimensionless variable $\rho = (r/r_0)$, where r_0 is the equilibrium internuclear distance, then:

$$V_{LJ} = -D_0\left(\frac{2}{\rho^6} - \frac{1}{\rho^{12}}\right) \tag{3.13}$$

$$V_M = -D_0 + D_0[1 - e^{-\alpha(\rho-1)}]^2 \tag{3.14}$$

$$V_{BE} = \frac{-D_0}{(1-6/\alpha)}\left[\frac{1}{\rho^6} - \frac{6e^{-\alpha(\rho-1)}}{\alpha}\right] \tag{3.15}$$

$\alpha = (r_0/d)$, where d is the so-called "range" of the potential. It is the distance in which the exponential will change by a factor of e. For the Morse function, the repulsive function has a range $= (d/2) \sim 0.2$ to 0.3 Å.

D_0 is the potential energy of interaction between the atoms when $r = r_0$ (i.e., $\rho = 1$) and differs from the DH_0° bond-dissociation energy at $0°K$ by the zero-point energy of the molecule E_0°. All of these curves have the same general shape shown in Figure 3.2. At $r = r_0$, $V = -D_0$, and this represents a minimum in the potential energy, hence:

$$\left(\frac{dV}{d\rho}\right)_{\rho=1} = 0 \tag{3.13}$$

is a relationship satisfied by all of the functions.

The additional parameter α can be related to the vibrational frequency ν of the diatomic molecule by the relation for a harmonic oscillator:

$$\nu = \frac{1}{2\pi}(k/\mu)^{1/2} \tag{3.14}$$

where μ is the reduced mass for the vibration and with the force constant k equal to the second derivative of the potential:

$$k = \frac{d^2V}{dr^2} \tag{3.15}$$

For the three potential functions we find:

$$k_{LJ} = \frac{72D_0}{r_0^2}$$

$$k_M = \frac{2\alpha^2 D_0}{r_0^2}$$

$$k_{BE} = \frac{6\alpha(7-\alpha)D_0}{(6-\alpha)r_0^2} \tag{3.16}$$

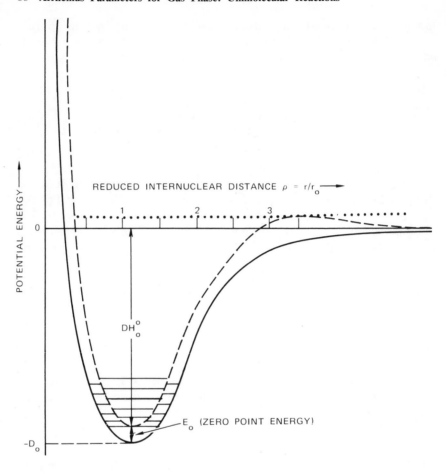

Figure 3.2   Potential energy function for a diatomic molecule. The solid curve represents
the potential energy of a nonrotating molecule. The horizontal lines designate the quantized vibrational levels. The dashed curve represents the potential energy of a rotating
molecule $V(\rho) + E_{rot}$ (see text). The dotted line shows the energy of a hypothetical molecule
with sufficient energy to dissociate.

For any molecule not at 0°K, in particular for one possessing rotational
energy, we must add to these potential functions a term expressing the
centrifugal force arising from the rotational energy E_{rot}:

$$E_{rot} = \frac{p_\theta^2}{2I_r} = \frac{p_\theta^2}{2\mu r^2} = \frac{E_{rot}^\circ}{\rho^2} \tag{3.17}$$

where p_θ is the angular momentum, which is constant for an isolated molecule. Note that the centrifugal force is always repulsive. E°_{rot} is the rotational energy at $\rho = 1$.

The centrifugal force term introduces a maximum in the potential energy function for diatomic molecules at large distances and gives us a natural position in phase space at which to locate the transition state. If we take the total potential and rotational energy expression, we can solve for the maximum and locate it:

$$V_{total} = V(\rho) + E_{rot}$$

$$\frac{\partial V_{total}}{\partial \rho} = 0 \qquad \text{at } \rho = \rho_{max}$$

The short-range experimental repulsive term in any of our potential functions falls off much more rapidly with distance than either the attractive or rotational terms and hence can be neglected at large distances. Thus taking the Lennard–Jones potential, we find:

$$V_{total} = -D_0\left(\frac{2}{\rho^6} - \frac{1}{\rho^{12}}\right) + \frac{E^\circ_{rot}}{\rho^2} \tag{3.18}$$

$$\xrightarrow[\rho > 1]{} -\frac{2D_0}{\rho^6} + \frac{E^\circ_{rot}}{\rho^2} \tag{3.19}$$

$$\left(\frac{\partial V_{total}}{\partial \rho}\right)_{\rho = \rho_{max}} = 0 = +\frac{12D_0}{\rho_{max}^7} - \frac{2E^\circ_{rot}}{\rho_{max}^3}$$

whence, solving for ρ_{max} we find:

$$\rho_{max}^4 = \left(\frac{6D_0}{E^\circ_{rot}}\right). \tag{3.20}$$

If now we assume that transition-state complexes at the top of the barrier will have average rotational energies $= RT$, then:

$$E_{rot}(\rho = \rho_{max}) = E^\circ_{rot}/\rho_{max}^2 = RT \tag{3.21}$$

and

$$E^\circ_{rot} = RT\rho_{max}^2. \tag{3.22}$$

Substituting this result in equation (3.20) and solving, we find:

$$\rho_{max} = \left(\frac{6D_0}{RT}\right)^{1/6}. \tag{3.23}$$

Because of the small fractional power dependence, this maximum is not very sensitive to either T or the precise value of D_0, and for most common values of both occurs in the range $\rho_{max} = 2.5$–3.0. We would find the same result for the Buckingham potential, and a very similar result for the Morse potential. The implication for bond-breaking reactions is that we may expect very large changes in bond lengths in simple bond-breaking reactions. This will result in an increased value for two of the rotational partition functions and a resultant increase in $\Delta S^{\ddagger}$.

If we use an r^{-6} attractive potential at long distances, then our model further predicts that at $\rho = \rho_{max}$, $V = V_{max} = \frac{2}{3}E^{\circ}_{rot\,max} = \frac{2}{3}RT$. If we further assume that the average relative kinetic energy along the line of centers is the classical $\frac{1}{2}RT$ for a single degree of freedom, then the separating pair will have a total relative translational kinetic energy of $\frac{7}{6}RT$ compared to RT for a random pair of molecules. Hence $E_{act} = \Delta E - \frac{1}{6}RT \approx \Delta E$ at $300°K.$[5]

For the hypothetical, first-order decomposition of a diatomic molecule, $\Delta S^{\ddagger}$ has contributions only from the rotational partition function and hence:

$$\Delta S^{\ddagger} = R \ln\left(\frac{I_r^{\ddagger}}{I_r}\right) = R \ln\left(\frac{\rho^{\ddagger}}{\rho_0}\right)^2$$

$$= 2R \ln\left(\frac{\rho^{\ddagger}}{\rho_0}\right). \tag{3.24}$$

If $\rho^{\ddagger} \cong 2.75$, then $\Delta S^{\ddagger} \approx 2R = 4.0$ eu. For such a diatomic molecule the reaction path degeneracy is unity, and at $300°K$ we would calculate:

$$A = \left(\frac{ekT_m}{h}\right)\exp\left(\Delta S^{\ddagger}/R\right) = 10^{14.1}\,s^{-1}$$

There is no way of evaluating this result, since diatomic molecules never undergo fission in a first-order process. Instead, their rates of fission are determined by the rates of energy transfer, a bimolecular process.

Transition states for reactions in which very large changes in bond lengths occur will be referred to as "loose" transition states. The experimental data indicate that most simple bond-fission reactions have this kind of transition state.

3.6 A-FACTORS FOR SIMPLE FISSION

It seems reasonable to expect that more complex molecules undergoing simple bond fission can be approximated by the simple potential energy function we have just employed for diatomics. The assumption implicit in such an approximation is that bond breaking is a localized process with very little effect on the neighboring or more distant atoms in the

---

[5] Essentially, rotational energy used in bond breaking is somewhat inefficient.

molecule. This may or may not be the case, and we shall have to examine each particular molecule to see if such an assumption is reasonable.

If we take the case of CH_4 as a simple example, we can consider the breaking of a single C—H bond. Current information on the resulting $\dot{C}H_3$ radical, even though somewhat sparse, suggests that there are no great changes in C—H bond lengths nor even in H–C–H angles in comparing CH_4 with $\dot{C}H_3$. The $\dot{C}H_3$ radical is nearly planar with an angle of ~117° for (H–C–H) compared to the corresponding tetrahedral angle of 109.5° in CH_4. We can thus take as our transition state for the bond rupture a molecule in which the CH_3 part is like the $\dot{C}H_3$ free radical, and the remaining C—H bond is extended to about 2.8 times its ground-state distance of 1.10 Å, or 3.08 Å (Figure 3.3). In the parent molecule this H-atom had three internal coordinates associated with it, one stretch and two bends. The stretch has become the reaction coordinate, and we need not consider it further. The two bending motions have become relatively large amplitude motions. It is unlikely in this exaggerated geometry that the restoring forces of normal bonds are operative, and we shall take an extremely simple model that assumes that the $C \cdots H$ bond can bend through an angle θ until the moving H atom makes van der Waal contact with its neighboring H atoms. The conical angle θ in Figure 3.3 represents this solid angle.

If we choose 1.3 Å as the van der Waal radius of nonbonded H atoms (i.e., the distance at which their potential of interaction is zero), then we can solve the geometrical problem and obtain a value of $\theta \sim 36°$ or 0.62

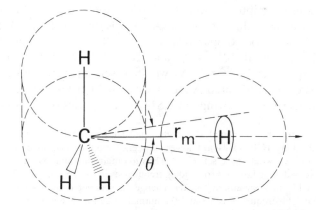

Figure 3.3 Free-volume model for the transition state of C—H bond fission in CH_4. CH_3 residue is assumed flat; stretched C—H bond is 2.8 times normal length. A van der Waal radius of 1.3 Å is drawn about each H atom. θ is the conical angle through which the H atom can bend.

radians. The two bending motions of the H atom relative to the CH_3 group thus have become a two-dimensional rotation, restricted to a relative solid angle $(\Delta\Omega/4\pi)$, given by:

$$\frac{\Delta\Omega}{4\pi} = \frac{1-\cos\theta}{2} \approx \frac{\theta^2}{4} \quad \text{(for } \theta < 1)$$

The entropy of a restricted two-dimensional rotor is given by (2.31):

$$S_{rot} = 6.9 + R\ln\left(\frac{I_r}{\sigma}\right) + R\ln(T/298) + R\ln\left(\frac{\Delta\Omega}{4\pi}\right) \tag{3.25}$$

Calculating I_r by the methods given earlier (Section 2.12), we have $I_r = I_1 \cdot I_2/(I_1 + I_2)$ with $I_1 \sim 1.5 r^2_{C-H}(\text{amu-Å}^2)$ and $I_2 \sim 7.8 r^2_{C-H}(\text{amu-Å}^2)$. I_1 is the moment of inertia of the flat CH_3 group about an axis through the C atom and perpendicular to the plane of Figure 3.3. I_2 is approximately the moment of the $C\cdots H$ long bond about this same C-atom axis. With $\sigma = 3$ we find $I_r/\sigma = 0.5$ amu-Å^2 and on substituting these values in (3.25), we find $S_{rot} \simeq 1.0$ eu at 298°K, a rather surprising result. We might have anticipated a greater contribution from such a large change in freedom of motion of the departing H atom. One reason for this small contribution arises from the fact that H has a very light mass, and the reduced moments are correspondingly small. A second reason is to be found in the symmetry factor for the internal rotation $\sigma = 3$. Our bookkeeping system applies this symmetry to the internal rotation but in compensation, the external symmetry of the TS is unity compared to 12 for CH_4.[6]

The various contributions to $\Delta S^{\ddagger}$ at 300°K are summarized in Table 3.1 and lead to a net value of 8.4 eu. At 300°K this gives an A factor of $10^{15.1}$ s^{-1}. This can be compared with values observed in higher temperature studies of about $10^{14.7}$ s^{-1},[7] and $10^{14.9}$ s^{-1},[8] both estimated to be uncertain by about $10^{\pm0.5}$ s^{-1}. As we shall see later, to make a proper comparison of our estimated A factor with the high-temperature data, we need to correct for the change in $\Delta S^{\ddagger}$ with temperature. This is done by

---

[6] If we had treated the TS complex as a distorted CH_4 molecule with the two bending motions of the long H bond as just lower frequency motions instead of restricted free internal rotations we would have set $\sigma^{\ddagger} = 3$. Then to obtain the same overall result for $\Delta S^{\ddagger}$ we would have had to assign the two bending frequencies in the TS values of 225 cm^{-1} each (see Table A.17) to compensate for the change in symmetry contribution to $\Delta S^{\ddagger}$ of $R\ln 3 = 2.2$ eu. This frequency is about $\frac{1}{6}$ the normal value for such a motion (Table A.13) and implies a 36-fold reduced force constant for the bending motion and a six-fold larger amplitude of motion for such a H atom.

[7] G. B. Skinner and A. A. Ruehwein, *J. Phys. Chem.*, **63**, 1736 (1959), shock-tube studies.

[8] H. B. Palmer and T. J. Hirt, *J. Phys. Chem.*, **67**, 709 (1963), flow studies.

computing $\Delta C_p^{\ddagger}$ and when it is done in the above case (Sec. 3.8), it leads to a value of $10^{14.5} \, s^{-1}$ at 1500°K.

It is instructive to make a comparison with the A factor expected for the fission of the tertiary C—H bond in isobutane, $(CH_3)_3C$—H. Now because the rest of the molecule is so heavy and large, we should expect that the extension of the C—H bond would have little effect on any of the three principal moments of inertia; hence we should omit the 2.4 eu

Table 3.1. Contributions to $\Delta S^{\ddagger}$ at 300°K for the Dissociation of the C—H Bond in CH_4

Degree of Freedom	Contribution to $\Delta S^{\ddagger}$
Translation (no change)	0
Rotation (2 moments of inertia each increase by a factor of about 3.3)	$R \ln 3.3 = 2.4$
External symmetry ($\sigma = 12; \sigma^{\ddagger} = 1$)	$R \ln 12 = 5.0$
Reaction coordinate C—H stretch $\sim 3000 \, cm^{-1} \rightarrow \nu^{\ddagger}$	0.0
2 H–C–H bends $\rightarrow$ restricted free rotations in two dimensions (see text) with $\sigma_{int}^{\ddagger} = 3$	1.0
Total	$\Delta S^{\ddagger} = 8.4$

contribution from this source which we listed for CH_4 (Table 3.1). Also the external symmetry change is from 3 to 1; hence we will have a contribution of only 2.2 eu from this source. However the reduced moment of the bending motion is about three-fold larger than in the case of CH_4, and this contributes $R \ln 3 = 2.2$ eu. If we now assume that the two-dimensional rotation of the H atom has about the same value of $(\Delta \Omega / 4\pi)$ as for the case of CH_4, we arrive at $\Delta S^{\ddagger}(300°K) = 6.3$ eu and $A = 10^{14.7} \, s^{-1}$, a factor of 3 lower than for CH_4. Most of this, of course, arises from the difference in reaction-path degeneracy.

When the two fragments composing the bond are comparable in moment of inertia, then we may expect larger changes in moments of inertia, correspondingly larger contributions to $\Delta S^{\ddagger}$ and thus larger values of A factors. A typical example would be the fission of CH_3I into $CH_3 + I$.

If we follow the prescription for a loose transition state, the $(C \cdots I)^{\ddagger}$ distance becomes $2.8r_{C-I} \simeq 5.9$ Å (see Table A.14). This is appreciably larger than the sum of the van der Waal radii of CH_3 and I^9 and suggests that the CH_3 group could be undergoing completely free internal rotation in the transition state.

In the rupture of the C—I bond in CH_3I, three modes of internal motion are converted to three translational degrees of freedom of the I atom. We can identify these degrees of freedom as the C—I stretch and two H—C—I bends.[10] Table 3.2 lists the various contributions to $\Delta S^{\ddagger}$ for the bond rupture in CH_3I. The total value $\Delta S^{\ddagger} = 10.5$ eu would then lead to an A factor of $10^{15.55}$ s^{-1} at 300°K. There are unfortunately no experimental data with which to compare this number.

When internal changes are coupled to the breaking bond, we must

Table 3.2.   Contribution to $\Delta S^{\ddagger}$ at 300°K for the Dissociation of the C—I Bond in CH_3I

Degree of Freedom[a]	Contribution to $\Delta S^{\ddagger}$
Translation (no change)	0
Rotation (two moments each increase by a factor of about $2.8^2 = 7.8$)	$R \ln 7.8 = 4.1$
External symmetry (no change) $(\sigma = 3; \sigma^{\ddagger} = 3)^b$	0
Reaction coordinate C–I stretch at 500 cm$^{-1} \rightarrow \nu^{\ddagger}$	-0.7
Two CH_3—I rocks (700 cm^{-1}) $\rightarrow$ 2 free rotations[b]	7.1
TOTAL	10.5

[a] For frequencies, see Table A.13; for corresponding entropies at 300°K, see Table A.17.

[b] The overall symmetry of a flat, symmetrical CH_3 is $3 \times 2 = 6$. We have chosen here to assign the factor of 3 to the rotation about an axis perpendicular to its plane. This rotation is then essentially the same as that of the CH_3I molecule about the C–I axis. The other factor of 2 is then assigned to the other two rotations about axes perpendicular to the C–I axis.

[9] As a rough rule, the van der Waals radius of an atom in a molecule can be taken as 0.95 Å larger than its covalent radius. This gives for carbon, 1.7 Å, for H, 1.3 Å, and for I, 2.3 Å. For CH_4 this gives a contact radius of 2.4 Å along a C—H bond, 2.2 Å in a CH_2 plane, and about 1.6 Å opposite a C—H bond for an average of about 2.1 Å.

[10] Actually, they correspond to two degenerate rocking motions of the methyl group relative to the C—I axis. When considering the internal motions, either stretches or bends, the laws of conservation of momentum require that the masses undergoing the largest amplitude motions will always correspond to the lightest atoms.

explore them in detail. For example, the reaction $\dot{C}_2H_5 \rightarrow C_2H_4 + H$ involves a considerable change in structure. We lose a hindered rotation in the C_2H_5 radical as the double bond begins to form. We may picture the transition state as

The contributions to $\Delta S^{\ddagger}$ are tabulated in Table 3.3

The result is $\Delta S^{\ddagger} = 0.4$ and $A(300°K) = 10^{13.4}\,s^{-1}$. This is much lower than the A factors for simple fissions we have considered earlier. The difference arises from the nature of the reaction. Atom–atom interactions extend out to large distances, as do atom–radical and radical–radical interactions; hence we anticipate loose transition states for the fission reactions of molecules where rupture leads to formation of two odd-electron species. The decomposition of the ethyl radical into a stable molecule plus an atom is different. We have no reason to anticipate interactions out to large distances when one of the fragments is a stable molecule, and this is in accord with the observation that reverse reaction, the addition of an H atom to C_2H_4 has a small activation energy and an A

Table 3.3. Contributions to Entropy of Activation at 300°K for $C_2H_5 \rightleftarrows (C_2H_5)^{\ddagger} \rightarrow C_2H_4 + H$

Degrees of Freedom		$\Delta S^{\ddagger}$
Translation	(no change)	0
Rotation	(neglect small changes)	0
Symmetry[a]	($\sigma = 6$; $\sigma^{\ddagger} = 1$)	$R \ln 6 = 3.6$
Spin	(no change)	0
Internal		
C—H stretch ($3100\,cm^{-1}$) $\rightarrow$ reaction coordinate		0.0
C—H stretch ($1000\,cm^{-1} \rightarrow 1300\,cm^{-1}$)		-0.1
H·C—C bend ($1150\,cm^{-1} \rightarrow 800\,cm^{-1}$)[b]		0.2
H·C—H bend ($1450\,cm^{-1} \rightarrow 1000\,cm^{-1}$)		0.1
CH$_2$ internal rotation (1.5 kcal barrier $\rightarrow$ 1/2 ethylene torsion $500\,cm^{-1}$)[c]		-3.4
Total		0.4

[a] A factor of 2 comes from the loss in rotation of the —CH$_2$ group. It is included here, rather than in the internal rotation correction.
[b] See Table A.13.
[c] See Tables A.18 and A.21.

factor that is in accord with our estimated A factor for the back reaction and the overall entropy change.[11]

We can anticipate at this point a conclusion from our later discussion, namely, that the transition states for radical–molecule reactions and molecule–molecule reactions will have what are termed "tight" transition states. These are transition states in which breaking or making bonds are within about 0.4 Å of their normal lengths. External rotations change very little in such transition states from the initial molecule, and only if there are large changes in internal torsions will $\Delta S^{\ddagger}$ be very different from zero.

We may expect similar effects to those described above for $\dot{C}_2H_5$ in other molecules where conjugative or resonance effects are to be expected. Thus the dissociation of H from propylene leads to a stiff allyl radical and a change of a nearly free rotation of the CH_3 group ($V_0 = 2$ kcal) to a CH_2 semi-torsion. A similar effect occurs for toluene when it loses an H atom and forms a stiff benzyl radical.

The loss of a secondary H atom from butene-1 is even more costly because here the formation of the stiff methyl allyl radical CH_3—$\dot{C}H{=}{=}CH{=}{=}\dot{C}H_2$ involves the freezing of the hindered rotation of two "heavy" groups, [ethyl-vinyl] in butene-1 and at 600°K is about 7. In compensation, however, we get back a semi-torsional motion, which we assign at half the frequency of the $300\ cm^{-1}$ associated with butene-2 or $150\ cm^{-1}$. It contributes about 3.0 eu at 300°K. At 600°K it contributes about 4.0. This appears to be a quite general behavior, which has the result that, no matter what the size of the group, we can assign a loss of about 3.5 ± 0.5 to changes from hindered rotation to semi-pi-bond torsion (see Table A.21). In addition, the terminal CH_2 in butene-1 loosens from a torsion at $\sim 750\ cm^{-1}$ to about half that, or $375\ cm^{-1}$, which represents a net gain of $+1.3$ at 600°K.

There are unfortunately no reliable quantitative data on simple fission of atoms with which to compare the estimates made above. The semi-quantitative data that do exist are, however, quite compatible with them.

3.7 SIMPLE FISSION INTO TWO LARGER GROUPS

The fission of a large polyatomic molecule AB into two smaller, but still polyatomic fragments A and B converts six internal modes of A—B into

[11] From data in the Appendix we can calculate $\Delta S^{\circ}_{300} = 21.5$ eu for $\dot{C}_2H_5 \rightarrow C_2H_4 + H$. This leads to an A factor of $10^{10.4}$ l/mole-s for the reverse addition reaction at 300°K, in excellent agreement with the observed value of $10^{10.4}$ l/mole-s reported in V. N. Kondratiev, Rate constants of gas-phase reactions, U.S. Govt. Printing Office, Washington, D.C. (1972). See also J. A. Kerr and M. J. Parsonage, *Evaluated Kinetic Data on Gas Phase Addition Reactions*, Butterworths, London, 1972.

three new translations and three new rotations of the products. These six internal modes correspond to: (a) one A—B bond stretch (reaction coordinate), (b) one A—B internal rotation, (c) two rocking modes or restricted rotation of A relative to A—B axis, and (d) two rocking modes or restricted rotation of B relative to A—B axis.

The fission of AB is expected to give rise to substantial increases in entropy in the transition state and hence to very large A factors. The main reason for this lies in the fact that any appreciable increase in the A—B bond length is expected to weaken considerably the force constants controlling the four rocking modes of A and B groups relative to the A—B axis. It is only in the past two decades that reasonably reliable kinetic data have become available for such processes. Table 3.4 summarizes some of the data for this category of reaction.

We notice that the A factors for the reactions described above range mainly between 10^{15} to 10^{17} sec^{-1}. They are not all completely comparable because they cover measurements from 330 to 950°K, and this is expected to have a small effect on the A factors.

Omitting the measurements involving ϕ-$\dot{C}H_2$, and with due respect to all of the errors implicit in the measurements of A factors, it is reasonable to represent the remaining group by a mean A factor of $10^{16\pm1}$ s^{-1}. This implies $\Delta S^{\ddagger}$ for these reactions of about 11.5 ± 5 gibbs/mole with about 8 gibbs/mole attributable to the changes in the four rocking frequencies and 3.5 gibbs/mole from increased rotational moment. This would imply an increase in each of the vibrational contributions from these modes of about 2.0 ± 1 gibbs/mole.

For rocking modes involving CH_3, NH_2, or OH groups, the initial frequencies are about 950 cm^{-1} (Table A.13) with a $S_{vib.}$ contribution at 600°K (Table A.17) of about 0.7 eu. For each of these to contribute 2.0 more to $\Delta S^{\ddagger}$ at 600°K requires a lowering of the rocking frequency to about 300 cm^{-1} in the transition state. This represents a decrease in the rocking force constants k_f of about a factor of 10, since $\nu \propto k_f^{1/2}$. It would imply a very low barrier to rocking in the transition state.[12]

However, such high A factors are precisely what we would expect for the loose transition states arising from the simple fission of bonds in larger molecules, and the restricted internal rotation model which we developed for atom–radical fission can account quite well for these low "bending" motions. Perhaps the best documented example is the pyrolysis of C_2H_6 into $2CH_3$ radicals and the reverse reaction, the recombination of CH_3 radicals. The rate constant for this recombination

---

[12] The structural implications of such low frequencies are discussed by S. W. Benson, *Advances in Photochemistry*, Interscience, New York, 1964, Vol. 2, p. 1.

Table 3.4 Arrhenius Parameters for Simple Fission of Molecules into Two Radicals

Reference[1]	Reaction	T_m (°K)	$\log A$ (s^{-1})	E (kcal/mole)
a	$C_2H_6 \rightarrow 2CH_3$	800	17.45	91.7
b		850	16.0	86.0
c	$C_4H_{10} \rightarrow 2C_2H_5$[2]	400	17.4	82
d	$C_2F_6 \rightarrow 2CF_3$[2]	350	17.4	88±3
e	$N_2O_5 \rightarrow NO_3 + NO_2$	330	14.8±1	21±2
f	$CH_3OOCH_3 \rightarrow 2CH_3O$	410	15.4	36.1
g	$t\text{-BuOO}t\text{-Bu} \rightarrow 2t\text{-BuO}$	410	15.6	37.4
h	$(CH_3CO)OO(CH_3CO) \rightarrow 2CH_3CO_2$	330	14.9	31
i	$C_2H_5ONO \rightarrow C_2H_5O + NO$	430	16.0	41.8
j	$C_2H_5ONO_2 \rightarrow C_2H_5O + NO_2$	420	16.85	41.2
k	$CH_3N{=}NCH_3 \rightarrow CH_3 + NNCH_3$	600	17.2	55.5
l	$CF_3N{=}NCF_3 \rightarrow CF_3 + NNCF_3$	600	16.17	55.2
m	$\Phi\text{-}CH_2CH_3 \rightarrow \Phi\text{-}CH_2 + CH_3$	950	14.6	70.1
	$\Phi\text{-}CH_2C_2H_5 \rightarrow \Phi\text{-}CH_2 + C_2H_5$	950	14.9	68.6
n	$C(NO_2)_4 \rightarrow \cdot C(NO_2)_3 + NO_2$	470	17.5	40.9
o	$(C_2H_5)_2Hg \rightarrow C_2H_5Hg + C_2H_5$	630	15.4	45.7
i	$t\text{-BuONO} \rightarrow t\text{-BuO} + NO$	430	16.3	40.3
p	$t\text{-BuNO} \rightarrow t\text{-Bu} + NO$	700	15.6	36.0

1. Bibliographic sources:
   a. C. P. Quinn, *Proc. Roy. Soc.* (London), **A275,** 190 (1963).
   b. M. C. Lin and M. H. Back, *Can. J. Chem.*, **44,** 505, 2357 (1966).
   c. A. Shepp and K. O. Kutschke, *J. Chem. Phys.*, **26,** 1020 (1957).
   d. P. B. Ayscough, *J. Chem. Phys.*, **24,** 944 (1956).
   e. R. L. Mills and H. S. Johnston, *J. Am. Chem. Soc.*, **73,** 938 (1951).
   f. P. L. Hanst and J. G. Calvert, *J. Phys. Chem.*, **63,** 104 (1959).
   g. L. Batt and S. W. Benson, *J. Chem. Phys.*, **36,** 895 (1962).
   h. O. J. Walker and G. L. E. Wild, *J. Am. Chem. Soc.*, **1132,** (1937).
   i. L. Batt and R. T. Milne, *Int. J. Chem. Kinet.*, **6,** 945 (1974).
   j. J. B. Levy, *J. Am. Chem. Soc.*, **76,** 3254, 3790 (1954).
   k. W. Forst and O. K. Rice, *Can. J. Chem.*, **41,** 562 (1963).
   l. E. Leventhal, C. R. Simonds, and C. Steel, *Can. J. Chem.*, **40,** 930 (1962).
   m. G. L. Esteban, J. A. Kerr, and A. F. Trotman-Dickenson, *J. Chem. Soc.*, 3873 (1963).
   n. J. M. Sullivan and A. E. Axworthy, Jr., *J. Phys. Chem.*, **70,** 3366 (1966).
   o. A. C. Lalonde and S. J. W. Price, *Can. J. Chem.*, **49,** 3368 (1971).
   p. K. Y. Choo, G. D. Mendenhall, D. M. Golden, and S. W. Benson, *Int. J. Chem. Kinet.*, **6,** 813 (1974).
2. Parameters calculated from data on rate of back reaction and overall thermochemistry (see Table 4.6).

has been measured by a variety of techniques and over a range of temperatures from 300 to 500°K. The observed value independent of temperature for the recombination is $10^{10.5 \pm 0.2}$ l./mole-s.[13] Combining this with the standard entropy change for the reaction at 300°K (see tabulated data in Appendix) of 37.9, we can calculate an A factor for the fission reaction of A (300°K) = $10^{17.0 \pm 0.2}$ s^{-1}. This corresponds to $\Delta S^{\ddagger}$

Table 3.5. Contributions to the Entropy of Activation at 300°K for the Fission $C_2H_6 \rightarrow 2CH_3$

Degrees of Freedom	$\Delta S^{\ddagger}$
Translation (no change)	0
Rotation (2 large moments of inertia each change by factor of 5.5)^a	$R \ln 5.5 = 3.4$
Symmetry ($\sigma = 18$; $\sigma^{\ddagger} = 18$; no change)	0
Spin (no change)	0
Barrier to internal rotation decreases from 3 kcal to 0 (Table A.20)	1.4
C—C stretch (1000 cm^{-1}) $\rightarrow \nu^{\ddagger}$ reaction coordinate	−0.1
Four CH$_3$ rocks at about 1000 cm^{-1} $\rightarrow$ 4 free rotations of CH$_3$ groups ($\sigma = 2$)^b	13.0
Total	17.7

^{a} Although the C-C distance is assumed to change by a factor of 2.8, the moment of inertia about the symmetry axis perpendicular to the C-C axis changes by only a factor of 5.5 if the CH$_3$ radicals in TS are assumed to be planar.

^{b} The four rocking motions split into two degenerate pairs, symmetric and antisymmetric, at 1150 cm^{-1} and 820 cm^{-1} each. The differences arise from the different reduced masses for the motions.

(300°K) = 17.2 ± 0.9 eu, in excellent agreement with the entropy changes computed in Table 3.5 for the loose transition state.

When the bond fission is coupled to other changes in the molecule due to pi-bond stabilization, the resultant transition state is stiffer by some amount and the A factor is thereby reduced. The extent of the loss in entropy attendant on such stiffening may be estimated as follows. Basically the change comes from the increase in the barrier to internal rotation. If the rotating groups are large ones, then the frequency describing the hindered torsion will be relatively low compared to (kT/h) at 300°K (i.e., 220 cm^{-1}), and the torsional partition function can be taken as $(kT/h\nu)$. The barrier change is usually from 2–3 kcal in the

[13] F. C. James and J. P. Simons, *Int. J. Chem. Kinet.*, **6**, 887 (1974).

molecule to 12–15 kcal in the transition state, a factor of about 6, and the frequency increases by about the square root. Hence, since $S_{vib} \approx R \ln \nu + constant$ for $\nu < kT/h$, $\Delta S^{\ddagger}_{vib} \sim R \ln 6^{1/2} = 1.8$ with a range of about ± 0.6.

In the fission of a benzyl compound $\phi CH_2 R \rightarrow \phi \dot{C} H_2 + R$, where R is a heavy group relative to H, the torsion about the ϕ—$CH_2 R$ bond is stiffened from about 2 kcal to 15 kcal, a factor of 7, and we may expect $\Delta S^{\ddagger}$ to be lower than for non-conjugated molecules by $R \ln (7)^{1/2} \approx 2.0$. This would represent a lowering of the A factor by a factor of nearly 3 compared to saturated molecules. As we note in Table 3.4, the A factor for fission of benzyl compounds is about a power of 10 lower than our standard $10^{16}\,s^{-1}$ reflecting, in part, the above stiffening. The rest comes from a much smaller increase in moment of inertia in the TS.

Compounds like bibenzyl or biallyl in which two such stiffenings take place would be expected to have a doubled effect and have A factors lower by about a power of 10. This is compatible with recent data on the recombination of allyl radicals[14,15] and the thermochemistry of the dissociation of biallyl which lead to an A factor of about $10^{15.5}\,s^{-1}$ for the dissociation.

3.8 TEMPERATURE COEFFICIENT OF RATE PARAMETERS

We have so far confined our attention to the A factors and activation energies at 300°K. Let us consider how to make calculations at any temperature. Since the Arrhenius A factor is equal to (ekT/h) $\exp\{\Delta S^{\ddagger}/R\}$, we can calculate $\Delta S^{\ddagger}_T$ from our value of $\Delta S^{\ddagger}_{300}$ by making a correction for $\Delta C^{\ddagger}_p$. In doing this we will calculate $C^{\ddagger}_p$ including the contribution of the reaction coordinate for which $C_p = R$. In this way we need make no extra allowance for the temperature variation of the (ekT/h) term. The same calculation then allows us to calculate $\Delta H^{\ddagger}$ including the contribution from the reaction coordinate. Table 3.5 shows such a calculation for the simple C—H bond fission in CH_4 (see also Table 3.1).

However, our restricted rotation model for the transition state will be sensitive to the assumed $C \cdots H$ distance and although this is not very temperature sensitive, it does vary with temperature. From (3.23) we can calculate:

$$\frac{\rho_T}{\rho_{300}} = \left(\frac{300}{T}\right)^{1/6} \tag{3.26}$$

[14] D. M. Golden, N. A. Gac, and S. W. Benson, *J. Amer. Chem. Soc.*, **91**, 2136 (1969).
[15] H. E. van den Bergh and A. B. Callear, *Trans. Faraday Soc.*, **66**, 2681 (1970).

At 1500°K we can calculate $(\rho_{1500}/\rho_{300}) = 1.31$, so that if $\rho_{300} = 2.8$, then $\rho_{1500} = 2.14$ and $r(C \cdots H) = 2.35$ Å. At this distance the nonbonded $H \cdots H$ distance is 2.60 Å which is just equal to our assumed van der Waal separation of 2.6 Å for H atoms, and the simple model breaks down. This reflects, in part, a peculiarity of H atoms and, in part, our ignorance of the precise potential function at large distances. We can avoid this problem by arbitrarily adopting $\rho_{max} = 3.0$ at 300°K for bonds involving H atoms.[16] This leads to $(\Delta\Omega/4\pi) \cong 0.16$ at 300°K about a factor of 2 larger than our previously calculated value (p. 93). At 1500°K with $\rho_{max} = 2.31$ and $r_{max} = 2.54$ Å, we obtain $(\Delta\Omega/4\pi) \sim 0.01$ $(\theta \sim 12°)$. This latter value combined with the $\Delta C_p^{\ddagger}$ changes in Table 3.6 and a revised estimate of A_{300} leads to a value of $A_{1500} = 10^{14.7} \text{ s}^{-1}$, in excellent agreement with the experimental results. The calculated value of the activation energy is also in good agreement with the best reported value of 103 kcal/mole.

Table 3.5. Contributions to $\Delta C_p^{\ddagger}$ for the Dissociation of the C—H Bond in CH_4

	Contribution to $\Delta C_p^{\ddagger}$ at T°K				
Degrees of Freedom	300	500	800	1000	1500
C—H Stretch (3000 cm^{-1}) → low frequency $\nu^{\ddagger}$ with $C_p = 2$ reaction coordinate	2.0	2.0	1.7	1.5	1.0
2 H—C—H bends in CH_4 (1450 cm^{-1}) → restricted free rotation in two dimensions with $C_p = 2.0$	1.8	0.8	−0.4	−0.8	−1.4
Total $\Delta C_{p_T}^{\ddagger}$	3.8	2.8	1.3	0.7	0.4
$\langle \Delta C_p \rangle_{T_1 - T_2}$		3.3	2.1	1.0	0.2

Average value of ΔC_p (300–1500°K) = 1.3 eu

∴ $\Delta S_{1500}^{\ddagger} = \Delta S_{300}^{\ddagger} + 1.3 \ln \frac{1500}{300} = \Delta S_{300}^{\ddagger} + 2.0$ eu

$\Delta H_{1500}^{\ddagger} = \Delta H_{300}^{\ddagger} + 1.3(1.200) = \Delta H_{300}^{\ddagger} + 1.6$ kcal/mole

$A_{1500} = A_{300}(10^{2.0/4.6}) = 10^{15.8} \text{ s}^{-1}$ (see p. 93)

$E_{1500} = \Delta E_{300} + 1.6$ kcal/mole = 105.3 kcal/mole[a]

[a] We are assuming that there is no activation energy for the recombination of $CH_3 + H$ at 300°K (for concentrations as standard states), so that $E_{act(300)} \cong \Delta E_{300}$ for the dissociation.

[16] A solution of the Lennard–Jones equation (3.23) with $E_{rot}^{\circ} = 600$ cal/mole at 300°K and with $D_0 = 112$ kcal (including 9 kcal of zero-point energy in the molecule) for CH_4 gives $\rho_{max}(300°K) = 3.22$ or $r_{max}(300°K) = 3.54$ Å. These values would lead to a maximum value of $(\Delta\Omega/4\pi) \sim 0.5$. However at 1500°K, $r_{max} = 2.71$ Å and $(\Delta\Omega/4\pi) \approx 0.014$.

What we can see from these calculations is that the restricted rotation model will be sensitive to the choice of potential function and also to the van der Waals radius. Uncertainties in these quantities introduce an uncertainty of about a factor of 4 in a priori calculations. Most important, however, is the strong decrease with temperature of the quantity $(\Delta\Omega/4\pi)$. The uncertainty in this effect is less sensitive to the potential parameters. Table 3.6 shows an estimate of the temperature parameters for the C—I bond fission in CH_3I while Table 3.7 does the same for the C_2H_6 decomposition. We see in Table 3.6 that until we get to temperatures where $r_{max} \lesssim 4.6$ Å, the sum of the van der Waals radii of I and CH_3, there is little effect of this quantity on the A factor. At $1000°K$, r_{max} has only decreased 18% from its value at $300°K$, which produces about a 36% change in the moment of inertia, and hence a decrease of this amount in the A factor. Hence $\Delta S^{\ddagger}_{1000} = 9.1 + R \ln (0.82)^2 = 8.3$ and $A_{1000} = 10^{15.1} \text{ s}^{-1}$ compared to $10^{15.45} \text{ s}^{-1}$ at $300°K$. From the overall thermochemistry of the reaction, these numbers yield values of $10^{11.0}$

Table 3.6.   Contributions to $\Delta C_p^{\ddagger}$ for the Dissociation of the C—I Bond in CH_3I[a]

Degrees of Freedom	$\Delta C_p^{\ddagger}$ at $T°K$			
	300	500	800	1000
Reaction coordinate:				
C—I stretch in CH_3I (500 cm^{-1} $\rightarrow$ reaction coordinate, $C_p = 2$)	0.7	0.3	0.1	0.0
$2CH_3$—I rocks (700 cm^{-1}) $\rightarrow$ two free rotations ($C_p = R$)	0.3	−0.9	−1.5	−1.7
Total $\Delta C_p^{\ddagger}$	1.0	−0.6	−1.4	−1.7
$\langle\Delta C_p^{\ddagger}\rangle_{T_1-T_2}$	0.2	−1.0	−1.6	
$\langle\Delta C_p^{\ddagger}\rangle_{300-1000} = -0.8$				
$\rho_{max(T)}/\rho_{max(300)}$	1.00	0.92	0.85	0.82
$r_{max(T)}$ Å[b]	5.88	5.41	5.00	4.82
$(\Delta\Omega/4\pi)$	1.0	1.0	1.0	1.0

$$\Delta S^{\ddagger}_{1000} = \Delta S^{\ddagger}_{300} - 0.8 \ln \left(\tfrac{1000}{300}\right) = 10.1 - 0.96 = 9.1$$
$$E_{act(1000)} = E_{act(300)} - 0.8(0.700) = 55.4 \text{ kcal}$$

[a] See Table 3.2 for symmetry and frequencies and tables in Appendix for other data.
[b] Estimated from equation (3.23) and the bond strength $DH°(CH_3$—I$) = 56$ kcal.

Table 3.7.   Contributions to $\Delta C_p^{\ddagger}$ for the C—C Bond Dissociation in C_2H_6

Degrees of Freedom	$\Delta C_p^{\ddagger}$ at $T°K$			
	300	500	800	1000
C—C stretch $(1000 \text{ cm}^{-1}) \to \nu^{\ddagger}$				
$(C_p = 2.0)$ reaction coordinate	+1.6	+1.0	+0.5	+0.3
Me Torsion barrier				
$3.0 \text{ kcal} \to 0 \text{ kcal}$ $(C_p = 1.0)$	−1.0	−0.9	−0.7	−0.4
4 CH_3 Rocks at 1000 cm^{-1}				
$\to 4$ free Me rotations $(C_p = 4.0)$	+2.4	0.0	−2.1	−2.7
$\Delta C_p^{\ddagger}$	+3.0	+0.1	−2.3	−2.8
$\langle \Delta C_p^{\ddagger} \rangle_{T_1 - T_2}$	1.60	−1.1	−2.6	
$\langle \Delta C_p \rangle_{300-1000} = -0.8 \text{ eu}$				
$r_{\text{max}(T)}$ (Å)	4.31	3.97	3.67	3.54
$(\Delta \Omega / 4\pi)^a$	0.85	0.74	0.64	0.24
for both methyls				

$$\Delta S_{1000}^{\ddagger} = \Delta S_{300}^{\ddagger} - 0.8 \ln \left(\tfrac{1000}{300}\right) + R \ln \left(\frac{\Delta \Omega_{1000}}{\Delta \Omega_{300}}\right) + R \ln \left(\frac{r_{1000}}{r_{300}}\right)^2$$

$$= 17.7 - 1.0 - 2.5 - 0.8 = 13.4 \text{ eu}$$

$$\Delta H_{1000}^{\ddagger} = \Delta H_{300}^{\ddagger} - 0.8(0.700)$$

$$= \Delta E_{300} - 0.6 - 0.56 = 88.2 - 1.2 = 87.0 \text{ kcal}$$

$$E_{\text{act}(1000)} = E_{\text{act}(300)} - 0.56 = 87.6 \text{ kcal}$$

$$A_{1000} = 10^{13.25} + \Delta S_{1000}^{\ddagger}/4.6 = 10^{16.2} \text{ s}^{-1}$$

^{a} See text for estimates. Contribution includes product of fraction of solid angle for each methyl.

l./mole-s for recombination of $CH_3 + I$ at 300°K and $10^{10.7}$ l./mole-s at 1000°K.

The extension of these methods for estimating a solid angle for restricted free rotation of two free radicals is much more difficult. The loss of symmetry in the relative rotation of one disk-like CH_3 radical near another one makes the estimation of free solid angle extremely difficult. One very crude way of doing this is to draw a sphere of $r = 2.2$ Å about each CH_3. If we consider the CH_3 radical as a flat disk with rounded edges, nearly spherical about each H atom, then the flat faces of this disk can move through a solid angle corresponding to about 40% of the sphere's surface. At $r \geq 4.4$ Å, each CH_3 can rotate completely freely. At $r = 2.6$ Å the rotation is completely inhibited. At $r = 3.5$ Å the overlap of the two spheres corresponds to about 10% of the surface of each. If this overlap zone falls entirely in the free area of one methyl, the other CH_3 is completely free in its rotation. If it falls entirely in the restricted area of the first CH_3, the second is completely inhibited.

Let us approximate this situation by reducing the area of each zone at its contact with the other zone by this same 10%. This intermediate area between free and restricted zones then amounts to about 40% of the total area in which we shall assume the rotation half free. Thus each methyl has a free rotation over about 40% of its solid angle. At $r = 4.30$ Å with about 1% overlap and a 90% freedom for rotation in the restricted position, we find a net free solid angle of about 93% for each methyl. The values of $(\Delta\Omega/4\pi)$ in Table 3.7 have been arrived at by such calculations and must be considered uncertain by about 50%. However, they do emphasize the sensitivity of the A factor to the decrease with temperature of the transition-state bond length which may, in fact, be overemphasized by the use of the Lennard–Jones potential. The Morse potential gives a much slower change of r_{max} with temperature, which would seem to be more reasonable. The absolute rate constant for C_2H_6 decomposition calculated by the foregoing is within a factor of 2 of the rate constants measured in the range 850–1050°K.[17]

3.9 CIS–TRANS ISOMERIZATIONS

One of the simplest types of unimolecular reaction occurs in geometrical isomerization about a double bond. The transition state seems relatively unambiguous in this reaction and corresponds to two bonded methylene groups with their planes at 90° to each other. The reaction coordinate is the internal torsional motion about what was originally a double bond. In a theory proposed by Benson, Golden, and Egger[18] it was suggested that because there can be no pi-bond overlap in the transition state, the transition state could be considered to be a biradical. For $CHD{=}CHD$, for example, this would correspond to the hypothetical species $\dot{C}HD{-}\dot{C}HD$. We would expect the C—C distance to be about 1.46 Å, as contrasted to 1.34 Å in the ground state. Although electronic triplet and singlet states might be expected to have nearly the same energy, it is likely that only singlet states are involved. Finally, because of the different groups attached, the transition state will have two optical isomers the entropy of mixing of which contributes a factor of 2 to $K^{\ddagger}$.

The symmetry numbers of the ground and transition states usually cancel each other. The A factor for such reactions can then be written as

$$A_{c-t} \sim 2\left(\frac{ekT_m}{h}\right)\frac{q_{elec}^{\ddagger}}{q_{tors}} \qquad (3.27)$$

[17] Shock-tube data (T. C. Clark et al., *J. Chem. Phys.*, **53**, 2982 (1970)) suggest a three-fold slower rate for CH_3 recombination at 1250°K than at 300°K.

[18] S. W. Benson, D. M. Golden, and K. Egger, *J. Am. Chem. Soc.*, **87**, 468 (1965).

where $q_{elec} = 1$, if triplet states are ruled out, and 4 if they participate. q_{tors} is the partition function corresponding to the reaction coordinate which is the torsional frequency about the double bond in the ground state. For cis-CHD=CHD, it is about 990 cm^{-1} [19] and at 800°K it contributes a factor of about $\frac{1}{2}$ to A_{c-t}. If we neglect triplet states, this leads to an A factor of 4×10^{13} s^{-1} for cis-CHD=CHD isomerization, which is in reasonable agreement with the reported value of 10^{13} s^{-1}.

As soon as we go to heavier substituents on the double bond, the torsion frequencies decrease considerably, to the extent that $kT > h\nu_{tors}$ and q_{tors} takes the classical form

$$q_{tors} = (1 - e^{-h\nu_{tors}/kT})^{-1}$$

$$\approx kT/h\nu_{tors} \tag{3.28}$$

The A factor then becomes

$$A_{c-t} = 2\nu_{tors}q_{elec}^{\ddagger} \rightarrow 2\nu_{tors} \tag{3.29}$$

Note that because of the cancellation of the (kT/h) term, the A factor has become independent of temperature; so $\Delta H^{\ddagger} = E$, and we drop the factor of e in (3.27).

When the groups attached to the double bond can interact with the biradical, we expect to witness a lowering of the activation energy of the reaction by an amount equal to the interaction or "resonance" energy, whereas the A factor will also decrease because of the stiffening of internal rotations accompanying this interaction. For such molecules, the simplified forms of (3.27) and (3.29) must be modified to take into account these changes in $\Delta S^{\ddagger}$.

Table 3.8 lists parameters for some homogeneous gas phase isomerizations together with A factors estimated from the foregoing discussion. With some obvious exceptions, the agreement is reasonably good. One is

$$CH_3CH=CHCN$$

for which no structural basis exists for the abnormally low A factor of 10^{11} s^{-1}. Comparing it to cis-butene-2 we note that the activation energy is decreased by 11.7 kcal, which would imply a conjugation energy of this amount plus any pi-bond interaction in the conjugated molecule, which may be anticipated to be about 2–3 kcal/mole for a total of about 14 kcal/mole. However recent kinetic data suggest that the radical stabilization by CN is about 7 kcal[20] and if we allow 2 kcal for the conjugation energy in the molecule, the expected activation energy relative to cis-butene-2 would be about 58 kcal. If we assume that the absolute value of the rate constant is correct, and increase the

[19] R. L. Arnett and B. L. Crawford, Jr., *J. Chem. Phys.*, **18**, 118 (1950).
[20] K. D. King and R. D. Goddard, *Int. J. Chem. Kinet.*, **7**, 109 (1975).

Table 3.8. Arrhenius Parameters for Some Thermal Cis-trans Isomerization

Reference[1]	Reactant	T_m °K	log A (sec^{-1})	E (kcal/mole)	log A (calc)
a	cis-CHD=CHD	770	13	65	13.6
b	cis-CH$_3$CH=CHCH$_3$	690	13.8	63	13.5
c		1100	14.62	66.2	
d	cis-CH$_3$CH=CHCN	700	11.0	51.3	13.2
e		1100	13.2	58.1	
e	cis-Pentadiene,1,3	1100	13.6	53.0	13.4
f		650	13.5	52.1	
g	cis-CH$_3$CH=CHCOOCH$_3$	660	13.2	57.8	12.8
h	cis-Stilbene	580	12.8	42.8	12.6
i	cis-Methyl cinnamate	610	10.5	41.6	12.9
j	cis-β-Cyanostyrene	610	11.6	46.0	12.6
k	cis-CF$_3$CF=CFCF$_3$	650	13.5	56.4	13.1
l		1100	14.3	58.8	
m	cis-CHCl=CHCl	825	12.8	56.0	13.2
l		1100	13.4	56.9	
l	cis-CHF=CHF	1100	13.4	60.7	13.4

1. Bibliographic sources:
   a. J. E. Douglas, B. S. Rabinovitch, and F. S. Looney, *J. Chem. Phys.*, **23**, 315 (1955).
   b. B. S. Rabinovitch and K. W. Michel, *J. Am. Chem. Soc.*, **81**, 5065 (1959); R. B. Cundall and T. F. Palmer, *Trans. Faraday Soc.*, **57**, 1936 (1961).
   c. S. H. Bauer and P. M. Jeffers, *Int. J. Chem. Kinet.*, **6**, 763 (1974).
   d. J. N. Butler and R. D. McAlpine, *Can. J. Chem.*, **41**, 2487 (1963).
   e. W. M. Marley and P. M. Jeffers, *J. Phys. Chem.* **79**, 2085 (1975).
   f. H. M. Frey, A. M. Lamant and R. Walsh, *J. Chem. Soc. (A)*, 2642 (1971).
   g. J. N. Butler and G. J. Small, *Can. J. Chem.*, **41**, 2492 (1963).
   h. G. B. Kistiakowsky and W. R. Smith, *J. Am. Chem. Soc.*, **56**, 638 (1934). Note that in the liquid state, log A = 10.4 and E = 36.7 kcal/mole. Data of T. W. T. Taylor and A. R. Murray, *J. Chem. Soc.*, 2078 (1938).
   i. G. B. Kistiakowsky and W. R. Smith, *J. Am. Chem. Soc.*, **57**, 269 (1935).
   j. G. B. Kistiakowsky and W. R. Smith, *J. Am. Chem. Soc.*, **58**, 2428 (1936).
   k. E. W. Schlag and E. W. Kaiser, Jr., *J. Am. Chem. Soc.*, **87**, 1171 (1965).
   l. P. M. Jeffers, *J. Phys. Chem.*, **78**, 1469 (1974).
   m. L. D. Hawton and G. P. Semeluk, *Can. J. Chem.*, **44**, 2143 (1966). This reaction is complicated by heterogeneous effects and radical reactions.

A factor to 10^{13} s^{-1}, it would boost E to 58.1 kcal at 700°K and reduce the net stabilization energy to 5 kcal, both of which changes seem quite reasonable.

Both of these changes in E_{act} would be compatible with the data on β-cyanostyrene. For the latter we would expect 11 kcal of net stabilization from the phenyl–radical interaction (13 kcal stabilization–2 kcal conjugation), plus 5 kcal net stabilization from the cyano–radical interaction for a total of 16 kcal. Subtracting this from the butene-2 value of $E_{act} = 63$ kcal/mole, we obtain $E = 47$ kcal, in good agreement with the observed 46 kcal. However the reported A factor, which is a power of 10 lower than that expected, suggests that the true activation energy might be closer to 49 kcal/mole.

In cis-stilbene, the nonplanar, strained molecule has the two phenyl groups tilted by about 45° with respect to the plane of the H—C≡C—H double-bond structure because of steric repulsion between these two phenyls. A crude calculation of the reduced moment of inertia for the torsion about the double bond gives $I_r \sim 65$ amu-Å^2, and a similar calculation for cis-butene-2 gives 7 amu-Å^2. Since the torsion frequencies ν_{tors} vary as $I_r^{-1/2}$, this suggests that for cis-stilbene $\nu_{tors} \sim \frac{1}{3}\nu_{tors}$(cis-butene-2) or about 130 cm^{-1} (Table A.21). This is sufficiently low compared to kT/h at 600°K ($\sim$420 cm^{-1}) that we can use eq (3.29) for the A factor, with suitable corrections for the significant structural changes. These latter are the increased barrier to rotation of the phenyl groups in the TS due to the conjugation energy. This latter is 13 kcal. To obtain the TS barrier we add the ground-state barrier height of about 2 kcal/mole estimated from ethyl benzene torsion to give a total barrier of about 15 kcal. In cis-stilbene we estimate the barrier to C-ϕ torsion at 2 kcal as in ethyl benzene. At 600°K we find from Table A.20 that this barrier change will lead to a net loss of $2 \times (2.5 - 0.3) = 4.4$ in going to TS. However there will be a significant change in the C—C stretch going from a double bond ($\sim$1650 cm^{-1}) to a single bond (1000 cm^{-1}) and a gain in $\Delta S^{\ddagger}$ of 0.5. In addition we must correct the ground state for the restricted rotation of the phenyl groups arising from their steric repulsion. A crude geometrical model leads to a factor of about $\frac{1}{2}$ for the relative motion of each phenyl in the ground state. For the two phenyl groups this suggests that the ground state entropy should be $2R \ln 2 = 2.8$ lower than a value we would calculate for stilbene from group values which make no allowances for such steric interference. These contributions are summarized in Table 3.9 and lead to $A_{600} = 10^{12.55}$ s^{-1}, in good agreement with the reported value.

Comparison of the experimental activation energies of cis-butene-2 and cis-stilbene yields a benzyl resonance energy of only 10 kcal, compared to an expected value of about 13 kcal. This is only apparent, however, because the pi-bond energies in the two olefins differ by 4 kcal. A 2 kcal discrepancy remains.

The data on the perfluorobutene-2 are interesting because by comparison with butene-2, they imply that the pi-bond energy in the fluorocarbon is about 6 kcal less than in the hydrocarbon.

It is possible to estimate the activation energies for the reactions mentioned above when the pi-bond energies are known. In the case of CHD═CHD, for example, the pi-bond energy at 300°K is 59.5 kcal. If we assume that the

Table 3.9.   Contributions to $\Delta S^{\ddagger}$ at 600°K in the cis–trans Isomerization of *cis*-Stilbene

Degree of Freedom	$\Delta S^{\ddagger}$
Reaction coordinate	−4.4
(130 cm^{-1} torsion about double bond)	
Optical isomerism and symmetry	1.4
($\sigma = 2$; $n = 1$; $\sigma^{\ddagger} = 2$; $n^{\ddagger} = 2$)	
Double bond → single bond	0.5
(1650 cm^{-1}) (1000 cm^{-1})	
2 Single bonds → 2–3 electron C—C bonds	−0.6
(1000 cm^{-1}) (1300 cm^{-1})	
Steric restriction of phenyl rotation	
in ground state	$2R \ln 2 =$ 2.8
Change in two phenyl torsion barriers	
2 kcal → 15 kcal	−4.4
	$\Delta S^{\ddagger} = -4.7$

$$A_{600} = \frac{ekT}{h} e^{\Delta S^{\ddagger}/R} = 10^{13.55 + \Delta S^{\ddagger}/4.6} = 10^{12.55} \text{ s}^{-1}$$

ground state of the biradical is planar, in accordance with the observations that sp^2, single-bonded, C atoms, such as those in butadiene, glyoxol, and styrene, form planar structures and that the perpendicular configuration has an activation energy of 3.6 kcal (the conjugation energy in butadiene), we estimate $\Delta H^{\ddagger}_{300} = 63.1$ kcal. At 800°K, using $\langle \Delta C_p^{\circ} \rangle = -0.5$ gibbs/mole, this becomes 62.8, and $E_{\text{act}} = \Delta H^{\ddagger} + RT = 64.6$ kcal at 800°K, in excellent agreement with the observed value. A similar calculation for butene-2 yields E_{act} (700°K) = 62.1 kcal, whereas for CHCl═CHCl, we estimate (at 800°K), 57.0 kcal/mole, all in excellent agreement with the data. The last value is based on the assumption that the C—H bond strength in CH₂Cl—CH₂Cl is 95 kcal/mole. This leads to a pi-bond strength of 55 kcal/mole in the olefin.

3.10   COMPLEX FISSIONS—CYCLIC TRANSITION STATES

A number of unimolecular reactions occur in which more than one bond in a molecule is broken. These vary in complexity over quite a range, but all have in common the formation of cyclic transition states. As examples, we may consider the elimination of HX from X-substituted hydrocarbons. These seem to involve four atoms in a ring transition state and are frequently referred to as four-center reactions. The pyrolysis of *i*-PrI provides an

example. The mechanism is

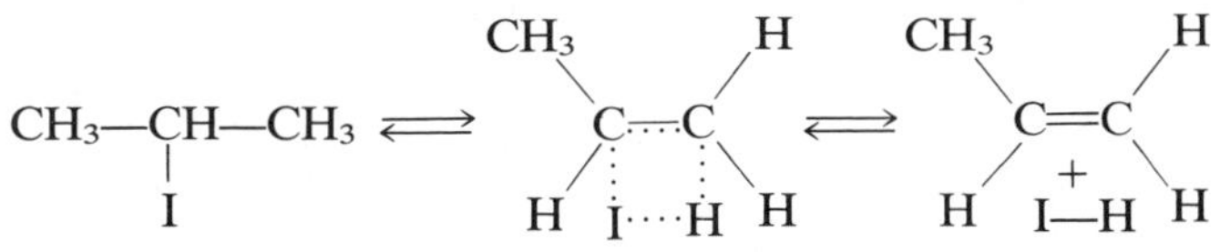

Cases are also found of three-center reactions:

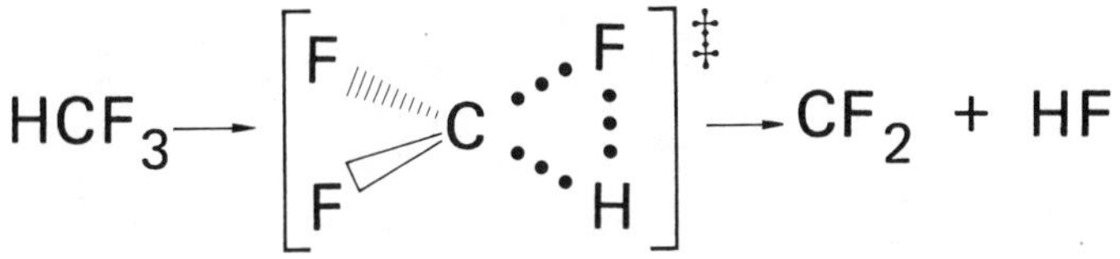

Nitroalkanes appear to follow a five-center path (*see footnote* 22, *p.* 115):

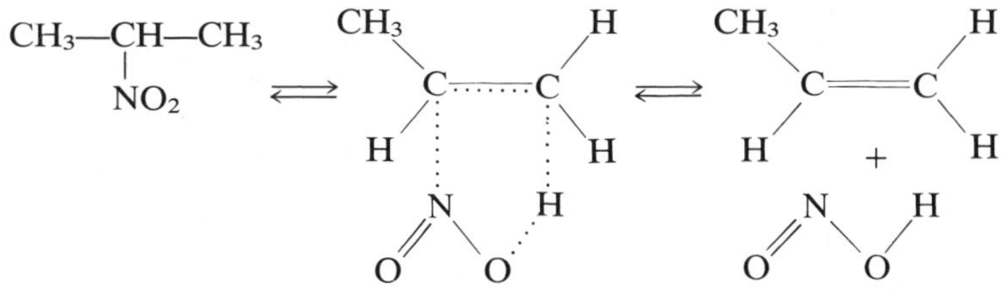

Alkyl esters pyrolyze by a six-center path:

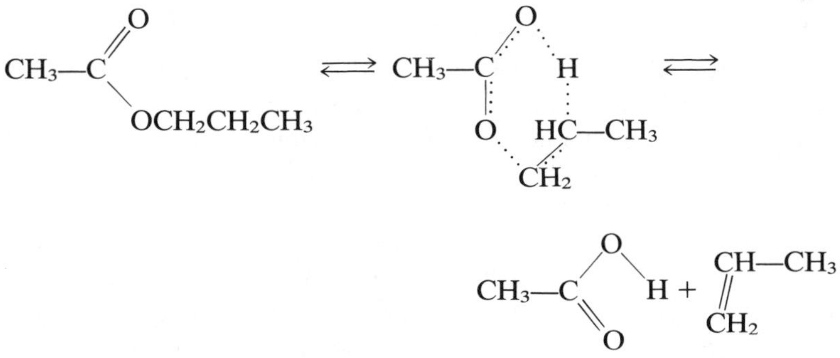

Retrograde (that is, reverse) Diels–Alder reactions have been proposed to follow a concerted six-center path, but there has been considerable controversy as to whether the alternate biradical path is not the correct one.

Decomposition of butadiene dimer illustrates the point:

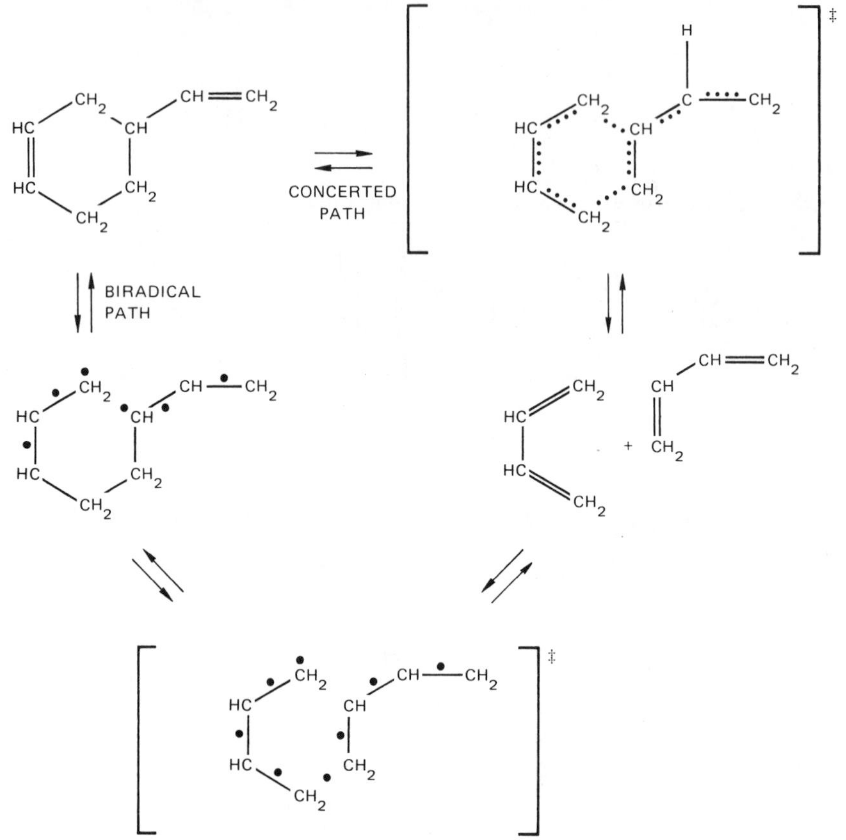

CONCERTED PATH

BIRADICAL PATH

Note that the biradical path involves the formation of a short-lived biradical intermediate, which may proceed on to fission to final products or recyclize to reactant. Thus there is another transition state between reactant and biradical that has not been drawn.

Depending on the relative rates of ring closure and biradical fission, the overall rate constant for the biradical path may involve either one or both transition states. Again, depending on various factors at present not completely understood, the reaction may follow the concerted path, the biradical path, or both. If the rate of biradical formation should be rate determining, the transition state for this path is about identical with that for the concerted process. We shall return to this reaction. In Table 3.10 are collected a number of rate constants for four-center fissions.

The four-center elimination reaction of HX from a substituted RX compound are all endothermic by from 12 to 20 kcal/mole, depending on X, and have activation energies that are very sensitive to structure. We discuss these in more detail later, but for the moment let us consider the A factors and for simplicity, start with C_2H_5Cl. The transition state involves the formation of a four-atom, cyclic ring, which, in principle, we expect to be fairly tight.

Table 3.10. Arrhenius Parameters for Three- and Four-Center Complex Fissions

Reference[1]	Reaction	T_m (°K)	log A (s^{-1})	E (kcal/mole)
a	$CH_3CH_2Cl \rightarrow C_2H_4 + HCl$	720	13.51	56.6
a	$n\text{-PrCl} \rightarrow C_3H_6 + HCl$	720	13.50	55.1
a	$n\text{-BuCl} \rightarrow n\text{-}C_4H_8 + HCl$	700	13.63	55.2
b	$sec\text{-BuCl} \rightarrow n\text{-}C_4H_8 + HCl$	630	13.62	49.6
c	$iso\text{-BuCl} \rightarrow i\text{-}C_4H_8 + HCl$	700	14.0(12.9)	56.8(53.3)
d	$t\text{-BuCl} \rightarrow i\text{-}C_4H_8 + HCl$	780	13.74	44.7
e	$EtBr \rightarrow C_2H_4 + HBr$	680	13.45	53.9
f	$n\text{-PrBr} \rightarrow C_3H_6 + HBr$	840	13.0	50.7
g	$n\text{-BuBr} \rightarrow 1\text{-}C_4H_8 + HBr$	670	13.2	50.9
h	$sec\text{-BuBr} \rightarrow 1\text{-}C_4H_8 + HBr$	600	13.53	46.5
i	$iso\text{-BuBr} \rightarrow i\text{-}C_4H_8 + HBr$	530	13.05	50.4
j	$t\text{-BuBr} \rightarrow i\text{-}C_4H_8 + HBr$	800	13.5	41.5
k	Cyclohexyl Br $\rightarrow$ cyclohexene + HBr	600	13.52	46.1
l	Cyclohexyl Cl $\rightarrow$ cyclohexene + HCl	600	13.0	50.0
m	Cyclopentyl Br $\rightarrow$ cyclopentene + HBr	600	12.84	43.7
j	$t\text{-BuOH} \rightarrow i\text{-}C_4H_8 + HOH$	1150	13.4	61.6
n	$t\text{-amyl OH} \rightarrow 2\text{-methyl butene-1} + HOH$	780	13.5	60.0
o	$EtI \rightarrow C_2H_4 + HI$	630	13.4	50.0
p	$i\text{-PrI} \rightarrow C_3H_8 + HI$	560	13.0(13.5)	43.5(45.0)
q	$t\text{-BuI} \rightarrow i\text{-}C_4H_8 + HI$	710	13.7	38.1
r	$CF_2HCl \rightarrow CF_2 + HCl$	890	13.84	55.8
s	$Al(Et)_3 \rightarrow Al(Et)_2H + C_2H_4$	430	10.9	30.1
t	$CH_3NHNH_2 \rightarrow CH_2{=}NH + NH_3$	1000	13.2	44
u	$t\text{-BuOMe} \rightarrow$ isobutene + MeOH	700	14.4(13.9)	61.5(59.0)
v	$CH_2{=}CHBr \rightarrow C_2H_2 + HBr$	1500	13.8	72.1
w	Cyclobutyl Cl $\rightarrow$ cyclobutene + HCl	700	13.6	55.2

Table 3.10 (*contd.*)

1. Bibliographic sources:

    a. H. Hartman, H. G. Bosche, and H. Heydtmann, *Zeit. Phys. Chem.*, NF **42**, 329 (1964).

    b. A. Maccoll and R. H. Stone, *J. Chem. Soc.*, 2756 (1961).

    c. K. E. Howlett, *J. Chem. Soc.*, 4487 (1962). Values in parentheses are "best" estimates by O'Neal and Benson (*loc. cit.*).

    d. W. Tsang, *J. Chem. Phys.*, **40**, 1171 (1964).

    e. P. J. Thomas, *J. Chem. Soc.*, 1192 (1959).

    f. A. T. Blades and G. W. Murphy, *J. Am. Chem. Soc.*, **74**, 6219 (1952).

    g. A. Maccoll and P. J. Thomas, *J. Chem. Soc.*, 5033 (1957).

    h. M. N. Kale, A. Maccoll, and P. J. Thomas, *J. Chem. Soc.*, 3016 (1958).

    i. G. P. Harden and A. Maccoll, *J. Chem. Soc.*, 1197 (1959).

    j. W. Tsang, *J. Chem. Phys.*, **40**, 1498 (1964).

    k. J. S. Green and A. Maccoll, *J. Chem. Soc.*, 2449 (1955).

    l. E. S. Swinbourne, *Aus. J. Chem.*, **11**, 314 (1958).

    m. M. N. Kale and A. Maccoll, *J. Chem. Soc.*, 5020 (1957).

    n. R. F. Schultz and G. B. Kistiakowsky, *J. Am. Chem. Soc.*, **56**, 395 (1934).

    o. S. W. Benson and A. N. Bose, *J. Chem. Phys.*, **37**, 2935 (1962).

    p. H. Teranishi and S. W. Benson, *J. Chem. Phys.*, **40**, 2946 (1964).

    q. W. Tsang, *J. Chem. Phys.*, **41**, 2487 (1964).

    r. J. W. Edwards and P. A. Small, *Nature*, **202**, 1329 (1964).

    s. A. T. Cocks and K. W. Egger, *J. C. S. Faraday I*, **68**, 423 (1972).

    t. D. M. Golden, R. K. Solly, N. A. Gac, and S. W. Benson, *J. Amer. Chem. Soc.*, **94**, 363 (1972); A-factor assumed.

    u. N. J. Daly and C. Wentrup, *Austr. J. Chem.*, **21**, 2711 (1968). Values in parentheses are K. Y. Choo, D. M. Golden, and S. W. Benson, *Int. J. Chem. Kinet.*, **6**, 631 (1974), from VLPP studies (890–1180°K).

    v. J. M. Simione and E. Tschuikow-Roux, *J. Phys. Chem.*, **74**, 4075 (1970).

    w. A. T. Cocks and H. M. Frey, *J. Amer. Chem. Soc.*, **91**, 7583 (1969). A competing side reaction is cleavage to $C_2H_4 + C_2H_3Cl$. At 700°K cyclobutene → butadiene.

The significant contributions to $\Delta S^{\ddagger}$ are a symmetry change of $+R \ln 3$, corresponding to the three equivalent H atoms, which can be eliminated with the Cl, and a loss of about 4 eu at 700°K, due to the stiffening of the CH_3 internal rotation. These would constitute a net contribution to $\Delta S^{\ddagger}$ of -1.8 eu, so that the A factor should become at 700°K, about $10^{13.2}$ s^{-1}. Despite some scatter, this seems to be in reasonable accord with the data for EtCl, EtBr, and EtI. Symmetry considerations would lead us to expect A factors of $10^{13.5}$ s^{-1} for i-PrX and $10^{13.7}$ s^{-1} for t-BuX elimination reactions, all again in reasonable agreement with the data. It is unfortunate that the uncertainty in the measured A factors correspond to factors of about $10^{0.5}$ s^{-1}, which is the order of the above differences.

A more systematic analysis than that performed above may be made as follows.[21] If we consider the formation of a four-membered ring, the four

---

[21] See H. E. O'Neal and S. W. Benson, *J. Phys. Chem.*, **71**, 2903 (1967) for more detailed discussion of A-factor prediction is concerted reactions.

atoms constituting the ring will have six frequencies associated with them. Of these, four can be taken as stretches, and two as deformations, one in plane and one out of plane. One of the stretches corresponds to the reaction coordinate. When one of the ring atoms is H or D, we can assign two of the stretches and the out-of-plane bend to it. For C_2H_5Cl, the $\Delta S^{\ddagger}$ contributions at 600°K are shown in Table 3.11.

In the n-PrX and n-BuX eliminations we lose the internal rotations of an ethyl and a propyl group, respectively. However these are not entirely lost because they are replaced by low torsional frequencies, about the incipient double bond, of about 350 to 300 cm^{-1}. Thus the net rotational loss in $\Delta S^{\ddagger}$ is probably about the same as for the CH_3 group. However there are only two reactive H atoms instead of three in EtX, so we may expect the A factor to be about $10^{13.1}$ s^{-1}, or not very different from the EtX compounds.

The data are consistent with the view submitted above but are not good enough to resolve the rather fine distinctions. The same will be true of the iso-butyl-X having only one reactive H atom.

When X is OH, we lose an additional rotation in the transition state and we may expect A factors for t-Bu—OH to be about $10^{13.2}$ s^{-1}. Again, the data are not good enough to discern such detail.

If the foregoing treatment is correct, we should expect that for the case when R is a cyclic structure, no loss in internal rotation occurs. Thus for cyclohexyl-X at high temperatures, the transition state has a symmetry contribution to $\Delta S^{\ddagger}$ of $R \ln 2$ arising from two available H atoms, which is however compensated by the two nearly equivalent ground state structures

Table 3.11. Activation Entropy for C_2H_5Cl Pyrolysis at 600°K^a

Degrees of Freedom	$\Delta S^{\ddagger}$
Symmetry	$+R \ln 3 = +2.2$
CH_3 internal rotation $\rightarrow$ CH$_2$—C$\overset{\text{CH}_3}{\underset{\text{H}}{}}$ torsion (400 cm^{-1})	-3.7
C—Cl stretch 650 cm^{-1} $\rightarrow$ (reaction coordinate)	-1.4
C—C—Cl bend $\rightarrow$ C—C·Cl bend (400 cm^{-1}) $\rightarrow$ (280 cm^{-1})	$+0.7$
H—C—H bend $\rightarrow$ H—C—C bend (1450 cm^{-1}) $\rightarrow$ (950 cm^{-1})	$+0.6$
Total	-1.6

^{a} Note that we have neglected the small contributions due to the ring H stretches.

Table 3.12. Arrhenius Parameters for Some Five- and Six-Center Complex Fissions

Reference[1]	Reaction	T_m (°K)	$\log A$ (s^{-1})	E (kcal/mole)
a, b	i-PrNO$_2$ → C$_3$H$_6$+HNO$_2$	570	11.0(11.3)	39.0(40)^2
b	EtNO$_2$ → C$_2$H$_4$+HNO$_2$	650	11.5(11.8)	41.5(43)^2
c, d	CH$_3$COOEt → C$_2$H$_4$+CH$_3$COOH	820	12.5	47.8
d	CH$_3$COOiPr → C$_3$H$_6$+CH$_3$COOH	760	13.0	45.0
e	CH$_3$COOt-Bu → i-C$_4$H$_8$+CH$_3$COOH	540	13.3	40.5
f	C$_2$H$_5$COOt-Bu → i-C$_4$H$_8$+C$_2$H$_5$COOH	540	12.8	39.2
g	CH$_3$COOCOCH$_3$ →			
	CH$_2$CO+CH$_3$COOH	600	12.0	34.5
h	CH$_3$—CH$_2$—O—CH=CH$_2$ →			
	CH$_3$CHO+C$_2$H$_4$	810	11.4	43.8
i	cis-Methylvinylcyclopropane →			
	hexadiene-1,4 (cis)	460	~11.0	31.0
j	3-Methyl hexadiene-1,5 ⇌ heptadiene-1,5	480	9.8(10.85)	32.5(35.0)^3
k	CH$_2$=CH—O—CH$_2$—CH=CH$_2$ →			
	CH$_2$=CH—CH$_2$—CH$_2$CHO	450	11.3	30.6
l	cis-Hexatriene-1,3,5-			
	cyclohexadiene-1,3	430	11.8	29.9
m	s-Trioxane → 3CH$_2$O	580	15.0	47.4
n	Cyclo [—CH(CH$_3$)—O—]$_3$ → 3CH$_3$CHO	510	15.1	44.2
n	Cyclo [—CH(nPr)—O—]$_3$ → 3nPr-CHO	480	14.4	42.0
n	Cyclo [—CH(iPr)—O—]$_3$ → 3iPr-CHO	480	14.5	42.0
o	Cyclohexene → C$_2$H$_4$+butadiene	860	15.2	66.2
p	Methyl cyclohexene-3 →			
	C$_3$H$_6$+butadiene	1020	15.1	66.6
p, q	Vinyl cyclohexene-3 → 2-butadiene	1000	15.2(15.7)	62.0(61.8)
r		670	11.5	47.0
s	CH$_2$ClCH=C(CH$_3$)$_2$ →			
	CH$_2$=CH—C(CH$_3$)=CH$_2$+HCl	500	12.0	38.3

1. Bibliographic sources:
   a. T. E. Smith and J. G. Calvert, *J. Phys. Chem.*, **63**, 1305 (1959).
   b. C. Fréjacques, *Compt. Rend.*, **231**, 1061 (1950).
   c. A. T. Blades and P. W. Gilderson, *Can. J. Chem.*, **38**, 1407 (1960).
   d. A. T. Blades, *Can. J. Chem.*, **32**, 366 (1954).
   e. E. V. Emovan and A. Maccoll, *J. Chem. Soc.*, 335 (1962).
   f. E. Warrick and P. Fugassi, *J. Phys. Chem.*, **52**, 1314 (1948).
   g. J. Murawski and M. Szwarc, *Trans. Faraday Soc.*, **47**, 269 (1951).
   h. A. T. Blades and G. W. Murphy, *J. Am. Chem. Soc.*, **74**, 1039 (1952).
   i. R. J. Ellis and H. M. Frey, *J. Chem. Soc.*, 5578 (1964).

j. A. Amano and M. Uchizama, *J. Phys. Chem.*, **69**, 1278 (1965).

k. F. W. Schuler and G. W. Murphy, *J. Am. Chem. Soc.*, **72**, 3155 (1950).

l. K. E. Lewis and H. Steiner, *J. Chem. Soc.*, 3080 (1964).

m. W. Hogg, D. M. McKinnon, and A. F. Trotman-Dickenson, *J. Chem. Soc.*, 1403 (1961).

n. C. C. Coffin, *Can. J. Res.*, **7**, 75 (1932); **9**, 603 (1933); N. A. D. Parlee, **18**, 223 (1940).

o. M. Uchizama, T. Tomioka, and A. Amano, *J. Phys. Chem.*, **68**, 1878 (1964).

p. W. Tsang, *J. Chem. Phys.*, **42**, 1805 (1965).

q. Values in parentheses are estimates from reverse reaction and K_{eq} by N. E. Duncan and J. E. Janz, *J. Chem. Phys.*, **20**, 1944 (1952).

r. A. G. London, A. Maccoll, and S. K. Wong, *J. Amer. Chem. Soc.*, **91**, 7577 (1969).

s. C. J. Harding, A. Maccoll, and R. A. Ross, *J. Chem. Soc.* (B), 634 (1969).

2. G. N. Spokes and S. W. Benson, *J. Am. Chem. Soc.*, **89**, 6030 (1967).

3. H. M. Frey and R. K. Solly, *Trans. Faraday Soc.*, **64**, 1858 (1968).

(axial and equatorial X). Thus we should find $A \sim (ekT_m/h)$, which is $10^{13.6}$ s^{-1} at 700°K. From Table 3.10 we note that this is in agreement with the data on cyclohexyl Br. The A factor for cyclopentyl Br, by contrast, seems too low. However the cyclopentyl rings have abnormally high entropies relative to six-membered rings because of a low-frequency, ring-puckering mode, the pseudorotation. Cyclobutane also has a low-frequency puckering mode. The A factors show very clearly the loss of such a mode in the transition state (see Section 2.20).

Data on the few five-center and larger number of six-center complex fissions are presented in Table 3.12. For purposes of comparison, we have included three isomerization reactions that also go through six-center complexes.

The HONO eliminations involve a transition state in which we have lost one CH_3 rotation and one NO_2 free rotation.[22] For EtNO$_2$ we expect a symmetry contribution of $R \ln 6$, an internal rotation loss of about -11, -1.0 from the reaction coordinate, yielding a net $\Delta S^{\ddagger} = -8.4$. Hence the A factor of 600°K should be $10^{11.8}$ s^{-1}, in reasonable accord with the data. For i-PrNO$_2$ the value should be about $10^{12.0}$ s^{-1}, the extra symmetry contribution being offset somewhat by the increased reduced moment of inertia around the C—N bond. The agreement here is poorer.

The ester pyrolyses involve the stiffening of two large internal rotations about the other O—C bond and the alkyl C—C bond. The acyl C—O bond has a large torsion barrier (Table A.19) and a frequency of ~ 250 cm^{-1}. These are replaced by two low-frequency torsions in the neighborhood of 250–400 cm^{-1}. At 700°K the net loss is about 2 eu per rotation or about 6.0 eu. Thus for CH_3COOEt we have $+R \ln 3$ from symmetry, -6.0 eu from rotation

---

[22] It is by no means certain that this is a five-center *TS*. The product may be the isomeric HN(O)O which is estimated to have only a slightly higher ΔH_f° than HONO. The A factor would still be about the same.

for a net $\Delta S^{\ddagger} \approx -3.8$ eu. Hence A (700°K) should be about $10^{12.8}$ s^{-1}, in good accord with the observed $10^{12.5}$ s^{-1}. The A factors for i-Pr and t-Bu esters should then be higher by factors of 2 and 3, respectively, or $10^{13.1}$ and $10^{13.3}$, in excellent agreement with the data. However the uncertainties in log A are about ±0.5, which appear when the data for t-Bu acetate and t-Bu propionate are compared. The rate constants have nearly identical values at 550°K, whereas the reported A factors differ by an unaccountable factor of 3. This is a case where group-additivity rules can be applied to ground states and transition states and would predict identical values for $\Delta S^{\ddagger}$.

The ethyl vinyl ether pyrolysis is isoelectronic with ethyl formate pyrolysis and the transition states are very similar (see Figure 3.6). The predicted A factor of $10^{11.6}$ s^{-1} is in good accord with the observed $10^{11.4}$ s^{-1}.

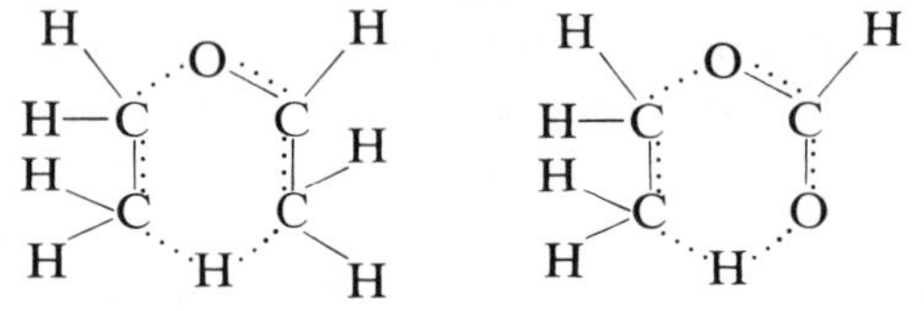

(a)   (b)

The lower A factor for the ether, compared with the ester, is due to the nearly free rotation about the vinyl C—O bond in the ether, compared with the stiff acyl C—O torsion in the ester (Table A.19). The ether loses about 2.3 gibbs/mole more in going to the transition state.

Similarly $(CH_3CO)_2O$ has an A factor very close to the predicted $10^{12.3}$ s^{-1}. The very similar transformation of cis-1-methyl, 2-vinyl-cyclopropane involves a loss of only two rotations:

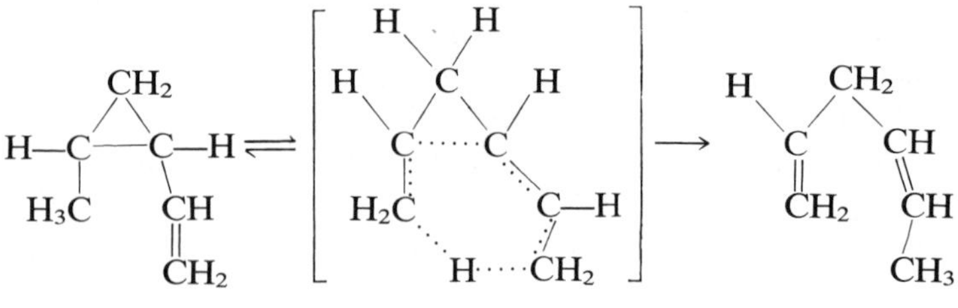

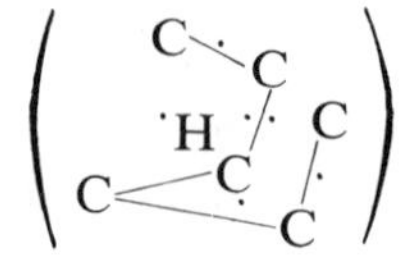

cis
conformation

However, the observed A factor is lower than that for the ether pyrolysis and lower than that anticipated $(10^{11.7})$ from the symmetry contribution of $R \ln 3$. This is due in part to the lower temperature, of 450°K, of the reaction, and hence, the lower contribution of the offsetting torsional modes.[23]

In contrast to all of the low A-factor reactions, which have involved the formation of six-membered rings for open chains, the pyrolysis of the symmetrical substituted trioxanes all have A factors about tenfold larger than the expected $10^{13.5} \, \text{s}^{-1}$. For these reactions $\Delta S^{\ddagger} \sim 5$, and the transition states must all involve a loosening of the ground state ring structure. This is a quite reasonable result and is predicted directly from the detailed scheme proposed by O'Neal and Benson.[24]

The reverse Diels–Alder reactions of cyclohexenes all have larger A factors than the trioxanes, despite the fact that the overall entropy change for them is smaller than that for the trioxanes. The trioxanes must certainly decompose by a concerted mechanism because the activation energy is so much lower than would be possible for any biradical path. We discuss these reactions later, from the point of view of the biradical intermediate, when we consider the reverse reaction of Diels–Alder additions to butadiene.

3.11 UNIMOLECULAR REACTIONS—BIRADICAL INTERMEDIATES

A large number of unimolecular reactions cannot be rationalized as occurring via a single transition state. They may involve a fairly complex mechanism involving the formation of one or more intermediate biradicals. The first and still controversial example is the isomerization of cyclopropane to propylene. Rabinowitch, Schlag, and Wiberg[25] showed that cis-1,2-dideutero-cyclopropane underwent geometrical isomerization to the trans form about twelve times faster than the structural isomerization to propylene. Benson[26] later proposed that such a result was kinetically and thermochemically compatible with a short-lived $(10^{-10} \, \text{s})$ trimethylene

---

[23] It is very surprising that the product is the cis, rather than the expected trans, olefin. This implies that the more crowded cis structure, with the vinyl group over the C_3 ring, is the favored form, and this could account for the low A factor. This is also the favored form for the trans isomerization (see Table 3.7).

[24] S. W. Benson and H. E. O'Neal, *loc. cit.*

[25] B. S. Rabinovitch, E. Schlag, and K. Wilberg, *J. Chem. Phys.*, **28**, 504 (1958); E. W. Schlag and B. S. Rabinovitch, *J. Amer. Chem. Soc.*, **82**, 5996 (1960).

[26] S. W. Benson, *J. Chem. Phys.*, **34**, 521 (1961).

biradical precursor common to both paths:

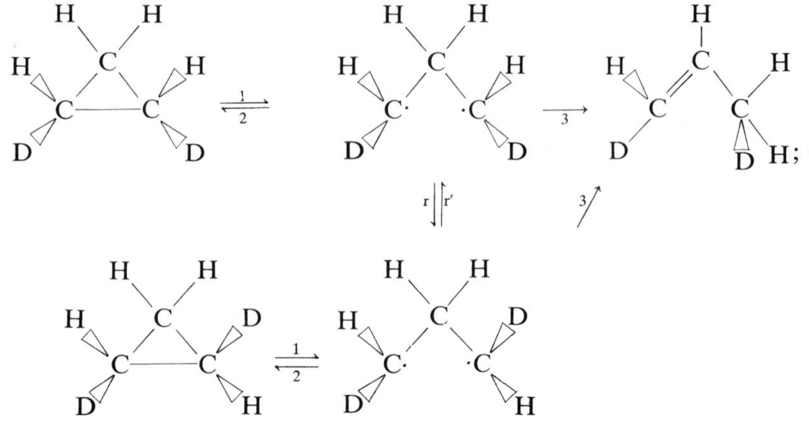

From assignable thermochemical properties of the biradical, it was possible to deduce Arrhenius parameters for each of the Steps 1, 2, and 3 in (3.28) above. The rotation steps are expected to take place so rapidly at the relevant temperatures that the biradical reaches rotational equilibrium, once it is formed; that is, $k_r \sim k'_r > (k_2 + k_3)$.

This simple picture has become more complicated since the proposal by Woodward and Hoffman[27] that the opening and closing of the ring bond in the cyclopropane molecule is not a random event but must occur concertedly with the rotations of the radical CH_2 end groups. These latter are perpendicular to the C—C—C plane in the stable ring and coplanar in the biradical (or the transition state). Orbital symmetry considerations suggest that the two CH_2 groups rotate in a conrotatory manner on bond opening. Detailed balancing then suggests that the ring closing of the biradical similarly takes place with conrotatory motion of the CH_2 radicals. The differences in the two points of view can be described in terms of a hypothetical potential energy surface for the reactions (see Figure 3.4). The concerted reaction approach considers the "biradical" to be the short-lived transition state at the top of a barrier separating two forms of the ring.

The biradical model proposes instead that there is a relatively stable biradical located in a potential trough between two distinct transition states.

The basis for proposing the biradical model lies in the observation that the activation energy for the reaction is greater than the energy estimated to open the ring bond. There is an assumption in making this estimate which is equivalent to saying that in the process of removing H atoms from each of the

[27] R. Hoffman, *Acc. Chem. Res.*, **1**, 17 (1968). An excellent review of work on "biradicals" is given by R. G. Bergmann in *Free Radicals*, J. K. Kochi (ed.), Vol. I, 191, Wiley, New York (1973).

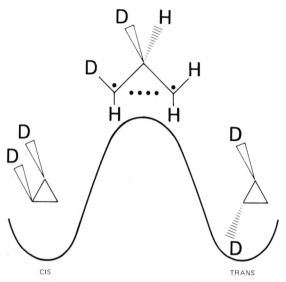

(a) Orbital symmetry model of the potential energy diagram of the *cis–trans* isomerization of 1,2-dideutero cyclopropane. Note that opening of the bond between CH*D* groups leads only to optical racemization.

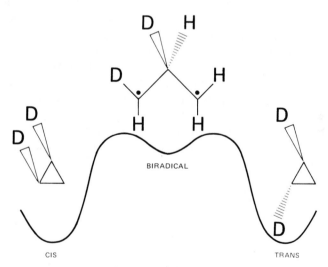

(b) Biradical model of the potential energy diagram of the *cis–trans* isomerization of 1,2-dideutero cyclopropane.

Figure 3.4.   Potential-energy diagrams for cyclopropane ring opening

end methyl groups of propane successively, the energy required is the same:

$$CH_3CH_2CH_3 \xrightarrow[\substack{98\,kcal \\ (observed)}]{} \dot{C}H_2CH_2CH_3 \xrightarrow[\substack{98\,kcal(?) \\ (assumed)}]{} \dot{C}H_2CH_2\dot{C}H_2$$

This assumption leads to an estimate of 54 kcal/mole to open the cyclopropane ring and thus a 10-kcal barrier to recyclizing the biradical. To eliminate the 10 kcal minimum for the biradical in energy space requires the proposal that the second C—H bond in the above process is 10 kcal stronger than the first, a very difficult interaction to understand since it implies a repulsion between the free electrons.[28] It should further be noted that in the hypothetical 1,3 biradical, the two C atoms are only 2.5 Å apart and if the CH_2 groups were perpendicular to the C—C—C plane, which would allow orbital overlap, they would still have about 8 kcal of sigma bonding. Only if one or both of the end CH_2 groups were twisted to lie in the plane of the ring would this sigma bonding be destroyed. It is presumed that pi-bond formation cannot exist at the large distance of 2.5 Å found in the biradical.

There is thus a mystery associated with the biradical, namely the origin of the barrier to recyclization. An equal mystery exists for the concerted reaction. If, indeed, the biradical is only the transition state at the top of a barrier, what is the nature of the forces which prevent the molecule which finds itself at the top of the barrier from migrating around the potential energy surface and exploring different molecular configurations (i.e., a metastable biradical) before returning to ground state? Some recent data by Berson[29] and his coworkers, on the relative rates of cis–trans isomerization versus racemization of the optically active forms of the trans isomer support the concerted model and suggest that there must be a barrier of at least 2 kcal separating the concerted path from that involving a randomly twisting biradical.[30]

An additional support for the biradical proposal comes from the data on the cis–trans isomerization of 1,2-dideutero spiropentane (Table 3.15). The weakest bond in this molecule is the one between the C_1—C_2 carbon atoms.

(cis)   (trans)

---

[28] This is, however, in accord with simple molecular orbital calculations, L. M. Stephenson, T. A. Gibson, and J. I. Brauman, *J. Amer. Chem. Soc.*, **95**, 2849 (1973).

[29] Note that in the concerted model, the breaking of the bond between CHD and CHD groups leads only to racemization while breaking of either of the other bonds leads only to cis–trans isomerization. See J. A. Berson, L. D. Pederson, and B. K. Carpenter, *J. Amer. Chem. Soc.*, **98**, 122 (1976).

[30] This is also in accord with more extended M.O. calculations by Y. Jean and L. Salem, *Chem. Comm.*, 382 (1971).

Orbital symmetry arguments would predict that the cleavage of this bond, followed by conrotatory (or disrotatory) rotation of the ĊHD groups could only lead to optical racemization, not to cis–trans isomerization. However, the latter reaction is the one observed. The biradical model would predict its Arrhenius parameters based on the cyclopropane model and taking into account the extra strain of spiropentane (8.8 kcal) as $k_{isom} = 10^{15.6-56.3/\theta}$. At 600°K this rate is three times slower than the observed rate which can be considered reasonably good agreement.

Despite these problems of interpretation, the biradical model provides a useful empirical tool for predicting Arrhenius parameters for substituted cyclopropanes from those of the unsubstituted.[31] In particular, the activation energies are found to be lowered by the expected amount if vinyl or phenyl groups are attached to the ring; namely, the stabilization energy of the conjugated radical. Nonpolar substituents are also found to have small, but predictable effects both on A factors and activation energies. Table 3.13 shows the predicted A factors for the biradical path while Table 3.14 shows the similar calculations for the concerted reaction. To compare with experiment, we should lower the A factor estimated on the biradical model by a factor of two giving A (cis $\rightleftarrows$ trans) = $10^{15.6}$ s^{-1}. The reason is that this model assumes that rotation in the metastable biradical is fast compared to ring reclosure so that each biradical has a 50% chance of closing to cis or to trans. Thus the two models have nearly the same A factors and both are, in fact, reasonably close to the experimental value of $10^{16.1}$ s^{-1}, which is probably uncertain by a factor of $10^{0.5}$ (Table 3.15).

Similar evidence exists for the pyrolysis of cyclobutane compounds via a biradical path. Geometrical isomerization of cis-1,2-dimethyl cyclobutane[32] occurs about four times more slowly than the major split into 2-propylene molecules. A minor split into C_2H_4 + butene-2 goes via a biradical, which can live long enough to rotate to produce some cis-butene-2 from the trans conformation before it splits to produce the favored trans-butene-2. The mechanism is

(cis)

(trans)

[31] H. E. O'Neal and S. W. Benson, *J. Phys. Chem.*, **72**, 1866 (1968); *Int. J. Chem. Kinet.*, **2**, 423 (1970).

[32] R. Gerberich and W. D. Walters, *J. Am. Chem. Soc.*, **83**, 3935; **84**, 4884 (1961).

Table 3.13.   Estimated Arrhenius A Factor for the cis–trans Isomerization of cis-1,2-Dideuterocyclopropane (Concerted Model)

Degrees of Freedom	$\Delta S^{\ddagger}_{300}$	$\Delta C^{\ddagger}_p$		
		300°K	500°K	800°K
Translation (no change)	0	0	0	0
Rotation (neglect small changes)	0	0	0	0
Symmetry ($\sigma = 1$, $n = 1 \rightarrow$ $\sigma^{\ddagger} = 1$, $n^{\ddagger} = 4$)[a]	$R \ln 4 = 2.8$	0	0	0
C—C Stretch ($1000\ cm^{-1}$) $\rightarrow$ C—C—C Bend ($350\ cm^{-1}$)	1.1	1.2	0.8	0.4
One CH_2 twist ($1100\ cm^{-1}$) $\rightarrow$ rotor ($V_0 = 2$ kcal) (reaction coordinate)[b]	−0.1	1.6	1.0	0.5
One CHD twist ($900\ cm^{-1}$) $\rightarrow$ rotor ($V_0 = 2$ kcal)	4.7	1.7	0.5	−0.3
Totals	8.5	4.5	2.3	0.6

$$\langle \Delta C^{\ddagger}_p \rangle = 2.3$$

$$\Delta S^{\ddagger}_{800} \approx \Delta S^{\ddagger}_{300} + \langle \Delta C_p \rangle \ln \left(\tfrac{800}{300}\right) = 8.5 + 2.3 = 10.8$$

$$A_{800} = A_{300} \exp \left\{ \frac{\Delta S^{\ddagger}_{800}}{R} \right\} = 10^{13.25 + 2.36} = 10^{15.6}\ s^{-1}$$

[a] TS is assumed to have both terminal methylene groups coplanar with the C—C—C plane. Note that in the concerted model, only the breaking of a CHD—CH_2 bond can lead to cis–trans isomerization.

[b] Since the $\dot{C}H_2$ and $\dot{C}HD$ ends rotate together, we should have taken an average of the two motions for the reaction coordinate. This would lower the A factor to $10^{15.4}\ s^{-1}$ (see Table 3.14).

Even more striking is the finding that in the deuterated compound, the CHD=CHD product was 50% cis and 50% trans as would be expected for a biradical mechanism.[33]

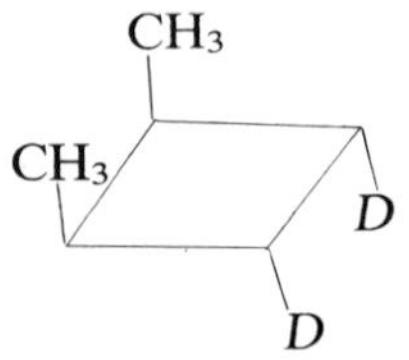

The Woodward–Hoffman rules indicate that concerted decompositions of cyclobutanes can only take place with geometrical inversion at one of the

[33] R. Srinivasan and J. N. C. Hsu, *J. Chem. Soc., Chem. Comm.*, 1213 (1972).

Table 3.14. Estimated Arrhenius A Factor for the cis–trans Isomerization of cis-1,2-Dideuterocyclopropane (Biradical Model)

Degree of Freedom	$\Delta S^{\ddagger}_{300}$	$\Delta C^{\ddagger}_p$		
		300°K	500°K	800°K
Translation (no change)	0	0	0	0
Rotation (neglect small changes)	0	0	0	0
Symmetry: $(\sigma = 1; n = 1) \rightarrow$ $(\sigma^{\ddagger} = 1; n^{\ddagger} = 3)^a$	$R \ln 3 = 2.2$	0	0	0
C—C Stretch $(1000\ cm^{-1}) \rightarrow$ C—C—C bend $(350\ cm^{-1})$ (reaction coordinate)^b	-0.1	1.6	1.0	0.5
One CH_2 Twist $(1100\ cm^{-1}) \rightarrow$ rotor $(V = 2.0\ kcal)^c$	3.0	1.9	0.8	-0.2
One CHD Twist $(900\ cm^{-1}) \rightarrow$ rotor $(V = 2.0\ kcal)^c$	4.7	1.7	0.5	-0.3
Totals	9.8	5.2	2.3	0
		$\langle \Delta C^{\ddagger}_p \rangle = 2.2$	$\langle \Delta C^{\ddagger}_p \rangle = 2.2$	

$$\Delta S^{\ddagger}_{800} \approx \Delta S^{\ddagger}_{300} + \langle \Delta C^{\ddagger}_p \rangle \ln \left(\tfrac{800}{300}\right) = 9.8 + 2.2 = 12.0\ \text{eu}$$

$$A_{800} = A_{300} \exp \left\{ \frac{\Delta S^{\ddagger}_{800}}{R} \right\} = 10^{13.25 + 2.62} \cong 10^{15.9}\ s^{-1}$$

^{a} Neglecting secondary isotope effects on bond strengths, this value of $n^{\ddagger}$ corresponds to the three different bonds being broken.
^{b} Note that $\Delta C^{\ddagger}_p$ includes the effect of the (kT/h) term whose $C_p = 2$.
^{c} Consult Tables A.1–A.23 for values. Note that when the bond between the two CHD groups is broken, we should have a higher $\Delta S^{\ddagger}$ contribution. This is largely compensated by the change in E_{act} arising from the secondary isotope effect.

breaking C—C bonds. Since such a path would involve appreciable distortion of the inverting center and would require a large increment of steric strain energy, the biradical pathway would seem to be favored. This has also been supported by recent quantum mechanical calculations of the energy of the tetramethylene biradical.[34] However the conjugation of the radical center with external pi bonds can relax the inversion requirement. The pyrolysis of acetyl cyclobutane would seem to qualify for such an exemption, and recent data indicates that it may decompose by a concerted pathway. The observed activation energy is indeed some 2–4 kcal lower than would be calculated by analogy with the alkyl cyclobutanes.[35,36]

[34] G. A. Segal, *J. Amer. Chem. Soc.*, **96**, 7892 (1974).
[35] R. K. Solly, D. M. Golden, and S. W. Benson, *Int. J. Chem. Kinet.*, **2**, 381 (1970).
[36] K. W. Egger, *J. Amer. Chem. Soc.*, **95**, 1745 (1973); A. T. Cocks and K. W. Egger, *J. C. S. Perkin II*, 2014 (1972).

Cyclobutene and alkyl cyclobutenes isomerize to butadienes with "normal" A factors of $10^{13.1}$ and an activation energy of 32.5 to 36 kcal/mole. The hypothetical biradical precursor in this case is indicated below. The structure of the parent molecule would seem to preclude fully developed allylic resonance in the transition state. The developing pi-electron orbitals in the breaking bond are at right angles to the double bond, so we would not expect

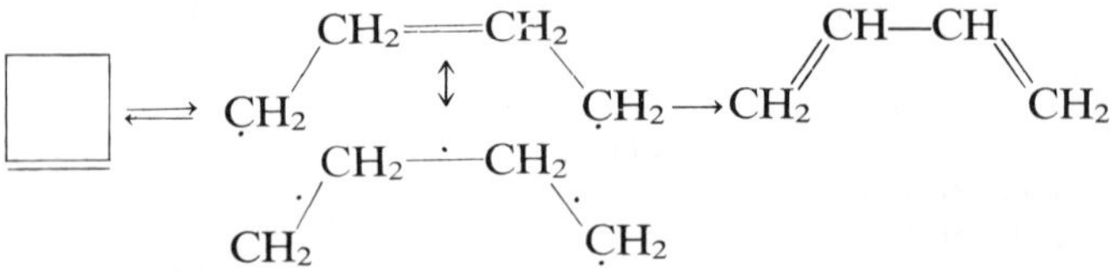

$$\square \rightleftharpoons \begin{array}{c} CH_2 \\ \dot{C}H_2 \end{array} \quad \updownarrow \quad \begin{array}{c} CH_2 = CH_2 \\ \dot{C}H_2 \rightarrow CH_2 \end{array} \quad \begin{array}{c} CH-CH \\ \\ CH_2 \end{array}$$

$$\begin{array}{c} CH_2 - CH_2 \\ \cdot \\ CH_2 \end{array} \quad \begin{array}{c} \\ \dot{C}H_2 \end{array}$$

extensive interaction. On this basis we could consider these reactions to proceed in a concerted fashion. The A factors predicted for either path are the same because the transition state must in either case be very rigid. There is only about 3.0 gibbs/mole entropy increase in the overall reaction.

The isomerization reactions of substituted cyclobutenes to butadienes are, in principle, stereospecific. When the single bond breaks and the terminal groups rotate, they do so in a conrotatory fashion, leading to the production of stereospecific olefins. Thus 3,4-dimethyl cyclobutene forms exclusively the *cis–trans* diene[37]:

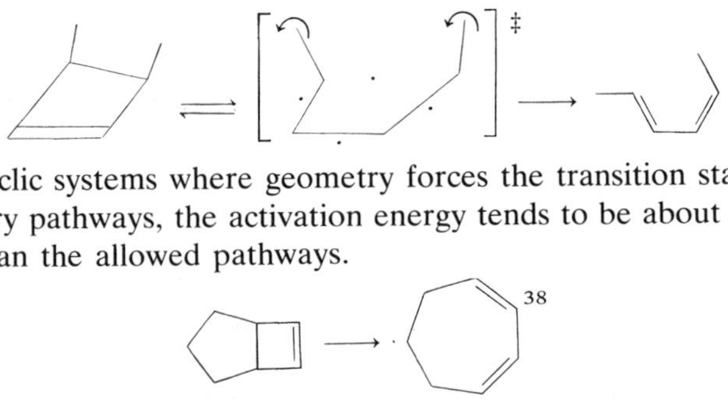

In bicyclic systems where geometry forces the transition state through disrotatory pathways, the activation energy tends to be about 10–15 kcal higher than the allowed pathways.

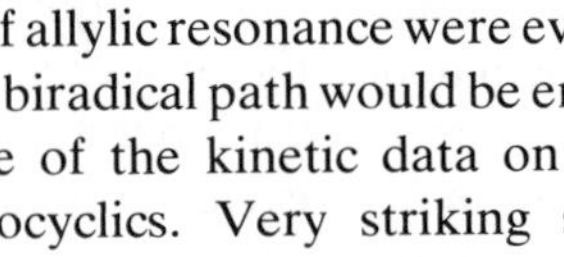

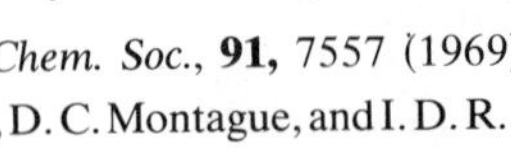

Note, however, that if allylic resonance were even two-thirds developed in the transition state, the biradical path would be energetically possible. Table 3.15 summarizes some of the kinetic data on these small ring systems, including some heterocyclics. Very striking support for the biradical

[37] R. Srinivasan, *J. Amer. Chem. Soc.*, **91**, 7557 (1969).

[38] G. R. Branton, H. M. Frey, D. C. Montague, and I. D. R. Stevens, *Trans. Faraday Soc.*, **62**, 659 (1969).

[39] D. M. Golden and J. I. Brauman, *Trans. Faraday Soc.*, **65**, 464 (1969).

hypothesis comes from the lowering of both the activation energy and the A factor observed when conjugating groups, such as vinyl, or phenyl, are introduced into these systems. We note that the pyrolysis of vinyl cyclo- propane produces as the major product, cyclopentene. The biradical mech- anism would be

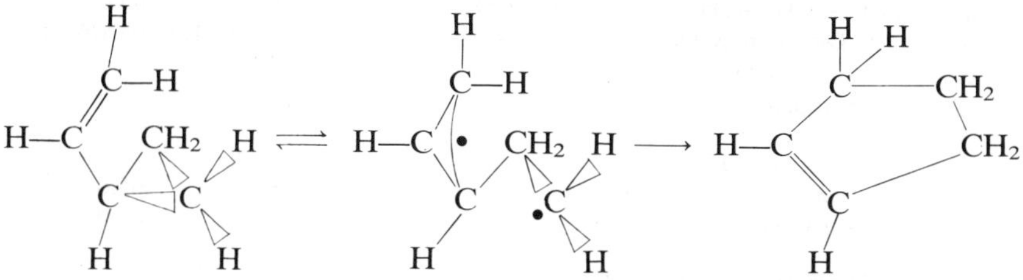

It should be pointed out that it is only when the vinyl group is in the unfavorable cis position to one of the C—C bonds in the ring that the reaction can occur with full allylic resonance. When the vinyl group is trans to the center of the ring, the geometry of the resulting, stiff allyl radical precludes five-membered ring formation. This is shown in the following diagrams:

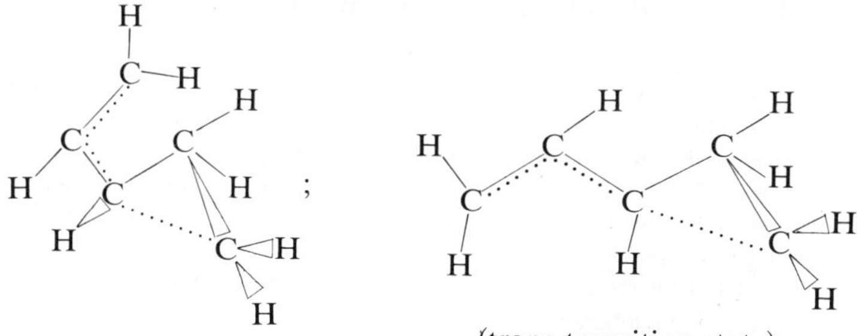

(cis transition state)   (trans transition state)

The trans form can, however, produce the pentadienes at an appreciably higher activation energy.

Comparison of any of the alkyl substituted cyclopropanes or cyclobutanes with the corresponding vinyl-substituted compounds shows an activation energy for the vinyl compound, lower by about 13 kcal and an A factor lower by about a factor of 10. The lowering of the activation energy by 13 kcal is precisely what we would expect if the rate determining step is the formation of the allylic biradical (3.39) because the allyl stabilization energy is 12.6 ± 1 kcal. By the same token, the lower A factor is explicable in terms of a net loss of hindered rotation of the vinyl group of about 4 eu.

Each substitution of a phenyl group should reduce $\Delta S^{\ddagger}$ by about 3–4 eu, as in the case of cis–trans isomerization, and E by 13 kcal, the benzyl- stabilization energy. Moreover, because of the much lower A factor for

Table 3.15. Arrhenius Parameters for the Pyrolysis of Small Ring Compounds

Reference	Reaction	T_m (°K)	log A (s⁻¹)	E (kcal/mole)
a	$\overline{CH_2\!-\!CH_2\!-\!CH_2}$ → $CH_3\!-\!CH\!=\!CH_2$	770	15.2	65.0
b	cis-Dideuterocyclopropane → trans	720	16.1(15.12)	65.1(65.4)
c	cis-1,2-Dideutero spiropentane → trans	600	14.5 (15.6)	51.5 (54.6)
d	$\overline{CH_3\!-\!CH\!-\!CH_2\!-\!CH_2}$			
	→ n-C₄H₈ (cis and trans)	740	15.40	65.0
	→ i-C₄H₈		14.62	66.0
e	$\overline{CH_2\!=\!CH\!-\!CH\!-\!CH_2\!-\!CH_2}$			
	→ $\overline{CH_2\!-\!CH\!=\!CH\!-\!CH_2\!-\!CH_2}$	640	13.6	49.7
	→ pentadiene-1,4		14.4	57.3
	→ pentadiene-1,3		13.0 (trans)	53.6
			13.9 (cis)	56.2
f	$\overline{CH_3\!-\!CH\!-\!CH_2\!-\!CH\!-\!CH_3}$ (cis) → trans	690	15.25	59.4
g	$\overline{\phi\text{-}CH\!-\!CH_2\!-\!CH\!-\!\phi}$ (cis) → trans (liquid)	460	11.2	33.5
h	Cyclobutane → 2C₂H₄	730	15.6	62.5
i	cis-Dimethyl cyclobutane → trans	690	14.81	60.1
	→ 2C₃H₆	680	15.48	60.4
	→ C₂H₄ + butene-2	680	15.57	63.4
j	Acetyl cyclobutane → CH₃COCH=CH₂ + C₂H₄	660	14.53	54.5
k	Methylene cyclobutane → CH₂=C=CH₂ + C₂H₄	710	15.7	63.0

1. Bibliographic sources:
   a. T. S. Chambers and G. B. Kistiakowsky, *J. Am. Chem. Soc.*, **56**, 399 (1934).
   b. E. W. Schlag and B. S. Rabinovitch, *J. Am. Chem. Soc.*, **82**, 5996 (1960). Values in parentheses are Arrhenius parameters for propylene formation from C₃H₄D₂. Note that because the mechanism proceeds via a biradical that can close with equal probability to yield *cis* or *trans*, the rate of ring opening is twice the rate of *cis–trans* isomerization.
   c. J. C. Gilbert, *Tetr.*, **25**, 1459 (1969). The values in parentheses are those estimated by assuming that the correct A factor is one third the A factor for cyclopropane isomerization (see reference b).
   d. D. W. Placzek and B. S. Rabinovitch, *J. Chem. Phys.*, **69**, 2141 (1965). Products include butene-1 and butene-2 (*cis* and *trans*) in ratios ~4:2:1.
   e. C. A. Wellington, *J. Phys. Chem.*, **66**, 1671 (1962).
   f. M. C. Flowers and H. M. Frey, *Proc. Roy. Soc.* (London), **A260**, 424 (1961); **A257**, 122 (1960). Collective production of butenes and pentenes is 100-fold slower.
   g. L. B. Rodewald and C. H. dePuy, *Tetrahedron Letters*, **40**, 2951 (1964).
   h. R. W. Carr, Jr. and W. D. Walters, *J. Phys. Chem.*, **67**, 1370 (1963).
   i. H. R. Gerberich and W. D. Walters, *J. Am. Chem. Soc.*, **83**, 3935, 4884 (1961).

Table 3.15 (*contd.*)

Reference	Reaction	T_m (°K)	log A (s^{-1})	E (kcal/mole)
l	CD$_2$=C—CH$_2$CH$_2$—CH$_2$ $\rightleftarrows$ CH$_2$=C—CH$_2$CH$_2$CD$_2$	650	14.5	50.0
m	Cyclo-C$_4$F$_8$ → 2C$_2$F$_4$	830	15.95	74.1
n	FC(Cl)—CF$_2$—CF$_2$—CFCl → 2CF$_2$=CFCl	740	15.4	65.3
	cis → *trans*	720	15.1	60.2
	trans → *cis*	720	14.9	60.2
o	CH$_2$CH$_2$O → CH$_3$CHO	670	14.2	57.0
p	CH$_2$CH$_2$CH$_2$O → C$_2$H$_4$ + CH$_2$O	710	14.8	60.0
q	CH$_2$CH$_2$CH$_2$—N=N → cyclopropane + N$_2$ + propylene	500	15.5	42.4
q, r	CH$_3$CHCH$_2$CH(CH$_3$)N=N (*cis* or *trans*) → dimethyl cyclopropane + N$_2$ + pentene-2 (*cis* and *trans*)	500	15.2	40.3
s	CF$_2$CF$_2$—O → CF$_2$O + C̈F$_2$	360	13.7	31.6
t	Cyclobutene → butadiene	390	13.1	32.6
u	3-Methyl cyclobutene → 1,3- pentadiene (*trans*)	390	13.5	31.6
v	1-Methyl cyclobutene → isoprene	420	13.8	35.1
w	(Perfluoro) butadiene → cyclobutene	500	12.1	35.4
x	Cyclobutylchloride → C$_2$H$_4$ + C$_2$H$_3$Cl	700	14.8	60.1

j. L. G. Paignault and W. D. Walters, *J. Am. Chem. Soc.*, **80**, 541 (1958).

k. J. P. Chesick, *J. Phys. Chem.*, **65**, 2170 (1961).

l. W. von E. Doering and J. C. Gilbert, *Tetra. Suppl.* **7**, 397 (1966). See also W. von E. Doering and W. R. Dolbier, Jr., *J. Amer. Chem. Soc.*, **89**, 4534 (1967).

m. B. Atkinson and A. B. Trenwith, *J. Chem. Soc.*, 2082 (1953).

n. B. Atkinson and M. Stedman, *J. Chem. Soc.*, 512 (1962).

o. M. L. Neufeld and A. T. Blades, *Can. J. Chem.*, **41**, 2956 (1963); S. W. Benson, *J. Chem. Phys.*, **40**, 105 (1964).

p. D. A. Bittker and W. D. Walters, *J. Am. Chem. Soc.*, **77**, 1429 (1955).

q. R. J. Crawford, R. T. Dummel, and A. Mishra, *J. Am. Chem. Soc.*, **87**, 3024 (1965).

r. R. J. Crawford and A. Mishra, *J. Am. Chem. Soc.*, **87**, 3768 (1965).

s. M. Lenze and A. Mele, *J. Chem. Phys.*, **43**, 1974 (1965).

t. W. Cooper and W. D. Walters, *J. Am. Chem. Soc.*, **80**, 4220 (1958).

u. H. M. Frey, *Trans. Faraday Soc.*, **60**, 83 (1964).

v. H. M. Frey, *Trans. Faraday Soc.*, **58**, 957 (1962).

w. E. W. Schlag and W. B. Peatman, *J. Am. Chem. Soc.*, **86**, 1676 (1964).

x. See references for Table 3.10.

benzyl torsion, compared to CH_2 torsion (due to higher reduced moment of inertia), it is probable that ring opening is no longer rate controlling in such compounds. If we assume that for cis–trans isomerization of 1,2-diphenyl cyclopropane, rotation in the biradical is rate determining, then we can anticipate that the entropy change going from the ring to the doubly conjugated biradical is practically zero. The A factor should then be essentially the A factor for torsion about the C-benzyl bond or about 10^{12} s^{-1}. A comparison of the A factor and E_{act} for cis–trans isomerization of the 1,2-diphenyl cyclopropane (Table 3.15) with those for 1,2-dimethyl cyclopropane shows a lowering of E by the expected 26 kcal and an A factor 10^4 smaller. The A factor is somewhat lower than the anticipated 10^{12} s^{-1} and this may reflect the effect of solvent since the reaction was carried out in the liquid phase.

3.12 POLYCYCLIC COMPOUNDS

The analysis of polycyclic systems is much more complex than for simple monocyclic compounds because more channels are available for reaction and also because the thermochemical data are less accurate. Using the methods described earlier for estimating entropies and heat capacities of polycyclic rings, there have been analyses[40] of a large number of unimolecular reactions. Both concerted and biradical paths occur. Thermochemical analysis can be definitive in demonstrating the incompatibility of a biradical path with observed Arrhenius parameters, but it cannot prove the occurrence of a biradical path. It can only demonstrate the compatibility of kinetic parameters with thermochemical data. Table 3.16 summarizes some of the data for these systems.

The cis–trans isomerization of 2-methyl-2,1,0-bicyclopentane affords an example of the biradical mechanism. The path is

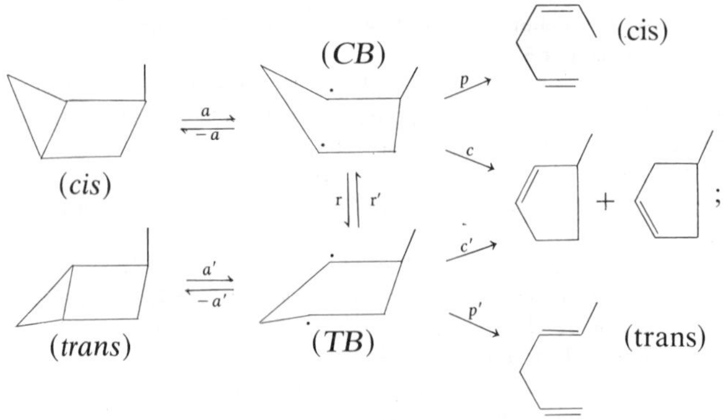

$$(3.30)$$

[40] H. E. O'Neal and S. W. Benson, *Int. J. Chem. Kinet.*, **2**, 423 (1970).

Table 3.16. Arrhenius Parameters for the Pyrolysis of Some Bicyclic Compounds

Reference[1]	Reaction	T_m (°K)	log A (s⁻¹)	E (kcal/mole)
a	[structure] → Cyclopentene	550	14.58(14.1)[2]	46.6(46.4)[2]
b	→ CH₂=CHCH₂CH=CH₂	600	14.4	52.3
c	[structure] (cis) → trans	450	14.45	38.9
b	[structure] → Hexadiene-1,5	450	13.4	36.0
d	[structure] → Hexadiene-1,5	630	15.17	55.0
e	[structure] → Cyclohexene	700	13.29	57.4
	→ 1-Methyl cyclopentane	730	13.89	61.17
f	[structure] → CH₂=CH(CH₃)—CH(CH₃)=CH₂	500	14.5	43.3
g	[structure] → C₂H₄ + Cyclopentadiene	500	13.78	42.8
h	Bicyclopentadiene → 2 cyclopentadiene	430	13.0	33.7
i	[structure] (β-Pinene) → (myrcene)	650	13.5	45.1
		650	15.9	49.9
j	Spiropentane → [structure]	660	15.9	57.6

1. Bibliographic sources:
   a. M. L. Halberstadt and J. P. Chesick, *J. Am. Chem. Soc.*, **84**, 2688 (1962).
   b. C. Steel, R. Zand, P. Hurwitz, and S. G. Cohen, *J. Am. Chem. Soc.*, **86**, 679 (1964).
   c. J. P. Chesick, *J. Am. Chem. Soc.*, **84**, 3250 (1962). $K_{eq} = (cis)/(trans) = 0.58$ at 220°C.
   d. R. Srinivasan and A. A. Levi, *J. Am. Chem. Soc.*, **85**, 3363 (1963).
   e. H. M. Frey and R. C. Smith, *Trans. Faraday Soc.*, **58**, 697 (1962).
   f. J. P. Chesick, *J. Phys. Chem.*, **68**, 2033 (1964).
   g. W. C. Herndon, W. B. Cooper, and M. J. Chambers, *J. Phys. Chem.*, **68**, 2016 (1964).
   h. J. B. Harkness, G. B. Kistiakowsky, and W. H. Mears, *J. Chem. Phys.*, **5**, 682 (1937).
   i. J. E. Hawkins and J. W. Vogh, *J. Phys. Chem.*, **57**, 902 (1955).
   j. M. C. Flowers and H. M. Frey, *J. Chem. Soc.*, 5550 (1961).

The activation energy of 39 kcal for the cis–trans isomerization is 7.7 kcal less than for the cyclopentene formation, which has been observed in the unsubstituted case. From the data on H migration in biradicals, we can deduce an activation energy of about 11 kcal for the 2–1 shift of an H atom in a 1,3 biradical. This implies here that both the ring inversion step (r) and the back reaction ($-a$) have 8 kcal less activation energy than the H atom shift.

From the known ΔH_f° of the bicyclopentane, it is possible to compute a ΔH_a for its conversion to the 1,3 biradical. At 298°K we find $\Delta H_a^\circ = 30$ kcal. The steady state rate of cis → trans isomerization is given by (ignoring back reaction and the other slower steps

$$\frac{-d\ (\text{cis})}{dt} = k'_{-a}(TB)$$

$$= \frac{k_r k_a k'_{-a}(\text{cis})}{k_r k'_{-a} + k_{-a} k_{-r} + k_{-a} k'_{-a}}, \qquad (3.31)$$

or for the observed first-order isomerization rate constant,

$$k_{\text{obs}} = \frac{1}{(\text{cis})} \frac{d\ (\text{cis})}{dt} = \frac{k_r K_a}{1 + k_r/k_{-a} + k_r/k'_{-a}} \qquad (3.32)$$

where $K_a = k_a/k_{-a}$.

If the ring inversion steps r and r' are slow compared to the ring-closing steps ($-a$) and ($-a'$), this reduces to Case I (slow inversion):

$$k_{\text{obs}} \to k_r K_a. \qquad (3.33)$$

If, in contrast, the converse is true, it reduces to Case II (fast inversion):

$$k_{\text{obs}} \to \frac{k_a}{1 + \dfrac{k'_r k_{-a}}{k_r k'_{-a}}} \sim \frac{k_a}{2}. \qquad (3.34)$$

In Case I we see that the observed activation energy E_{obs} is $\Delta H_a + E_r$, so that $E_r = 8.9$ kcal. In Case II $E_{\text{obs}} = E_a$, so that $E_{-a} = 8.9$ kcal. This latter value is comparable to the 10 kcal we find for the activation energy for 1,3 biradicals needed to close the cyclopropane rings; thus it is a quite reasonable result. In order for Case II to apply we would require that the barrier to inversion E_r would be less than 8.9 kcal. Nuclear magnetic resonance data on the inversion barrier in cyclohexanes yields a value of about 10 kcal,[41] so that E_r for cyclopentanes must be appreciably less because they have rapid pseudorotation that leads to ring inversion. It thus suggests that Case II applies.

If we make the same steady state calculation of the rate of formation of cyclopentene, we find that the activation energy (E_c) for H migration in the

---

[41] An excellent review is by V. M. Griffins, E. Wyn-Jones, and R. F. M. White, *Internal Rotation in Molecules*, W. J Orville-Thomas (ed.), Ch. 12, Wiley, New York, 1974.

biradical (Step c) is 16.6 kcal rather than the 11 kcal of the trimethylene biradical. We can rationalize this in terms of the fact that the transition state for this step requires that four carbon atoms in the ring be planar and thus includes the barrier to ring inversion, as well as the normal activation energy to H migration. This would suggest a 5 kcal barrier to ring inversion which seems quite reasonable.

The biradical mechanism for small ring pyrolysis has been analyzed quantitatively[42] in terms of a "loose" transition state and shown to give A-factor estimates to within a factor of $10^{\pm 0.3}$ s^{-1} and activation energies to within ±2 kcal. What is implied by "loose" is that the transition state for ring opening looks very much like the final state, the free biradical. This is not surprising because the data on radical–radical recombination are all in accord with a "loose" transition state for recombination.

The reaction coordinate may change considerably, however, depending on ring size. Looked at from the reverse direction, ring closing of the biradical, the reaction coordinate for closing of $\dot{C}H_2$—CH_2—$\dot{C}H_2$ to cyclopropane is the low-frequency (350 cm^{-1}) C—C—C bending mode. For the tetramethylene biradical, however, the reaction coordinate is the rotation of the two $\dot{C}H_2$—CH_2— groups about the central C—C bond.

A good example of a concerted reaction is provided by the reverse Diels–Alder reaction of norbornene to give cyclopentadiene plus C_2H_4:[43]

The observed rate constant (Table 3.16) is $k_1 = 10^{13.78-42.8/\theta}$ s^{-1}, measured near 500°K. The A factor indicates that $\Delta S^{\ddagger}$ is very small and that we are dealing with a tight transition state. The plausible biradical route would would be:[44]

| ΔH_f° (kcal/mole) | (22.3) | (71) | (32.4) | (12.5) |

However, $\Delta H_a^\circ \simeq 48$ kcal/mole which is 5 kcal in excess of the observed activation energy. The anticipated activation energy for the biradical Step a

[42] S. W. Benson and H. E. O'Neal, *J. Phys. Chem.*, **72**, 1866 (1968).

[43] R. Walsh and J. M. Wells, *Int. J. Chem. Kinet.*, **7**, 319 (1975).

[44] An energetically more favorable route would be opening to the cyclohexenyl methyl radical leading to a cycloheptadiene or a methyl cyclohexadiene as products.

would be $48 + E_{-a}$ where E_{-a} is the activation energy for closing the bicyclic 5-membered ring. E_{-a} is expected to be about 6–8 kcal, making $E_a \geq 54$ kcal, far in excess of the measured value. Thus the biradical mechanism is ruled out and this decomposition must be concerted. Since $\Delta S^{\ddagger} \approx 1.4$ eu, we can conclude that the TS must be very tight with a structure not very different from the starting compound.

We should not close our discussion of biradicals without some mention of some of the stereochemical data. The pyrolysis of cyclic azo compounds can provide a biradical whether or not the C—N bonds are broken consecutively or simultaneously. The pyrolysis of substituted pyrazolines lead to the formation of the same products as are obtained from trimethylene biradical, namely substituted cyclopropanes and propylenes (Table 3.15).

Surprisingly, the *cis*-pyrazoline[45] gives a 2:1 ratio of trans/cis. It appears that single inversion of structure is favored in the products. The significance of these results is still uncertain[46] and the uncertainty is complicated by the fact that biradicals are formed with considerable excess internal energy and while path 1′ seems most probable, it is not certain to how much extent path 1″ occurs.

3.13 STERIC INHIBITION OF RESONANCE IN TRANSITION STATES

Perhaps the most interesting feature of the bicyclopentane (2, 1, 0) pyrolysis is the slow rate at which the ring is opened by the biradical to form the diene. From our analysis of the cyclobutane system we would conclude that the activation energy for a 1,4 biradical to split into two olefins is about 6 kcal:

---

[45] R. J. Crawford and A. Mishra, *J. Amer. Chem. Soc.,* **88,** 3963 (1966) and later papers.
[46] For a detailed discussion, see R. G. Bergman, *loc. cit.* If path 1′ were predominant, then we could expect almost five- to 10-fold more olefins relative to ring compounds based on our experience with cyclopropanes. This suggests that path 1″ may be very important or else that peculiar dynamical effects are coming into play (p. 118). See also P. Dervan and T. Uyehara, *J. Amer. Chem. Soc.* **98,** 1262 (1976).

This is not the case for the cyclopentane biradical, where the activation energy to open the ring is about 22 kcal (Table 3.16). The cause of this large activation energy must be ascribed to steric inhibition of pi bonding. In order for the cyclopentadiyl-1,3 biradical

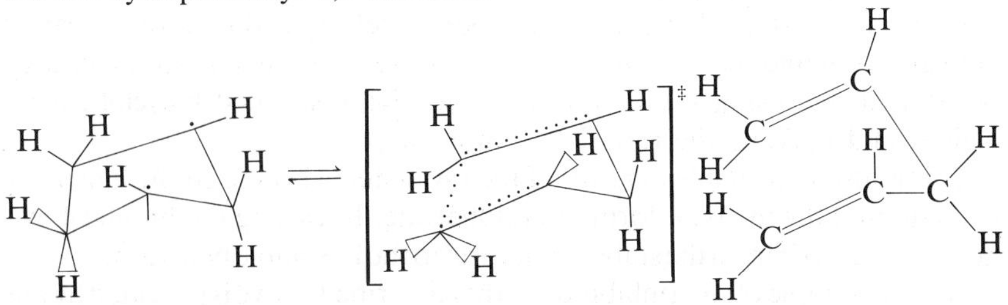

to make the transition to diene, the apex of the ring must be bent almost at right angles, so that the electrons in the breaking C—C bond can interact to form the pi bond in the emergent olefin.

A very similar barrier to that described above has been observed[47] in the decomposition of the cyclopentyl radical to form the pentene-1-yl-5 radical:

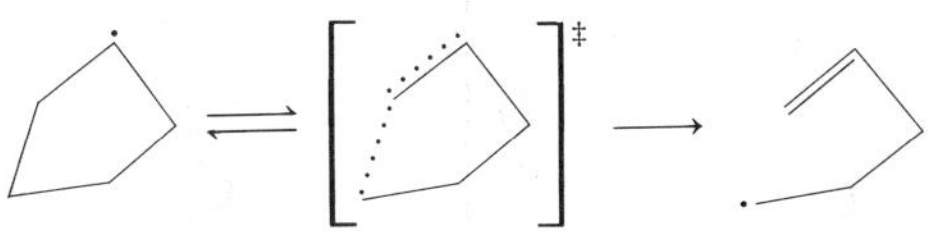

Despite the fact that the reaction above is endothermic by only 17 kcal, the observed activation energy is 37 kcal so that the "excess" activation energy is about 20 kcal, very close to the 22 kcal just estimated for the biradical.

A striking contrast to the situation just described is provided by the pyrolysis of bicyclohexane (2, 2, 0), which goes to hexadiene-1,5 via a cyclohexadiyl-1,4 biradical:

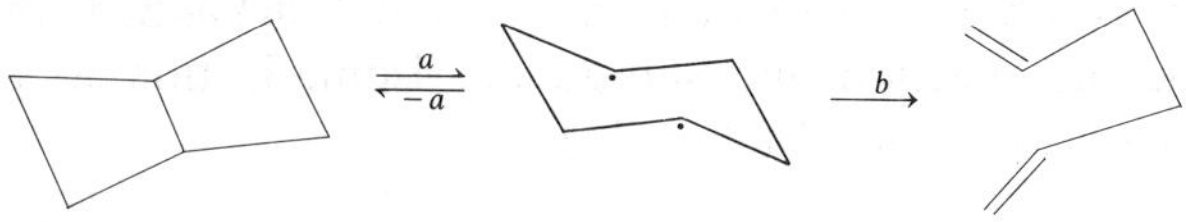

The strain free chair form of the C_6 ring permits the development of the pi bond between the electrons in the breaking bond and the odd electrons without severe distortion of the transition state. We can estimate the ΔH_f° of the bicyclohexane (2, 2, 0) as 28 kcal/mole by assuming double the strain energy of a C_4 ring. With the biradical $\Delta H_f^\circ \sim 56$ kcal, we find $\Delta H_a = 28$ kcal, compared to an observed activation energy of 36.0 kcal, so that E_b, the

[47] H. E. Gunning and R. L. Stock, *Can. J. Chem.*, **42,** 357 (1964); A. S. Gordon, *Can. J. Chem.*, **43,** 570 (1965). See also R. Walsh, *Int. J. Chem. Kinet.*, **2,** 71 (1970); W. P. L. Carter and D. C. Tardy, *J. Phys. Chem.*, **78,** 1573 (1974). General radical isomerization is reviewed by J. W. Wilt in *Free Radicals*, Vol. I, Ch. 8, J. K. Kochi (ed.), Wiley, New York, 1973.

activation energy for splitting the ring to yield diene, is now ≤ 8 kcal/mole. This can be compared to 22 kcal/mole for opening the cyclopentyl biradical and 4–6 kcal/mole for opening the tetramethylene biradical, showing that there is perhaps 2 kcal of strain involved in the C_6 ring transition state. It is, however, more likely that the extra 8 kcal of activation should be assigned to E_{-a}, as representing a strain energy developed in closing the biradical to the highly strained bicyclohexane. This would make a the rate-determining step.

This is also reasonable in terms of the estimated 6 kcal of strain which has been quoted for the boat form of cyclohexane. In closing the biradical, we need to convert it from the stable chair form to the less stable boat form.

A final example of such inhibition as that described above is provided by the case of the cyclobutyl radical. The activation energy for its opening to form butene-1-yl-4 radical is reported as 18 kcal,[48] despite the fact that the overall reaction is 3 kcal exothermic:

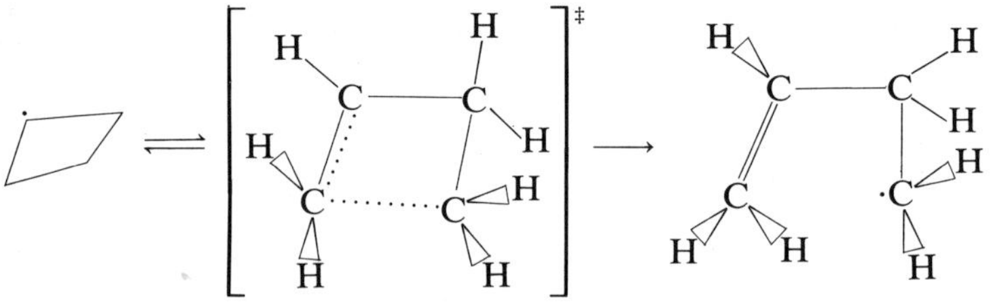

$$(3.35)$$

As we see from the structure, the breaking bond is at right angles to the emerging pi bond and hence there must be very small interaction of pi electrons in the transition state.

In apparent contrast to these cases is the case of 1,2-dideuteromethylene cyclobutane, for which an activation energy of 50 kcal has been reported for the ring-opening reaction and subsequent isomerization to methylene, 1,2-dideuterocyclobutane (Table 3.15):

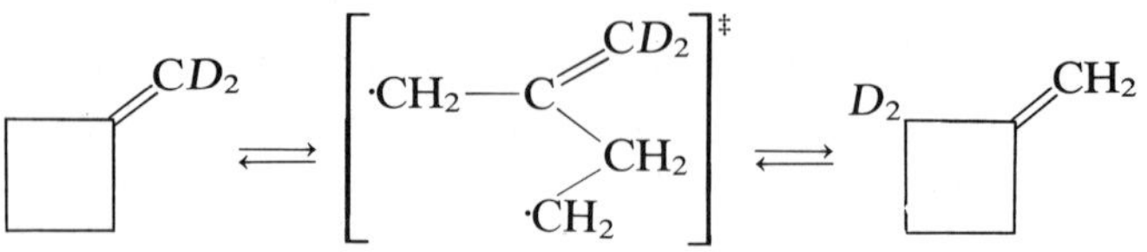

This value of 50 kcal suggests that the full allylic stabilization energy is available in the transition state for the ring opening so that there is not more

[48] A. S. Gordon, S. Ruthven-Smith, and C. M. Drew, *J. Chem. Phys.*, **36,** 824 (1963). See also R. Walsh, *loc. cit.* The competing reaction to recyclize the cyclopropyl methyl radical is much more rapid.

than about 2 kcal of steric inhibition in this case. However, if we consider that the transition state is in this case very close in geometry to the biradical, and thus not very different in geometry from tetramethylene, the excess strain relative to cyclobutane of 2 kcal seems quite reasonable.

3.14 ARRHENIUS A FACTORS FOR RING PYROLYSES

The A factors for bicyclic reactions can only be estimated with less accuracy than those for other reactions because of the absence of data on their entropies. However, if we estimate the entropies of the bicyclic species by the empirical methods suggested in the earlier section, we find reasonable agreement between the estimated and observed A factors.

EXAMPLE

Estimate the A factor for the pyrolysis of bicyclopentane $(2, 1, 0)$, BCP, to form cyclopentene at 550°K.

From the preceding discussion, we see that the biradical mechanism predicts that the opening of the ring to form cyclopentadiyl biradical is not rate determining because ring closing is fast compared to H-atom migration. Thus we can assume that the transition state involving H migration is rate determining and calculate $\Delta S^{\ddagger}$ on this basis. The mechanism is

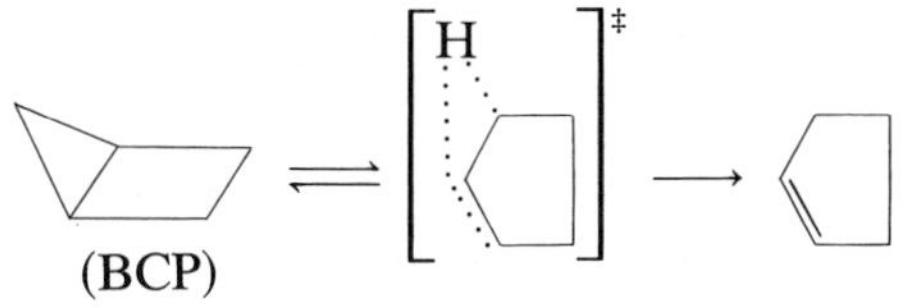

(BCP)

We must first calculate S^0 (BCP). If we assume that is has a stiff, four-membered ring like cyclobutene rather than cyclobutane (no pseudorotation), we can use the methods of Section 2.16. Starting with dimethyl cyclopropane $[S^0_{int(298)} = 78.2$ eu from groups], we subtract 9.3 eu (Table 2.13) to close to the four-membered ring. This gives 68.9 eu for S^0_{298} (BCP).

For the transition state, we can assume $S^0_{298\,(int)}$ to be very close to cyclopentene. We correct its entropy by +1.4 eu because it has optical isomers, to obtain $S^{0\ddagger}_{298} = 72.0$ eu. The reaction coordinate is the C—H stretch at 3100 cm^{-1} (Table A.13), and its entropy at 550°K (Table A.17) is zero eu. Neglecting any corrections from ΔC_{pm}, which is expected to be small, we thus calculate $\Delta S^{\ddagger} = 3.1$ eu.

At 550K°, with the Arrhenius A factor given by

$$A = \left(\frac{ekT_m}{h}\right)e^{\Delta S^{\ddagger}/R},$$

we find $A = 10^{14.2}$ s^{-1}. This is in excellent agreement with the two reported values of $10^{14.1}$ and $10^{14.6}$ s^{-1} (Table 3.16).

3.15 REVERSE DIELS–ALDER REACTIONS

We have already noted that the experimental evidence is fairly compelling that many C_3 and C_4 rings pyrolyze by a biradical mechanism. Recent evidence is just as convincing that the same is true for C_6 rings. The pyrolysis of cyclooctadiene-1,5 has been shown to produce 4-vinyl cyclohexene at low temperatures and butadiene at higher temperature[49,50] ($\sim 600°K$).

The inference is very strong from these results that both products come from a common precursor because 4-vinyl cyclohexene does not give butadiene at 600°K. The only consistent mechanism is one involving the allylic biradical

$$\text{(CO)} \quad \underset{-a}{\overset{a}{\rightleftharpoons}} \quad \underset{-c}{\overset{c}{\rightleftharpoons}} \quad \text{(VC)} \qquad \underset{-b}{\overset{b}{\rightleftharpoons}} \quad + \quad \text{(B)} \tag{3.36}$$

Steady-state analysis yields

$$\frac{d(\text{VC})}{dt} = \frac{k_c k_a (\text{CO})}{k_b + k_c + k_{-a}} = k_{\text{VC}}(\text{CO}) \tag{3.37}$$

$$\frac{d(\text{B})}{dt} = \frac{2 k_b k_a (\text{CO})}{k_b + k_c + k_{-a}} = 2 k_{\text{B}}(\text{CO}). \tag{3.38}$$

The best first-order rate constants (Doering et al.) are:

$$k_{\text{VC}} = 10^{15.6 - 52.4/\theta} \text{ s}^{-1}$$
$$k_{\text{B}} = 10^{16.3 - 55.9/\theta} \text{ s}^{-1}$$

with $k_{\text{B}/k_{\text{VC}}} = k_b/k_c = 10^{0.7 - 3.5/\theta} \text{ s}^{-1}$. In the experimental range of interest k_b/k_c varies from about $\frac{1}{3}$ to $\frac{1}{4}$. Furthermore, the strain energy of CO (~ 7 to 8 kcal) is appreciably greater than that of VC (~ 1.5 kcal). Since much of the former arises from eclipsing of H atoms and repulsion of the pi bonds, it seems reasonable to suppose that there may be more activation energy involved in closing the C_8 than the C_6 ring. The intrinsic A factors for these two closures are expected to be about the same, while symmetry and optical isomerism will favor C_6 ring closing by a factor of 4. Hence we may guess that $k_c \gg k_{-a}$. Thus we conclude that $k_{\text{VC}} \sim k_a$.

[49] R. Srinivason and A. A. Levi, *J. Am. Chem. Soc.*, **86**, 3756 (1964).
[50] W. von E. Doering, M. Franck-Neumann, D. Hasselmann, and R. L. Kaye, *J. Amer. Chem. Soc.*, **94**, 3833 (1972).

We can now use group additivity to estimate for the biradical (*cis, cis*) that $\Delta H^{\circ}_{f\,300} = 73.3 \pm 2$ kcal/mole and $S^{\circ}_{300} = 103.2 \pm 1.5$ eu. For CO we estimate $\Delta H^{\circ}_{f\,300} = 23.5$ kcal/mole and $S^{\circ}_{300} = 85.4 \pm 1.0$ eu. This then leads to the value (neglecting $\Delta C^{\ddagger}_{p}$):

$$k_{-a} = 10^{11.7-2.6/\theta}\, \text{s}^{-1}$$

An a priori estimate of k_{-a}, assuming that torsion about one of the central bonds is the reaction coordinate, leads to $A_{600} \sim 10^{11.6}\,\text{s}^{-1}$ in excellent agreement. We can similarly estimate the activation energy at about 3 kcal arising from the eclipsing of H atoms and C atoms about the central C—C bond as the cis–cis diradical approaches the transition state.

The A factor for k_c is estimated by our usual methods as $A_{600} \sim 10^{12.0}\,\text{s}^{-1}$ with an activation energy of about 1.0 kcal/mole. Since $\Delta H^{\circ}_f(\text{VC}) = 17.6$ kcal/mole and $S^{\circ}_{300}(\text{VC}) = 92.7$ eu, we can deduce:

$$k_c \approx 10^{12.0-1.0/\theta}\, \text{s}^{-1}$$

$$k_{-c} \approx 10^{14.3-56.7/\theta}\, \text{s}^{-1}$$

The directly measured ratio k_B/k_{VC} differs slightly (Doering et al.) from the independently measured ratio of k_B/k_{VC}, namely:

$$k_B/k_{VC} \simeq k_b/k_c = 10^{1.2-5.0/\theta}\, \text{s}^{-1}$$

Using the known ΔH°_f and S° for butadiene, we can then estimate:

$$k_b = 10^{13.2-6.0/\theta}\, \text{s}^{-1}$$

$$k_{-b} = 10^{8.5-27.3/\theta}\, \text{s}^{-1}$$

These values are now subject to some independent tests. First we can take the measured rate of decomposition of 4-vinyl cyclohexene to butadiene, for which the rate law is:

$$\frac{-d(\text{VC})}{(\text{VC})\,dt} = \frac{k_{-c}k_b}{k_b+k_c} = \frac{K_{-c}k_b}{1+k_b/k_c}$$
$$= k_{\text{exp VC}}\,(800^{\circ}\text{K}) \approx 0.6 k_b K_{-c}$$
$$= 10^{15.3-61.7/\theta}\, \text{s}^{-1}$$

This is again in good agreement with the value measured in shock-tube studies[51] of $10^{15.2-62/\theta}\,\text{s}^{-1}$. A further self-consistency check comes from the Diels–Alder dimerization of butadiene to VC, for which the rate law is:

$$\frac{-d(B)}{(B)^2\,dt} = \frac{2k_{-b}k_c}{k_b+k_c} = 2k_{\text{exp B}}$$

[51] W. Tsang, *Int. J. Chem. Kinet.*, **2**, 311 (1970).

or

$$k_{\exp B} = \frac{k_{-b}}{1 + k_b/k_c} \approx 0.8 k_{-b} \ (500°K)$$

$$= 10^{8.4-27.3/\theta} \ \text{l./mole-s.}$$

This is in excellent agreement with the experimental value[52] of $10^{8.2-26.8/\theta}$ l./mole-s, but serves only as a check on the self consistency of the rate data for the reversible Diels–Alder reaction and the thermochemical data for the reaction.

Doering et al. (*loc. cit.*) have made some interesting measurements on the rates of racemization and also vinyl exchange of an optically active deuterium labelled ($\diagdown CH{=}CD_2$) vinyl cyclohexene. The mechanism is indicated by:

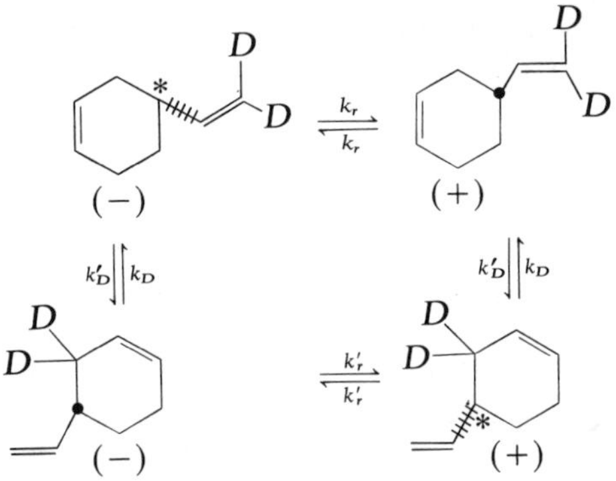

The absolute rates are very similar but the reported Arrhenius parameters differ markedly from each other and from what we might have anticipated from the biradical model:

$$k_\alpha \approx k'_\alpha = 10^{12.1-49.7/\theta} \ \text{s}^{-1}$$

$$k_D \approx k'_D = 10^{13.1-52.1/\theta} \ \text{s}^{-1}$$

Since the temperature range was quite small (45°C) and the rates very slow, the Arrhenius parameters are subject to large absolute uncertainty. If we use the biradical mechanism and assume that bond opening is rate determining for D exchange with the ring, we can then predict:

$$k_D \approx k_{-c} = 10^{14.3-56.7/\theta} \ \text{s}^{-1}$$

At 630°K this rate is a factor of 2 slower than the reported value. It is only 50% faster than the reported k_α. However, the latter is twice the k_r shown in the mechanism,[53] so that k_{-c} is about 50% faster than the

---

[52] D. Rowley and H. Steiner, *Disc. Faraday Soc.*, **10**, 198 (1951).
[53] The author calculated k_α by the equation, $k_\alpha = (1/t) \ln (\alpha_0/\alpha_t)$, where α_0 and α_t are the specific rotations at times $t = t_0$ and $t = t$, respectively. The mechanistic $k_r \approx k'_r = (k_\alpha/2)$.

mechanistic k_r. Thus the racemization and exchange measurements may be considered to be in agreement with the biradical model.

A final study giving very strong support to this biradical mechanism is that of the pyrolysis, racemization, and isomerization of *trans*-1,2-divinyl cyclobutane[54] in nonpolar solution.

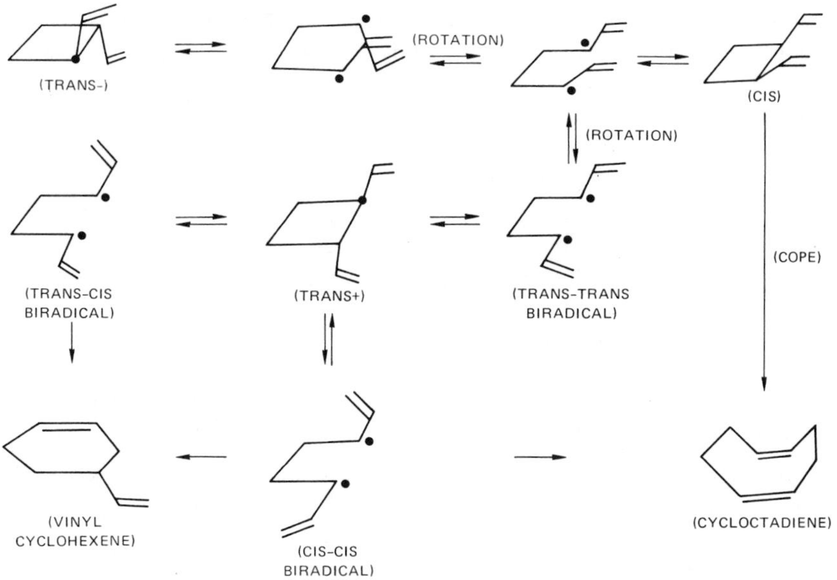

The mechanism and rate parameters are in very good agreement with the values estimated from thermochemical data and the parameters from the cyclooctadiene reaction study. At 180°C the fastest rate constant is for 4-vinyl-cyclohexene production while only slightly slower are k_r for racemization and k for cyclooctadiene formation. Butadiene formation is about 10% of VC production as might have been estimated from our earlier discussion.

Further light on the 1,8-octadienyl biradical is found in the recent work of Stephenson et al.,[55] who followed the thermal reaction of 1,4-*cis*–*cis*-dideutero-butadiene to produce tetradeuterovinyl-cyclohexene. It might be expected in such a system that a trans–trans diradical formed from two butadienes would live long enough before reacting to rotate about the 3–4 or 5–6 bonds and undergo cis–trans isomerization. If they then cleaved back into butadiene, one would expect some cis–trans dideuterobutadiene. None was found. This requires that the rotation rates are less than 1 to 10% of the

[54] G. S. Hammond and C. D. deBoer, *J. Amer. Chem. Soc.*, **86,** 899 (1964).

[55] L. M. Stephenson, R. V. Gemmer, and S. Current, *J. Amer. Chem. Soc.*, **97,** 5909 (1975). See J. A. Berson and R. Malherbe, ibid., 5910 (1975), for similar results on *trans, trans*-penta-1,3-diene-1-*d* dimerization. Compare with data in W. M. Marley and P. M. Jeffers, *J. Phys. Chem.* **79,** 2085 (1975).

competing rates in the biradical. For the trans–trans biradical, the competing rates are closure to divinyl cyclobutane or reopening to butadiene. The former is about 10-fold faster (190°C) than the later (k_b), which is itself about 20-fold faster than the rotation so the results are really not unexpected.

In contrast to the butadiene system, it is found that the reported rate parameters for the reverse Diels–Alder reaction of cyclohexene to butadiene + ethylene is only compatible with a concerted process. If, however, the parameters are raised, the biradical process is possible. It would be required, however, to raise them from $10^{15.2-66.2/\theta}\,\mathrm{s}^{-1}$ (Table 3.12) to about $10^{16.5-72/\theta}\,\mathrm{s}^{-1}$. This would permit a 5-kcal activation energy for the addition of C_2H_4 to butadiene to form the biradical.

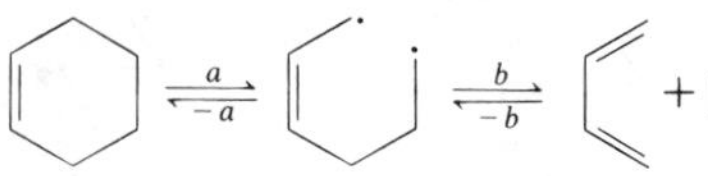

We find that $\Delta H_a^\circ = 66.9$ kcal, which is to be compared to an observed activation energy of 66.2 kcal. This implies that $k_b > k_{-a}$ at 900°K. If the activation energies differ by only 4 kcal, this would be consistent with a priori estimates of the relative A factors.[56]

[56] These conclusions are in excellent agreement with an extensive *ab initio* calculation by R. E. Townshend et al., *J. Amer. Chem. Soc.*, in press (1976).

4

Arrhenius Parameters for Bimolecular and Termolecular Reactions

4.1 COLLISION THEORY OF BIMOLECULAR REACTIONS

Historically, bimolecular reactions were interpreted in terms of a collision model derived from the kinetic theory of gases. The model assumed that the rate of a bimolecular reaction between A and B in the gas phase was given by some fraction α of the rate at which A and B collide. The fraction α was further broken down into two factors. First was the Boltzmann factor $e^{-E/RT}$, which represents the fraction of collision pairs $[A \cdots B]$ that have the necessary activation energy E_{AB} localized in the proper internal coordinates so that chemical reaction may take place. The second factor was the steric factor p, which made provision for the possibility that only a fraction $p \leq 1$ of properly energized collisions would have the geometry suitable for reaction.

From the collision point of view the bimolecular rate constant k_{AB} could be written as

$$k_{AB} = p_{AB} Z_{AB} e^{-E_{AB}/RT} \tag{4.1}$$

where Z_{AB} is the frequency of collisions between A and B expressed in some standard concentration units (for example atmospheres or moles/liter).

The kinetic theory of gases gives Z_{AB} as

$$Z_{AB} = \pi r_{AB}{}^2 \left(\frac{8RT}{\pi \mu_{AB}} \right)^{1/2} \tag{4.2}$$

141

where r_{AB} is the distance between the centers of mass of A and B in the collision complex, and $\mu_{AB} = M_A M_B/(M_A + M_B)$ is the reduced molecular weight of the colliding pair.

If all the quantities above are in cgs units, Z_{AB} will have units of cc/molecule-s. To transform to more usual units of liter/mole-s, we must multiply by $N_0/1000 = 6.02 \times 10^{20}$, where $N_0 =$ Avogadro's number $= 6.02 \times 10^{23}$ molecules/mole.

We can now compare (4.2) with the transition-state formula for a bimolecular reaction

$$k_{AB} = \frac{kT}{h} K_{AB}^{\ddagger} = \frac{kT}{h} \frac{Q_{AB}^{\ddagger}}{Q_A Q_B} \exp\left(\frac{-\Delta E_0^{\circ\ddagger}}{RT}\right) \qquad (4.3)$$

where $Q_{AB}^{\ddagger}$, Q_A, and Q_B are the molecular partition functions for the species $AB^{\ddagger}$, A, and B respectively, and $\Delta E_0^{\circ\ddagger}$ is the energy of formation of the $AB^{\ddagger}$ at $0°K$ from A and B. If we factor each partition function into its translational, rotational, vibrational, and electronic contributions, this last equation becomes

$$k_{AB} = \frac{kT}{h}\left(\frac{Q_{AB}^{\ddagger}}{Q_A \cdot Q_B}\right)_{\text{tran}} Q_{\text{rot}\|AB}^{\ddagger}$$

$$\times \left[\frac{Q_{\text{vib}AB}^{\ddagger} \cdot Q_{\text{rot}\perp AB}^{\ddagger} \cdot Q_{\text{elec}AB}^{\ddagger}}{Q_{\text{vib}A} \cdot Q_{\text{vib}B} \cdot Q_{\text{rot}A} \cdot Q_{\text{rot}B} \cdot Q_{\text{elec}A} \cdot Q_{\text{elec}B}}\right] \qquad (4.4)$$

$$\times \exp\left(\frac{-\Delta E_0^{\circ\ddagger}}{RT}\right).$$

By $Q_{\text{rot}\|AB}^{\ddagger}$ we mean the apparent rotational partition function for $AB^{\ddagger}$ considered as a hypothetical diatomic molecule of masses M_A and M_B separated by the distance r_{AB}. The quantity $Q_{\text{rot}\perp AB}^{\ddagger}$ is then the rotational partition function for the remaining degree of rotational freedom around the AB axis. This is an approximate description. More exactly $Q_{\text{rot}\perp AB}^{\ddagger} = (Q_{\text{rot}AB}^{\ddagger}/Q_{\text{rot}\|AB}^{\ddagger})$, where $Q_{\text{rot}AB}^{\ddagger}$ is the total precise rotational partition function for $AB^{\ddagger}$.

If we substitute the proper molecular quantities for the Q_{tran} it can be shown that[1]

$$Z_{AB} = \left(\frac{kT}{h}\right)\left(\frac{Q_{AB}^{\ddagger}}{Q_A \cdot Q_B}\right)_{\text{tran}} [Q_{\text{rot}\|AB}^{\ddagger}]. \qquad (4.5)$$

Thus if we compare the collision theories, from (4.2), and the transition-state theories, from (4.3), we see that we can make a structural analysis of

[1] See, for example, S. W. Benson, *Foundations of Chemical Kinetics*, McGraw-Hill, New York, 1960, p. 273.

the steric factor p_{AB} as

$$p_{AB} = \frac{Q_{vibAB^\ddagger} Q_{rot\perp AB} Q_{elecAB} \exp{-(\Delta E_0^{o\ddagger} - E_{AB})/RT}}{Q_{vibA} \cdot Q_{vibB} \cdot Q_{rotA} \cdot Q_{rotB} \cdot Q_{elecA} \cdot Q_{elecB}}. \tag{4.6}$$

From the point of view taken above we see that collision theory is a zeroth-order approximation to transition-state theory. In the simple hypothetical case when A and B are atoms all the terms in (4.6) become unity, and $p_{AB} = 1$. This is a somewhat artificial case because there are few chemical reactions that two atoms can undergo. Some of these few examples are radiative recombination, associative ionization, electronic excitation, and the inverse process of deexcitation:

$$A + B \rightleftarrows AB + h\nu$$
$$A + B \rightleftarrows A - B^+ + e^- \tag{4.7}$$
$$A + B \rightleftarrows A^* + B$$

4.2   STRUCTURAL ANALYSIS OF STERIC FACTORS

When A and B are not atoms but are more complex, it is invariably the case that the denominator in (4.6) is larger than the numerator. The reason for this is that in the process of transforming polyatomic A plus polyatomic B into polyatomic $AB^\ddagger$ we have changed six external rotations and six translations of A and B into three translations, three rotations, and six internal motions of $AB^\ddagger$. One of the latter is, of course, the reaction coordinate.

The factor Z_{AB} in (4.5) already accounts for the translational partition functions and two of the rotations of $AB^\ddagger$. Hence p_{AB} reflects the remaining terms, which include one rotation of $AB^\ddagger$ against six rotations for A and B and $(3n - 7)$ internal vibrations in $AB^\ddagger$ against $(3n - 12)$ in $A + B$.

The ratio of electronic partition functions is almost always ≤ 1. Because vibrational partition functions are generally in the range of 1–10, whereas rotational partition functions are in the range 10–100 per rotational degree of freedom, we expect p_{AB} to be in the range 1–10^{-5} depending on the complexity of A and B and the "tightness" of the transition state $AB^\ddagger$. This latter is of essential importance, for among the $(3n - 7)$ internal vibrations of $AB^\ddagger$ we will have one nearly free internal rotation (A against B) about the $A \cdots B$ axis and up to four rocking modes of A and B with respect to the $A \cdots B$ axis. If these latter are very loose, they can just about cancel the rotational partition functions of A and B, and the result is that p will be very close to unity. This we shall see occurs in the recombination of alkyl radicals, for example $2CH_3 \rightarrow C_2H_6$.

Occasionally we see claims of steric factors in excess of unity. These are rationalized by claiming that the Boltzmann factor $f_B = e^{-E_{AB}/RT}$ in (4.2) is not correct and that one should use instead the factor (for a classical system)

$$f_B = \frac{1}{(S-1)!} \left(\frac{E_{AB}}{RT}\right)^{S-1} e^{-E_{AB}/RT} \qquad (4.8)$$

to take into account the fact that the activation energy E_{AB} can be distributed among all possible S internal degrees of freedom of the collision complex. For a quantized system with a geometrical mean vibration frequency $\nu_m = (\nu_1 \cdot \nu_2 \cdot \nu_3 \cdots \nu_s)^{1/s}$ the factor is

$$f_B \sim \frac{1}{(S-1)!} \left(\frac{E_{AB}}{h\nu_m}\right)^{S-1} e^{-E_{AB}/RT}.$$

This would be correct if each such distribution has a unit probability of contributing to chemical reaction. However, only those collision complexes that have the proper localization of activation energy in the necessary bonds can react (that is, are in the transition state). All the others will redissociate.[2] Equation (4.8) is thus improper for a bimolecular reaction, and we can take it as a good critique of experimental rate data that when their steric factors exceed unity by any significant amount, they are most likely in error.

Two important exceptions exist to this last rule, and both are apparent rather than real. The first concerns the bimolecular process of energy transfer, indicated by

$$A + M \rightleftarrows A^* + M. \qquad (4.9)$$

Here a collision between A and M with energy distributed between the two becomes localized in the internal modes of A^*. In actuality, such a process undoubtedly takes place from the nearby energy states of A just below A^* through a series of "lucky" collisions, and each with a "normal" or even small steric factor.

However, by convention, the overall process described above is usually referred to the ground state of A. Thus at substantial energies E^* in A^* there is a large cumulative degeneracy factor, which reflects the large number of ways of distributing this energy among the S internal degrees of freedom of A.[3] We can see this for the very first step when E^* is small.

---

[2] S. W. Benson and A. E. Axworthy, Jr., *J. Chem. Phys.*, **21**, 428 (1953).
[3] This is the preexponential factor in (4.8).

Let us assume a molecule A with twelve internal oscillators all of very nearly the same frequency, so that if A has one vibrational quantum, it may exist with equal probability in any of the twelve modes. The degeneracy of the first vibrational state, $A^{(1)}$ is 12, and for the process of collisional excitation of the first and second vibrational levels of A, we can write the hypothetical kinetic mechanism

$$A^{(0)}+M \underset{-1}{\overset{1}{\rightleftarrows}} M+A^{(1)};$$

$$A^{(1)}+M \underset{-2}{\overset{2}{\rightleftarrows}} M+A^{(2)}; \tag{4.10}$$

$$\frac{d(A^{(2)})}{dt}=k_2(M)(A^{(1)})-k_{-2}(M)(A^{(2)}). \tag{4.11}$$

The degeneracy of the $A^{(2)}$ state is $(S+1)(S)/2=78$, so that $(k_2/k_{-2})=(78/12)e^{-h\nu/RT}$ and $(k_1/k_{-1})=12e^{-h\nu/RT}$. If $A^{(1)}$ is at or near equilibrium with $A^{(0)}$ and we are looking at the initial rate of formation of $A^{(2)}$ (i.e., we ignore deactivation of $A^{(2)}$), $(A^{(1)})=12(A^{(0)})e^{-E°/RT}$ where $E°=h\nu$ and

$$\frac{d(A^{(2)})}{dt}=12k_2(M)(A^{(0)})e^{-E°/RT}$$

$$=78k_{-2}(M)(A^{(0)})e^{-2E°/RT}. \tag{4.12}$$

We see that even if the steric factor of k_{-2} was 0.1, the apparent steric factor for the overall process would be 7.8 and would exceed unity. It is easy to extrapolate to higher levels and see that if we are dealing with a large number of quanta $(E^*\gg h\nu)$, the statistical factor can become very large.

The second process applied to apparently large steric factors is dissociative recombination of ions + electrons:

$$AB^++e^-\rightarrow A+B+\text{energy}. \tag{4.13}$$

Steric factors for these reactions can be as large as 10^5. This is accounted for by the fact that we are using the wrong collision frequency. Electrons, being about 10^4 lighter than most molecules, have 10^2 higher average thermal velocities and thus have 100-fold greater collision rates. The remaining factor of 10^3 comes from the fact that at a 100 Å separation the coulombic interaction of the two ions (E^2/r) is 3.3 kcal, which is much larger than RT (0.6 kcal) at 25°C. Hence ion collisions at low temperatures have a much larger effective diameter than they do for neutral particles. Because this is about 20–30 times larger than collision diameters for neutrals, and the collision frequency goes as the square of the effective collision diameter, we find the remaining factor of about 10^3 in the "abnormal" collision cross section.

4.3 BIMOLECULAR REACTIONS—CATEGORIES

Transition-state theory yields for the A factor of a bimolecular reaction the value $(ekT/h) \exp(\Delta S_p^{\ddagger}/R)$, where $\Delta S_p^{\ddagger}$ is the entropy change (always negative) for forming a mole of transition state complex from the two reactants.[4]

$$A + B \rightleftarrows (AB)^{\ddagger} \rightarrow \text{products}.$$

Three categories of reactions may be distinguished for bimolecular reactions. The first is a metathesis in which an atom is transferred from A to B. An example is the well-known chain step:

$$H + Br_2 \rightleftarrows (H \cdots Br \cdots Br)^{\ddagger} \rightleftarrows HBr + Br.$$

The second is a displacement or exchange reaction, quite frequent in the chemistry of unsaturated molecules:

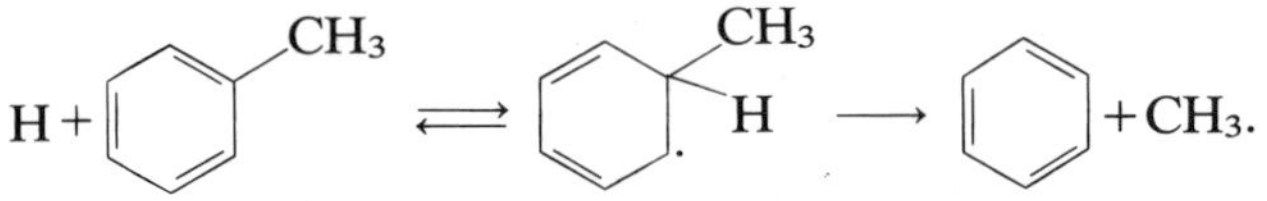

The last category is the association reaction. A typical example is the recombination of radicals:

$$2 \cdot CH_3 \rightleftarrows (CH_3 \cdots CH_3)^{\ddagger} \rightarrow C_2H_6.$$

It should be noted that the displacement reaction is a complex process involving the formation of a true intermediate by an association reaction. Because these reactions are generally exothermic, the adduct is usually a vibrationally excited species, which in the gas phase may have a very short lifetime. Such reactions can thus involve two discrete transition states.

Unless the vibrationally excited species formed in an association reaction loses its energy by a collision, it will inevitably dissociate back into original or into more stable species. Where the rate of this dissociation is rapid compared to collision frequencies, we find that the reaction rate is controlled by the rate of these deenergizing collisions and the kinetics

---

[4] This depends again on standard states. The discussion above is correct for pressure units. With $\Delta S_c^{\ddagger}$, based on concentration units, the activation energy is higher by RT, and we get a factor of e^2. The same is true for the Arrhenius equation for second-order reactions.

tend toward third order. Such processes are called energy transfer processes because the rate limiting step is the transfer of energy between species.

Note that in every case of the formation of a transition state from two reactants the reverse process is a unimolecular dissociation, the rate constant of which is derivable from the considerations presented in Chapter 3.

4.4 LOOSE AND TIGHT TRANSITION STATES

We have already seen in our discussion of fission processes that the "looseness or tightness" of the transition state, and hence the value of $\Delta S^{\ddagger}$, depends intimately on the nature of the potential interaction between the two fragments separating. The same reasoning must apply to the two species taking part in a bimolecular reaction. Let us summarize our semiquantitative analysis to this point.

We have described as a "loose" transition state, one in which the fragments are separated by distances between the bonding centers in each of them, which are about 2.9 ± 0.2 times the stable bond distance. For the elements C, N, O, and F, this will involve separations of about 4–4.5 Å. At these distances the separating species will be almost independent of each other. Their mutual interactions will be of the order of RT in energy in the transition state, and we expect no activation energy in the reverse reaction, the combination of the two species.

At the other extreme from this loose transition state will be the "tight" transition state. An empirical rule based on the difference in bond lengths of 0.3 Å in H_2 (two-electron bond) and H_2^+ (one-electron bond) suggests that the extension of a single bond by 0.3–0.5 Å reduces the bond strength by about a factor of two. We will assume that tight transition states involve distances between bonded atoms that may be extended by about this amount. Thus in the metathesis reaction of $H + D_2 \rightarrow DH + D$, we would anticipate that if the transition state $(H \cdots D \cdots D)^{\ddagger}$ does not correspond to a stable species, that it will be characterized by a tight transition state in which the $H \cdots D$ and $D \cdots D$ distances are about 1.0 to 1.1 Å. Moreover since the TS is unstable, relative to either products and reactants, we shall expect a nontrivial activation energy for the reaction. In fact all of these expectations are confirmed, and the activation energy is about 8 kcal despite the fact that the reaction is nearly thermoneutral.

In contrast to this, the exothermic, metathesis reaction $O + OH \rightarrow O_2 + H$ will proceed through a stable adduct HO_2 relative to reactants (as well as products), and so we expect the transition state to be loose corresponding to one appropriate for radical recombination (or for the reverse reaction, simple bond fission). This is also confirmed, and we shall find, in addition, that there is no activation energy for the reaction. This type of metathesis is really then an association reaction. How to distinguish the two in the absence of direct experimental information is not always simple, and we shall return to this question later. Let us consider for the moment, the tight transition states.

4.5 LOWER LIMITS TO BIMOLECULAR A FACTORS

The calculation of $\Delta S^{\ddagger}$ for a bimolecular process requires the knowledge of the structure of the $AB^{\ddagger}$ complex. Some useful information in this regard can be obtained very quickly by fixing a lower limit to $\Delta S^{\ddagger}$, which is always negative, by assuming that $AB^{\ddagger}$ is a very tight complex. In such

Table 4.1. Arrhenius Parameters for Some Metathesis Reactions Involving Atoms

Reference[1]	Reaction	T_m° (°K)	log A (l./mole-s)	E (kcal/mole)
a	$Br + H_2 \rightarrow HBr + H$	620	10.8(11.4)	18.2(19.7)
b	$I + H_2 \rightarrow HI + H$	680	11.4	34.1
a	$Cl + H_2 \rightarrow HCl + H$	500	10.9(10.7)[8]	5.5(5.3)[8]
c	$O + O_3 \rightarrow 2O_2$	380	10.5	5.7
a	$Br + C_2H_6 \rightarrow HBr + C_2H_5$	500	10.8	13.5
a	$Cl + C_2H_6 \rightarrow HCl + C_2H_5$	500	11.0	1.0
d	$I + CH_4 \rightarrow HI + CH_3$	630	11.7	33.5
e	$H + C_2H_6 \rightarrow H_2 + C_2H_5$	300–1100	11.1	9.7
f	$H + D_2 \rightarrow HD + D$	900	10.7	9.4
d	$I + CH_3I \rightarrow I_2 + CH_3$	600	11.4	20.5
g	$O + NO_2 \rightarrow O_2 + NO$	330	10.3	1.0
h	$O + H_2 \rightarrow OH + H$	600	10.50(11.34)	10.2(13.7)
i	$H + HOH \rightarrow H_2 + OH$	(300–2500)	10.9	20.5
j	$H + N_2O \rightarrow HO + N_2$	(450–1487)	10.7	13
k	$N + O_2 \rightarrow NO + O$	300	9.3	6.3
k	$N + NO \rightarrow N_2 + O$	(300–5000)	10.2	0
l	$K + Br_2 \rightarrow KBr + Br$	(beam)	11.2	−0.4
k	$O + C_2H_6 \rightarrow HO + C_2H_5$	(300–650)	10.4	6.4
i	$O + OH \rightarrow O_2 + H$	300	10.3	0
m	$F + H_2 \rightarrow HF + F$	300	11.1	1.7
m	$Na + CCl_4 \rightarrow NaCl + CCl_3$	550	10.3	3.5

1. Bibliographic sources:
    a. G. C. Fettis and J. H. Knox, *Progress in Reaction Kinetics*, McMillan, New York, 1964, Vol. 2, p. 26. Also see p. 17 for $Cl + H_2$. Values in parentheses are "best" estimates by F and K.

cases, lower limits of $AB^\ddagger$ can be quickly fixed by comparison with known molecules. Thus, in the metathesis reaction

$$H + C_2H_6 \longrightarrow \left(H \cdots H \cdots \underset{\underset{H}{|}}{\overset{\overset{H}{|}}{C}}{-}CH_3 \right)^{\ddagger} \longrightarrow H_2 + C_2H_5.$$

We can assign $S^\ddagger(C_2H_7)$ a value by comparison with C_2H_6. We note that its translation and rotational entropy should not be seriously increased over C_2H_6 by the additional light H atom, so that the chief changes will be in symmetry and spin. Hence

$$S^{\circ\ddagger}(C_2H_7) \geq S^\circ(C_2H_6) + R \ln 6 + R \ln 2,$$

and

$$\Delta S^\ddagger = S^\ddagger(\dot{C}_2H_7) - S^\circ(C_2H_6) - S^\circ(H)$$
$$\geq -S^\circ(H) + R \ln 12.$$

At 400°K, $\Delta S^\ddagger(400) \geq -24.1$ with a standard state of 1 atm. Transforming to molarity by adding $R \ln(RT) = 7.0$ gibbs/mole, we find $\Delta S_c^\ddagger(400) \geq -17.1$ gibbs/mole, so that $A = e^2(kT/h) \exp(\Delta S_c^\ddagger/R) \geq 10^{10.1}$ l./mole-s. This is reasonably close to most of the A factors reported for this reaction, which are in the neighborhood of 10^{11} (Table 4.1). To account

(8) S. W. Benson, F. R. Cruikshank, and R. Shaw, *Int. J. Chem. Kinet.* **1**, 29 (1968).

b. J. H. Sullivan, *J. Chem. Phys.*, **30**, 1292 (1959).

c. S. W. Benson and A. E. Axworthy, Jr., *J. Chem. Phys.*, **42**, 2614 (1965).

d. M. C. Flowers and S. W. Benson, *J. Chem. Phys.*, **38**, 882 (1963). See corrections noted by D. M. Golden, R. Walsh, and S. W. Benson, *J. Am. Chem. Soc.*, **87**, 4053 (1965).

e. R. R. Baldwin and A. J. Melvin, *J. Chem. Soc.*, 1785 (1964).

f. A. A. Westenberg and H. deHaas, *J. Chem. Phys.*, **47**, 1393 (1967).

g. F. S. Klein and J. T. Herron, *J. Chem. Phys.*, **41**, 1285 (1964).

h. A. A. Westenberg and N. deHaas, *J. Chem. Phys.*, **50**, 2512 (1969). Values in parenthesis from shock tube studies (1400–2000°K) are consistent and demonstrate effect of $\Delta C_v^\ddagger \sim 3.1$ eu. G. L. Schott, R. W. Getzinger, and W. A. Seitz, *Int. J. Chem. Kinet.*, **6**, 921 (1974).

i. D. L. Baulch et al., *Evaluated Data for High temperature Reactions*, Vol. I, Butterworths, London, 1972.

j. G. Dixon-Lewis, M. M. Sutton and A. Williams, *J. Chem. Soc.* 5724 (1965).

k. D. Garvin and R. F. Hampson, *Chemical Kinetics Data Survey*, VII, NBSIR 74-430, Govt. Printing Office, Washington, D.C. (1974).

l. T. T. Warnock, R. B. Bernstein, and A. E. Grosser, *J. Chem. Phys.*, **46**, 1685 (1967).

m. V. N. Kondratiev, *Rate Constants of Gas-Phase Reactions*, L. J. Holtschlag (trans.) and R. M. Fristrom (USNBS) (ed.), National Technical Information Service, Springfield, Va. (1972).

for the difference, we should have to make the quite reasonable assumption of two H·H—C bends with frequencies at about $400\,\text{cm}^{-1}$ in the transition state.

For a reaction of a heavy atom, we can take as an example

$$I + CH_4 \rightleftarrows (I\cdots H\cdots CH_3)^{\ddagger} \rightarrow HI + CH_3.$$

Here $S^{\ddagger} \geq S^{\circ}(CH_3I) + R \ln 2$. To improve the comparison, we should add a correction for the increased rotational entropy of CH_3I due to a long $C\cdots I$ distance. This latter can be estimated as discussed earlier (page 147) by adding 0.6 Å to the sum of the normal C—H and H—I covalent lengths. This yields $r(C\cdots I) = 3.4$ Å, compared to a normal $r(C—I) = 2.10$ Å. Because the entropy goes as $\frac{1}{2}R \ln(I_A I_B I_C)$, where I_A, I_B, and I_C are the principal moments of inertia and vary as $\Sigma\, M_i r_i^2$, this changes two of these moments each by a factor of nearly 2.6. Hence we should add $R \ln 2.6$ to our estimate:

$$S^{\ddagger} \geq S^{\circ}(CH_3I) + R \ln 5.2;$$

$$\Delta S^{\ddagger} \geq S^{\circ}(CH_3I) - S^{\circ}(CH_4) - S^{\circ}(I) + 3.3 = -24.7$$

at 300°K. $\Delta S_c^{\ddagger} > -18.3$ gibbs/mole, so that $A > 10^{9.7}$ liter/mole-s.

The observed value of $10^{11.7}$ is considerably above this (Table 4.1), indicating the presence of very low frequency bending modes in the transition state. At 700°K, these would have to be two degenerate rocking vibrations at $\sim 200\,\text{cm}^{-1}$ each to account for the difference, again a quite reasonable result. An equivalent explanation is an internal rotation of CH_3 in a nonlinear complex plus a $200\,\text{cm}^{-1}$ C—H—I bend.

For a reaction such as $CH_3 + I_2 \rightarrow CH_3 + I$ let us use, as the closest molecular analogue, CH_2I_2 which has one less H atom and an $I\cdots I$ distance not much greater (3.4 Å) than that anticipated for I_2 in the transition state (2.66 Å + 0.5 Å). This leads at 300°K to:

$$\Delta S_c^{\ddagger} = S_{300}^{\ddagger}(CH_3\cdots I_2) - S_{300}^{\circ}(CH_3) - S_{300}^{\circ}(I_2)$$

$$\geq S_{300p}^{\ddagger}(CH_2I_2) + 1.4(\text{spin}) - 0.8(\text{symmetry}) - 46.4 - 62.3$$

$$+ 6.4(\text{atm} \rightarrow M/L)$$

$$\geq 74.0 + 7.0 - 108.7 = -27.7\,\text{eu}.$$

We should further correct this for the reasonable expectation that the *TS* will have bent $C\cdots I\cdots I$ bonds and an internal $CH_3—I_2$ torsion with zero barrier. At 300°K (Table A.18) this gives an additional 5.8 eu so that $\Delta S_c^{\ddagger} \geq -21.9\,\text{eu}$, which leads to $A_{300} \geq 10^{8.9}$ l./mole-s. Again this is considerably below the reported values, which range from $10^{9.6}$ to $10^{10.1}$ l./mole-s, suggesting that the transition state is much looser than our tight model. If we consider that our calculation lacks a correction for the reaction coordinate which must correspond to the I—I stretch at about 450 cm^{-1} ($S_{300}^0 = 0.9$ eu), we see that the actual transition state has about 5.0 eu more entropy than the tight transition state. It would take two nearly free internal rotations of the CH_3 group at right angles to the $C \cdots I_2$ bond to contribute this much entropy.

In both of these latter examples, we see that the tight transition state gives a considerably lower A factor than the experimentally observed value, so much so (factor of 15–100) in fact, that we are led to conclude that the transition state for both of these reactions must be loose! There is no way of anticipating such a finding except by analogy.[5]

Consider the reaction $CH_3 + I_2$, which is exothermic by about 20 kcal and has zero, or nearly zero, activation energy. The fact that the activation energy is so small is an indication that the transition state cannot be very tight, namely that there cannot be a very close interaction of $CH_3 + I_2$.

A final example comes from the attack of CH_3 on C_2H_6 to abstract H. Correcting only for symmetry and spin,

$$S^{\ddagger\circ}(CH_3 \cdots H \cdots CH_2\text{—}CH_3) \geq S^\circ(C_3H_8) + R \ln 4;$$

$$\Delta S_c^{\ddagger} \geq S^\circ(C_3H_8) + 11.2 - S^\circ(C_2H_6) - S^\circ(CH_3)$$

$$\geq -25.4 \text{ gibbs/mole}$$

At 400°K,

$$A_c \geq 10^{8.0} \text{ liter/mole-s},$$

which is reasonably close to the reported value of $10^{8.5}$, indicating that the transition state is fairly stiff. In fact, the reasonable corrections for one free CH_3 internal rotation and increased moment of inertia give $A_c = 10^{8.5}$ in excellent agreement.

4.6 METATHESIS REACTIONS—ATOMS

With the exception of the reactions involving I atoms for which the A factors tend to be higher than the others by a factor of about 10, the method of minimum A factors tends to give good agreement with the observed values, depending on whether the transition state is assumed to have making and breaking bonds colinear or bent. Thus, in general, a tight transition state seems applicable to these atom-metathesis reactions.

---

[5] Recent reports by Y. T. Lee in fact show that HI_2 is a stable compound and by analogy CH_3I_2, as well.

To make the analyses quantitative, let us choose specific rules for constructing the *TS*. In the reaction:

$$A + B—C \rightarrow [A \cdot B \cdot \dot{C}]^{\ddagger} \rightarrow A—B + C$$

we shall assume that the transition state has a tight structure in which we have the equivalent of one electron bonds between A and B and between B and C, as shown. In that case, these bonds will be about 0.4 Å larger than their normal single bond length.[6] This will permit us to calculate moments of inertia. We will further assume that bending and stretching frequencies associated with these bonds are about 0.7 of their normal values.

Let us apply these rules to the $H + D_2 \rightarrow HD + D$ reaction. If we assume a linear complex ($r_{H \cdot D} = r_{D \cdot D} = 1.11$ Å), we can estimate two stretching modes at 2100 cm^{-1} (H—D) and 1700 cm^{-1} (D—D) (making allowance for reduced masses). For the two degenerate bending modes we can estimate 1000 cm^{-1}. One of the stretches, the asymmetric stretch, will be the reaction coordinate. If now we reference this transition state to its nearest structural analog D_2, for which $S_{300}^{\circ} = 34.6$ eu. we can estimate $S^{\ddagger}(H \cdot D \cdot D)$ as follows:

$$S_{300}^{\circ}(H \cdot D \cdot D) = S_{300}^{\circ}(D_2) + 1.8(\text{translation}) + 1.4(\text{spin})$$

$$+ 1.4(\text{symmetry}) + 3.9(\text{rotational moment})$$
$$+ 0.2(\text{vibration})$$
$$= 43.3 \text{ eu}.$$

Hence $\Delta S_{300}^{\ddagger} = -18.7$ eu and $A_{300} = 10^{11.0}$ l./mole-s. This, while slightly in excess of the observed value (Table 4.1) of $10^{10.7 \pm 0.5}$ l./mole-s, can be considered in good agreement. Note that A is not very sensitive to the distances or frequencies chosen. If we had chosen a nonlinear molecule with an H—D—D angle of about 140°, we would have obtained a $\Delta S^{\ddagger}$ value more positive by about 3.0 eu, because of the extra rotational degree of freedom. If we accept the experimental A factor as correct, then our calculation rules out a nonlinear *TS*, in agreement with theoretical calculations based on molecular orbitals.

If we apply the same method to the reaction:

$$H + H_2O \rightleftarrows [H \cdot H \cdot \dot{O}—H]^{\ddagger} \rightarrow H_2 + OH$$

we find:

$$S_{300}^{\ddagger}(H \cdot H \cdot OH) = S_{300}^{\circ}(H_2O) + 0.2(\text{translation}) + 1.4(\text{spin})$$
$$+ 1.4(\text{symmetry}) + 3.3(\text{rotational moments}) + 0.2(\text{vibration})$$

---

[6] Based on H_2 and H_2^+ bond lengths we have chosen $\Delta r = 0.32$ Å for a one-electron bond. For larger bonds we should probably scale this to the bond length, so that for $I \cdot I$ we might expect Δr to be about 1.0 Å (see pg. 147).

or

$$\Delta S^{\ddagger}_{300} = -S^{\circ}(H) + 6.5 = -20.9 \text{ eu.}$$

This yields $A_{300} = 10^{10.5}$ l/mole-s, in reasonable agreement with the reported value of $10^{10.9}$ (Table 4.1). A TS nonlinear in the $H \cdot H \cdot O$ bonds would yield for the extra internal rotation an A factor of about $10^{11.2}$, in slightly better agreement with the reported value. The data are not sufficiently accurate to permit discrimination between the two choices of TS.

The reaction $N + O_2 \rightarrow NO + O$ yields for a linear complex based on N_2O as a model:

$$S^{\ddagger}_{300}(N \cdot O \cdot O) = S^{\circ}_{300}(N_2O) + 0.3(\text{translation}) + 1.4(\text{rotation})$$
$$+ 1.4(\text{spin—assuming tight } TS) + 1.0(\text{vibration})^7$$
$$= 56.6 \text{ eu.}$$

Hence $\Delta S^{\ddagger}_{300} = -29.0$ eu and $A_{300} = 10^{8.8}$ l./mole-s, in reasonable agreement with the observed value of $10^{9.3}$ l./mole-s. A bent transition state using NO_2 as a model would have $S^{\ddagger}_{300} = 60.0$ eu and $A_{300} = 10^{9.4}$ l./mole-s, in much better agreement. In this case the data favor the nonlinear TS, but is really not accurate enough to be decisive. Valency rules would favor the bent TS.

An interesting contrast to this reaction is provided by the structurally very similar one:

$$N + NO \rightleftarrows (N \cdot N \cdot O)^{\ddagger} \rightarrow N_2 + O$$

which has an A factor 10-fold higher than that for $N + O_2$ and has zero activation energy (Table 4.1). The only significant difference in $\Delta S^{\ddagger}$ for the two reactions lies in the entropy differences of NO (50.3 eu) and O_2 (49.0 eu). The TS models will be very much the same and we find:

$$A_{300}(\text{linear } TS) = 10^{8.5} \text{ l./mole-s}$$
$$A_{300}(\text{nonlinear } TS) = 10^{9.1} \text{ l./mole-s}$$

Both of these are so much lower than the observed value, $A = 10^{10.2}$ l./mole-s, that we are forced to conclude that we must in this case be dealing with a loose TS. As we shall see later, this is, in fact, the case and the $N + NO$ reaction is a radical recombination, rather than a metathesis. The same is also true of such reactions as metal atoms plus halogens of which $K + Br_2 \rightarrow KBr + Br$ is a typical example. With an A factor of $10^{11.0}$ l./mole-s and a slightly negative activation energy, we can estimate that every collision at 300°K with an impact parameter of 6 Å leads to reaction. This is, of course, characteristic of a recombination

---

[7] We assume two bends at 400 cm^{-1} each.

Table 4.2.   Estimation of the A Factor for the Reaction $H + C_2H_6 \rightarrow H_2 + \dot{C}_2H_5$

Degree of Freedom	$\Delta S^{\ddagger}_{300}$	Contributions to $\Delta C^{\ddagger}_v$	
		300°K	800°K
Reference reaction			
($TS = C_2H_6$)	−27.4	−3.0	−3.0
Symmetry	3.6	—	—
Spin	1.4	—	—
Translation	0.2	—	—
External rotation[a]	1.7		
Reaction coordinate			
(3000 cm^{-1} C—H stretch $\rightarrow \nu^{\ddagger}$)	0	2.0	1.7
Internal rotation			
(I_r changes by factor of 1.5)	0.3	—	—
New $H \cdot H$ stretch (2800 cm^{-1})	0	0	0.3
Two degenerate $H \cdot H \cdot C$ bends			
at 1000 cm^{-1}	0.2(3.4)[b]	0.8(1.4)[b]	3.1(2.5)[b]
Two H—C—H bends (1400 cm^{-1})			
$\rightarrow H \cdot C$—H (1000 cm^{-1})	0.2	0.6	0.6
Totals	−19.8(−16.6)	0.4(1.0)	2.7(2.1)

$$\langle \Delta C^{\ddagger}_v \rangle_{800} = 1.6(1.6)$$

$A_{300} = 10^{10.7}$ l./mole-s (linear TS)

$\quad\quad = 10^{11.4}$ l./mole-s (nonlinear TS)

$A_{800} = 10^{11.1}$ l./mole-s (linear TS); $10^{11.8}$ (nonlinear TS)

[a] Two major moments each change by factor of 2; third moment changes by factor of 1.5.
[b] Values in parentheses correspond to nonlinear complex with $H \cdot H \cdot C$ angle of about 150° and assumed free rotation about the axis. This free rotation replaces one of the degenerate bending modes in the linear complex.

reaction, the species ($K^+Br_2^-$) being a stable, ionic intermediate. The interaction has been described by a "harpooning" mechanism in which at distances closer than 6 Å the electron finds it energetically more stable to "jump" from K to Br_2 producing a coulombically bound ion pair, $K^+Br_2^-$.[8]

As final examples of our methods of estimation of $\Delta S^{\ddagger}$, let us consider the case of H atom and O atom metathesis with C_2H_6 at 800°K. The reactions are:

$$H + C_2H_6 \rightleftarrows [H \cdot H \cdot CH_2\text{—}CH_3]^{\ddagger} \rightarrow H_2 + C_2H_5$$

$$O + C_2H_6 \rightleftarrows [O \cdot H \cdot CH_2\text{—}CH_3]^{\ddagger} \rightarrow OH + C_2H_5.$$

[8] This model was first introduced by J. Magee, *J. Chem. Phys.*, **8**, 687 (1940). For a recent discussion, see J. L. Kinsey, *Chemical Kinetics*, Ch. 6, J. C. Polanyi (ed.), Butterworth, London, 1972.

When we add an atom to a polyatomic molecule, we add three new internal degrees of freedom. In addition, in a metathesis reaction, we perturb the three degrees of freedom of the atom being transferred. Thus, in general, we must consider additions and/or changes in at least six internal degrees of freedom.

The data and calculations are shown in Tables 4.2 and 4.3. For the first reaction we reference our transition state to C_2H_6 and for the second, to C_2H_5F. In calculating the temperature coefficient of the Arrhenius parameters, we use $\Delta C_v^{\ddagger}$ and not $\Delta C_p^{\ddagger}$, since we will want to express our rates in units of moles/liter rather than atmospheres. In the case of unimolecular reactions since $\Delta n^{\ddagger} = 0$, $\Delta C_p^{\ddagger}$, and $\Delta C_v^{\ddagger}$ are the same. For bimolecular reactions, $\Delta n^{\ddagger} = -1$ and $\Delta C_p^{\ddagger} = \Delta C_v^{\ddagger} - R$.

Table 4.3.   Estimation of A Factors for the Reaction $O + C_2H_6 \rightarrow OH + C_2H_5$

Degree of Freedom	$\Delta S_{300}^{\ddagger}$	Contributions to $\Delta C_v^{\ddagger}$ 300°K	800°K
Reference reaction			
($TS = C_2H_5F$)	−30.1	−1.8	−2.3
Spin (triplet TS)	2.2	—	—
Symmetry	0	—	—
Translation	−0.3	—	—
External rotation[a]	1.7	—	—
Reaction coordinate			
($1100\ cm^{-1}$ C—F stretch $\rightarrow \nu^{\ddagger}$)	−0.1	1.7	0.6
Internal rotation			
(no significant change)	0	—	—
New O·H stretch ($2100\ cm^{-1}$)	0	0	0.7
Two degenerate O·H·C bends			
at $600\ cm^{-1}$	$1.0(3.7)^{b}$	$2.0(2.0)^{b}$	$3.6(2.8)^{b}$
F—C—C bend ($420\ cm^{-1}$)			
$\rightarrow$ (O·H)·C bend ($280\ cm^{-1}$)	0.6	0.2	0.0
F—C—(H)$_2$ bend ($1000\ cm^{-1}$)			
$\rightarrow$ (O·H)·C(H)$_2$ bend ($700\ cm^{-1}$)	0.2	0.4	0.2
Totals	−24.8(−22.1)	2.5(2.5)	2.8(2.0)
		$\langle \Delta C_v^{\ddagger} \rangle_{800} = 2.7(2.3)$	

$A_{300} = 10^{9.7}$ l./mole-s (linear TS); $10^{10.3}$ (nonlinear TS)
$A_{800} = 10^{10.2}$ (linear TS); $10^{11.0}$ (nonlinear TS)

[a] Two major moments increased by about a factor of 2; third by a factor of 1.5.
[b] Values in parentheses correspond to nonlinear O·H·C complex with O·H·C angle about 150° and assumed free rotation about the C·H axis.

We note that there is a small variation in Arrhenius parameters with temperature. The experimental A factor for $H + C_2H_6$ is $10^{11.1}$ (Table 4.1), and since it is actually weighted for higher temperatures, would favor the linear TS.

For the reaction of $O + C_2H_6$ we note a more pronounced variation of the A factors with temperature, and the experimental value of $10^{10.4}$ favors the nonlinear TS.

A number of metathesis reactions between atoms and molecules are shown in Table 4.1, from which we see that the bulk of these reactions have A factors that may be represented as $10^{10.5 \pm 0.5}$ liter/mole-s. This type of representation is probably as accurate as any given by the more complex treatments that have been done. The uncertainty in A is in most cases also of this order.

4.7 METATHESIS REACTIONS OF RADICALS

Compared to atom–molecule metathesis, the metathesis of diatomic and polyatomic radicals with molecules involves the conversion of more rotational degrees of freedom into internal coordinates with lower entropies, and we expect lower A factors for these reactions. This turns out generally to be the case.

Table 4.4 shows the Arrhenius parameters for some atom-transfer reactions between two molecules or radicals and molecules. Again, about the best generalization that we can make for these types of reactions is that, for the radical–molecule reactions, most A factors are $10^{8.5 \pm 0.5}$ liter/mole-s. Two exceptions are noted to this rule, and both are probably in error. In the case of the $CH_3 +$ benzene abstraction, the data are not compatible with the reverse reaction $C_6H_5 + CH_4$. The entropy change in the reaction

$$\cdot CH_3 + C_6H_6 \underset{-b}{\overset{b}{\rightleftarrows}} \cdot C_6H_5 + CH_4$$

is 3.2 gibbs/mole, so that if the back reaction has an A factor of $10^{8.6}$ l./mole-s, the forward must be $10^{9.3}$, not the $10^{7.5}$ indicated. It is more likely that both are in error, more reasonable values for both being $A_b = 10^{8.9}$ and $A_{-b} = 10^{8.2}$. Also $\Delta H_b = 7$ kcal so that $E_b = 18.1$ kcal if $E_{-b} = 11$.

The disproportionation of $2C_2H_5$ radicals to $C_2H_4 + C_2H_6$ has what appears to be a normal A factor. However it has zero activation energy and if we compute an A factor by correcting $S^{\ddagger}$ relative to n-butane as a reference substance, we estimate $A_{300} \gtrsim 10^{7.4}$ l./mole-s, appreciably lower than the observed value of about $10^{8.3 \pm 0.5}$. However if we now use the

Table 4.4. Arrhenius Parameters for Some Metathesis Reactions not Involving Atoms

Reference	Reaction	T_m (°K)	log A (l./mole-s)	E (kcal/mole)
a	$CH_3 + \overset{*}{C}H_4 \rightarrow CH_4 + \overset{*}{C}H_3$	500	8.8	14.6
b	$CH_3 + C_2H_6 \rightarrow CH_4 + \cdot C_2H_5$	420	8.5	10.8
	$CH_3 + C(CH_3)_4 \rightarrow CH_4 + \cdot CH_2C(CH_3)_3$	420	8.5	10.4
	$CH_3 + benzene \rightarrow CH_4 + phenyl$	450	7.5[c]	9.6[c]
d	$C_2H_5 + C_2H_5COEt \rightarrow C_2H_6 + \dot{C}_2H_4COEt$	450	8.0	7.8
e	$2C_2H_5 \rightarrow C_2H_4 + C_2H_6$	450	9.6(8.3)[f]	0
g	$C_2H_6 + C_2H_4 \rightarrow 2C_2H_5$	450	11.3(10.4)[n]	60.0
i	$2C_2H_4 \rightarrow C_2H_5 + C_2H_3$	1300	11.1(10.4)[j]	62
k	$2NO_2 \rightarrow (sym)\ NO_3 + NO$	800	9.7	23.6
l	$CH_3 + CCl_4 \rightarrow CH_3Cl + CCl_3$	400	8.6	9.1
m	$CF_3 + CCl_4 \rightarrow CF_3Cl + CCl_3$	450	8.5	10.4
n	$CF_3 + CHD_3 \rightarrow CF_3H + CD_3$	420	8.1	10.5
	$\rightarrow CF_3D + CHD_2$		8.5	12.7
o	$C_6H_5 + CH_4 \rightarrow C_6H_6 + CH_3$	450	8.6	11.1
p	$CF_3 + CH_3Br \rightarrow CF_3Br + CH_3$	420	7.5	8.1
	$+ CH_3Cl \rightarrow CF_3Cl + CH_3$	420	—	>17.0
	$+ CH_3I \rightarrow CF_3I + CH_3$	420	6.8	3.3

a. F. S. Dainton, K. J. Irvin, and F. Wilkinson, *Trans. Faraday Soc.*, **55**, 929 (1959); F. S. Dainton and D. E. McElcheran, *Trans. Faraday Soc.*, **51**, 657 (1955).

b. A. F. Trotman-Dickinson and E. W. R. Steacie, *J. Chem. Phys.*, **19**, 329 (1951).

c. It is doubtful if these are the correct parameters of this reaction.

d. M. H. J. Wijnen and E. W. R. Steacie, *J. Chem. Phys.*, **20**, 205 (1952).

e. A. Shepp and K. O. Kutschke, *J. Chem. Phys.*, **26**, 1020 (1957).

f. R. Hiatt and S. W. Benson, *J. Amer. Chem. Soc.*, **94**, 25 (1972); **94**, 6886 (1972). These values in parentheses are calculated from a combination of rate data on recombination together with thermochemical estimates of the differences in $S°$ and $\Delta H_f°$ of the alkyl radicals.

g Calculated from reference e and thermochemical data.

h Revised estimates, see reference i.

i. Estimates by S. W. Benson and G. R. Haugen, *J. Phys. Chem.*, **71**, 1735, (1935) (1967), from data on hydrogenation of C_2H_4.

j. Revised estimates, see reference f.

k. P. G. Ashmore and M. G. Burnett, *Trans. Faraday Soc.*, **58**, 253 (1962).

l. J. Currie, H. Sidebottom, and J. Tedder, *Int. J. Chem. Kinet.*, **6**, 481 (1974).

m. W. G. Alcock and E. Whittle, *Trans. Faraday Soc.*, **62**, 139, 664 (1966).

n. T. E. Sharp and H. S. Johnston, *J. Chem. Phys.*, **37**, 1541 (1962).

o. F. J. Duncan and A. F. Trotman-Dickson, *J. Chem. Soc.*, 4672 (1962).

p. W. G. Alcock and E. Whittle, *Trans. Faraday Soc.*, **61**, 244 (1965).

above rules for the tight TS associated with metathesis and assume an extra internal rotation around the nonlinear, C·H·C central bonds, we then may estimate $10^{8.6}$ l./mole-s in reasonable agreement with the observed data.

As an example of the estimation methods for the tight transition state, let us examine the A factor for the attack of CF_3 on CH_3Br to abstract Br.

$$CF_3 + CH_3Br \rightleftarrows [F_3C·Br·\dot{C}H_3]^{\ddagger} \rightarrow CF_3Br + CH_3$$

Table 4.5 summarizes the calculation using CF_3SCH_3 as a model for the transition state. We find a value of $A_{300} = 10^{7.7}$ l./mole-s with a large temperature coefficient rendering it $10^{7.8}$ at 400°K, in good agreement with the experimental value of $10^{7.5 \pm 0.5}$ (Table 4.4). We would anticipate a similar value for $CF_3 + CH_3I$, and the reported value is so much lower than this that we would suspect it to be in error. Even a linear TS would

Table 4.5. Estimation of the A-Factor for the Reaction $CF_3 + BrCH_3 \rightarrow$ $CF_3Br + CH_3$

Degree of Freedom	$\Delta S^{\ddagger}_{300}$	Contributions to $\Delta C^{\ddagger}_v$	
		300°K	600°K
Reference reaction ($TS = CF_3SCH_3$)	−41.2	0.3	5.0
Spin	1.4	—	—
Symmetry (no change)	0	—	—
Translation	1.0	—	—
External rotation (increase in moments)	1.5	—	—
Two reduced barriers for internal rotation ($V_0 = 2$ kcal $\rightarrow 0$)	1.8	−2.1	−2.1
Reaction Coordinate C—S stretch $(60\ cm^{-1}) \rightarrow \nu^{\ddagger}$	−0.5	1.0	0.3
Four rocking modes (CF_3 and CH_3) decrease to 70% $(400\ cm^{-1} \rightarrow 280)$ $(900\ cm^{-1} \rightarrow 630)$	1.8	1.3	1.0
C—S—C bend $(400\ cm^{-1} \rightarrow 280)$	0.6	0.2	0.1
Totals	−33.6	0.7	4.3

$$\langle \Delta C^{\ddagger}_v \rangle_{600} = 2.5$$

$$A_{300} = 10^{7.7}\ \text{l./mole-a (note bent } TS)$$
$$A_{600} = 10^{8.1}\ \text{l./mole-s}$$

not yield a value as low as $10^{6.8}$ l./mole-s. In the metathesis of an atom between two polyatomic fragments we add three new frequencies to the molecule being attacked, one stretching mode and two rocking modes. We also change the three frequencies associated with the atom being transferred. Thus, three old and six new frequencies must be considered, one of them being the reaction coordinate.

The large temperature coefficient observed (Table 4.5) for the metathesis reaction, $CF_3 + BrCH_3$, is not unusual. In forming the transition state in such a metathesis we transform a net of three translations and three rotations of the reactants into the reaction coordinate (kT/h), plus five internal coordinates, usually low frequency. These six latter modes have higher heat capacities than the former, and so it will usually be the case that for radical–molecule bimolecular reactions with tight transition states, $\Delta C_v^{\ddagger}$ may be as large as 6 cal/mole-°K. This is of particular importance in comparing reactions at low and at high temperatures where the usual Arrhenius behavior will be badly in error. Table 4.6, for the metathesis $CH_3 + C_2H_6 \rightarrow CH_4 + C_2H_5$, illustrates this.

A few generalizations are worth noting. The activation energies for H-atom abstractions by organic radicals from organic molecules in exothermic reactions are all about 8 ± 2 kcal and are not too sensitive to the overall exothermicity of the reaction. The same appears to be true for Cl

Table 4.6. $\Delta C_v^{\ddagger}$ for the Reaction $CH_3 + C_2H_6 \rightarrow CH_4 + C_2H_5{}^a$

Degree of Freedom	$\Delta C_v^{\ddagger}(T)$ (cal/mole-°K)			
	300	600	900	1200
Model reaction[a]	−1.1	−0.1	−0.2	−0.4
2C·H·C bends (700 cm^{-1})[b]	1.7	3.2	3.6	3.7
1C·H·C (symmetrical stretch at 700 cm^{-1} replaces C—C stretch at 1000 cm^{-1})	0.5	0.4	0.2	0.2
1C·H·C (asymmetrical stretch at 2000 cm^{-1} becomes reaction coordinate $\nu^{\ddagger}$)[b,c]	2.0	2.0	2.0	2.0
1CH$_3$ torsion, $V_0 = 3.0$ kcal $\rightarrow 0$	−1.3	−0.9	−0.6	−0.3
1C—C—C bend (400 cm^{-1}) $\rightarrow$ (C·H)·C—C bend (280 cm^{-1})	0.2	0.1	0	0
2CH$_3$ rocks (900 cm^{-1}) $\rightarrow$ 600 cm^{-1}	1.0	0.6	0.4	0.2
1C$_2$H$_5$ rock (700 cm^{-1}) $\rightarrow$ 500 cm^{-1}	0.4	0.2	0.1	0
Totals	3.4	5.5	5.5	5.4

[a] For transition state $S°(CH_3 \cdots H \cdots C_2H_5)^{\ddagger}$ use $S°(C_3H_8) +$ corrections.
[b] These are the three frequencies assigned to the "extra" H atom.
[c] This is contribution of the (kT/h) term.

atoms. F atoms appear never to be abstracted or to shift in cyclopropane pyrolyses, and we can estimate a lower limit for such abstractions of about 16 kcal/mole. Activation energies for Br atom abstraction appear to be about 6 kcal, although very little quantitative data are available. From a few unpublished experiments in the author's laboratories, I-atom abstractions may have activation energies anywhere from zero to about 6 kcal mole (see Table 4.4).

The activation energies presented above follow the order of the H—X bond strength in that they are lowest for the weakest bonds (for example, C—I) and highest for the strongest C—F bonds. HF has the highest bond strength (136 kcal); H—Cl and H—H are about equal (103, 104), and $DH°$ (H—Br) = 87, whereas $DH°$ (H—I) = 71. These numbers also follow the ionization potentials of the atoms involved. This latter is understandable if the atom abstraction goes through a polar state involving electron donation by the atom

$$
\begin{matrix}
R \\ R-\overset{|}{\underset{R}{C}}\cdot + X : \overset{S}{\underset{S}{C}} -S
\end{matrix}
\rightleftharpoons
\left[\overset{(-1/2)(+1/2)}{-\overset{|}{C}\cdot X\cdot \overset{|}{C}-} \right]^{\ddagger}
\rightleftharpoons
\begin{matrix}
R \\ R-\overset{|}{\underset{R}{C}} : X + \overset{S}{\underset{S}{C}} -S
\end{matrix}
$$

$$
\left(\cdot X \cdot \overset{S}{\underset{S}{C}} -S \right)
\qquad
\left[-\overset{|}{C}\cdot X \cdot \overset{|}{\underset{}{C}}- \right]^{\ddagger} {}^{(+1/2)(-1/2)}
\qquad
\left(-\overset{|}{\underset{}{C}}\cdot X \right)
$$

The fact that there is a measured activation energy for halogen-atom abstractions can be considered strong evidence against the existence of weakly stable complexes of the type $R—X—R$. Such complexes have been postulated to exist as stable bridged species in α-halogen radicals:[9]

4.8 ASSOCIATION REACTIONS—RECOMBINATION

Association reactions that involve molecules represent the inverse processes to the complex dissociation reactions we have discussed above. Because we have indicated some very simple rules for estimating the dissociation reaction parameters, the association reaction parameters can

[9] For a more complete discussion and some contrary evidence, see P. S. Skell and K. F. Shea, *Free Radicals*, Vol. II, J. K. Kochi, (ed.), Wiley, New York, 1973, p. 809.

be obtained via the equilibrium constant

$$A + B \underset{-a}{\overset{a}{\rightleftarrows}} AB;$$

$$K_a = \frac{k_a}{k_{-a}};$$ (4.14)

$$\frac{A_a}{A_{-a}} = e^{\Delta S_a/R};$$

$$\Delta H_a = E_a - E_{-a}.$$

Note that when k_a is measured in concentration units, we must correct ΔH_a from pressure to concentration units by adding RT (see Section 1.5):

$$\Delta H_a(p) = \Delta H_2(c) + RT(\Delta n) = \Delta H_a(c) - RT.$$

We shall not discuss these further, as far as the molecule–molecule reactions are concerned.

Two other categories involving radicals are worth discussing. The first of these is the association of radicals. If we used the parent hydrocarbon for estimating the A factors, we would adopt values of about $10^{8.5}$ liter/mole-s for the recombination reactions. Thus for the reaction

$$2CH_3 \rightarrow C_2H_6$$

$\Delta S° = -37.9$ gibbs/mole and $\Delta S°(c)$ at $300°K \sim -31.5$ gibbs/mole. This would yield a lower limit for the A factor of 10^7 liter/mole-s, compared to the observed value of $10^{10.5}$ liter/mole-s. The difference corresponds to about 16 gibbs/mole and at $300°K$ as we have already noted (Table 3.5), would require a TS in which four very loose rocking modes would be assigned frequencies of about 140 cm^{-1} each. However this is a completely artificial number since it assigns to the methyl rocking modes almost as much entropy as they would have if they were rotating freely at $300°K$. At $900°K$, since the value of C_p for two weak vibrations is $2R$ (4 eu) and that for a two-dimensional rotor is only R, the 140 cm^{-1} would give to each CH_3 in the TS more entropy than they would have if they were rotating freely! The free rotor with restricted amplitude gives a much better representation of the experimental data both for recombination and dissociation (p. 91) and is, in addition, a physically more acceptable model.

Although earlier photochemical, sector methods gave very high values to rate constants for radical recombination, very much like the value for CH_3, more recent measurements, together with thermochemical analysis, have tended to support lower values although the experimental uncertainty is still relatively high.

For rapid association reactions, which proceed with little or no activation energy, an upper limit to the A factor is provided by collision theory. The collision rate constant is given by

$$k_z = \frac{\pi d_{AB}^2}{\sigma_{AB}} \left(\frac{8RT}{\pi \mu_{AB}} \right)^{1/2} \text{(cc/molecule-s)} \quad (4.15)$$

$$= \frac{\pi d_{AB}^2}{\sigma_{AB}} \left(\frac{8RT}{\pi \mu_{AB}} \right)^{1/2} \frac{N_0}{1000} \text{(l./mole-s)}$$

where d_{AB} is the "collision" diameter of the colliding pair. $\mu_{AB} = M_A M_B / (M_A + M_B)$, and σ_{AB} is the symmetry number for the pair (1 for $A \neq B$ and 2 for $A = B$). Using $d_{AB} = 4.0 \times 10^{-8}$ cm, we find $k_z(CH_3)_{400} = 10^{11.1}$ liter/mole-s. If we further assume that only that fraction of all collisions ($\frac{1}{4}$) leading to singlet C_2H_6 are to be counted (i.e., triplet C_2H_6 is repulsive), we find $10^{10.5}$ liter/mole-s as the upper limit to the CH_3 recombination. The number can be raised by boosting the collision diameter to 4.0 or 4.5 Å, or else allowing some triplet recombination. The latter assumption is somewhat dubious, and the former reasonable, but the implication is, in any case, clear: that *every* gas-kinetic collision of two CH_3 radicals in the singlet state leads to C_2H_6 formation. Even if the measured recombination rate constant is too high by a factor of 2, the result is very unexpected. It implies that at distances of ~4.0 Å, the interaction of two CH_3 radicals is sufficiently large, compared to RT (0.6 kcal), that the methyl groups have time to rotate into a proper relative orientation, which will lead to C_2H_6 formation before they separate. If we assume that the CH_3—CH_3 bond energy can be represented by a Lennard–Jones potential, at large distances only the attractive term $-2V_0(r_0/r)^6$ is important. The repulsion is provided by the centrifugal energy, which can be written as $(p_\theta^2/2\mu r^2)$, where p_θ is the constant angular momentum of a colliding pair; $\mu = M_{CH_3}/2$; r is the internuclear distance; $r_0 = 1.54$ Å, the C—C bond distance is C_2H_6; and V_0 is the C—C bond-dissociation energy in C_2H_6 corrected for zero-point energy. This latter can be estimated at about 6 kcal, and yields a value of $V_0 = 94$ kcal.

The formalism developed earlier (3.23) can now be used to calculate a value of r_m, the maximum separation for a CH_3—CH_3 pair which can be deactivated to C_2H_6 on collision.

$$\left(\frac{r_m}{r_0} \right) = \left(\frac{6V_0}{P_\theta^2/2\mu r_m^2} \right)^{1/6} = \left(\frac{6V_0}{E_{rot}} \right)^{1/6} \quad (4.16)$$

The significance of r_m is that, for a given value of E_{rot}, all molecules with impact parameters (that is, the distance between projected trajectories) greater than r_m cannot come into range of the attractive forces, and only collisions with impact parameter $<r_m$ leads to effective contact. Thus r_m is a collision diameter. If we assume that $E_{rot} = RT = 0.8$ kcal at 400°K, $(r_m/r_0) \cong 3.0$ and is not very sensitive to temperature. Other potential curves, such as the Morse curve, lead to equivalent results with slightly smaller values. This calculation forms the basis for assigning r_m as a collision diameter to the radical collision.

If we apply the same analysis we made above to the other alkyl radicals, such as ethyl, propyl, and butyl, we see that because V_0 changes negligibly even for the isomeric radicals, (r_m/r_0) is nearly a constant, and the only difference between them arises from a decreasing k_Z due to increasing mass. Thus for Et we estimate $k_Z(Et) \sim 0.71 k_Z(Me)$ and $k(n\text{-Bu}) \sim 0.5 k_Z(Me)$. However the actual recombination rate constants for these radicals fall below these upper limit collision frequencies (Table 4.7), indicating an orientation requirement on collision. This orientation requirement is best discussed quantitatively in terms of the "loose" transition state model which we have already considered in some detail (Secs. 4.3, 4.4) and which we shall not elaborate on further here.

There is one further aspect of radical recombination which bears inspection and that has to do with very polar species which we can take as species whose dipole moments exceed 1.4 Debye. The formula for dipole–dipole interaction is given by:

$$V = \frac{a\mu_1\mu_2}{r^3} (2 \cos \theta_1 \cos \theta_2 - \sin \theta_1 \sin \theta_2 \cos \phi) \qquad (4.17)$$

where r is the distance between the centers of the two dipoles μ_1 and μ_2 considered as point dipoles making angles of θ_1 and θ_2, respectively, with the line joining their centers. The angle ϕ is the dihedral angle between the two planes, each containing one dipole and the line through their centers.

The energy V varies between a maximum repulsion of $(2a\mu_1\mu_2/r^3)$ when $\theta_1 = \theta_2 = 0°$ or 180° (head-to-head or tail-to-tail) and $(-2a\mu_1\mu_2/r^3)$ (attraction) when $\theta_1 = \theta_2 + 180° = 0°$ or 180° (head-to-tail alignment). When μ_1, μ_2 are expressed in Debye units (esu-Å) and r is in Å, then V will be in kcal/mole when the constant a is set equal to 14.4 kcal-Å/mole. Thus at 3.0 Å separation, two 1.4 Debye dipoles will have a maximum interaction of 2.1 kcal/mole.

If we consider species such as $\dot{C}F_3$ radicals or $\dot{C}Cl_3$ radicals which are quite polar, we can estimate their dipoles as at least as large and probably larger than their related saturated parents CHF_3 and $CHCl_3$ for which $\mu = 1.65$ and 1.01 Debye, respectively. Unlike their hydrocarbon analogs $\dot{C}H_3$ which are nearly flat, these polar radicals are close to tetrahedral in shape. Thus if two $\dot{C}F_3$ or two $\dot{C}Cl_3$ radicals collide along their halide

Table 4.7. Arrhenius Parameters Parameters for Some Association Reactions Involving Radicals

Reference[1]	Reaction	$T_m(°K)$	log A (l./mole-s)	E (kcal/mole)
a	$2CH_3 \rightarrow C_2H_6$	300–400	10.5 ± 0.2	0
b	$2C_2H_5 \rightarrow C_4H_{10}$	400	10.4	0
c		800	10.0 ± 0.5	0
d		350–415	9.3 ± 0.5	0
f	$CH_3 + C_2H_5 \rightarrow C_3H_8$	400	10.7	0
d		350–415	10.2 ± 0.5	0
e		1100	9.4 ± 0.5	0
g	$2i\text{-}C_3H_7 \rightarrow (i\text{-}Pr)_2$	400	8.6	0
g		750	9.5 ± 0.3	0
h	$2t\text{-}Bu \rightarrow (t\text{-}Bu)_2$	373	9.5	0
j		420	8.2 ± 1	0 (?)
k		700	8.8 ± 0.3	0 (?)
l		300	9.1	0
m	$2tBuO \rightarrow (t\text{-}BuO)_2$	420	8.8	0
n	$2CF_3 \rightarrow C_2F_6$	300	9.5 ± 0.3	0
o	$2CCl_3 \rightarrow C_2Cl_6$	385–460	9.6	0 (?)
p	2 Allyl $\rightarrow$ biallyl	300	9.9	0
q		1000	9.8	
r	$CH_3 + O_2 \rightarrow CH_3O_2\cdot$	300	9.5	0 (3rd order)
s	$C_2H_5 + O_2 \rightarrow C_2H_5O_2\cdot$	400	9.6	0
t	$CH_3 + NO \rightarrow CH_3NO$	300	8.8	0
u		300	9.4	(?)
v	$2CF_2 \rightarrow C_2F_4$	300	8.4	1.6

1. Bibliographic sources:
   a. G. B. Kistiakowsky and E. K. Roberts, *J. Chem. Phys.*, **21**, 1637 (1953); see resume of results in F. C. James and J. P. Simons, *Int. J. Chem. Kinet.*, **6**, 887 (1974).
   b. A. Shepp and K. D. Kutschke, *J. Chem. Phys.*, **26**, 1020 (1957).
   c. D. M. Golden, et al., *Int. J. Chem. Kinet.*, **8**, 381 (1976).
   d. R. Hiatt and S. W. Benson, *J. Amer. Chem. Soc.*, **94**, 25, 6886 (1972). Values listed here reflect a lower assigned $S°(C_2H_5)$.
   e. A. Lifchitz and M. Frenklach, *J. Phys. Chem.*, **79**, 686 (1975).
   f. J. C. J. Thynne, *Trans. Faraday Soc.*, **58**, 676 (1962).
   g. R. Hiatt and S. W. Benson, *J. Amer. Chem. Soc.*, **94**, 25 (1972).
   h. D. M. Golden, et al., *J. Amer. Chem. Soc.*, **96**, 7645 (1974).
   i. E. L. Metcalfe, *J. Chem. Soc.*, 3560 (1963).
   j. R. Hiatt and S. W. Benson, *Int. J. Chem. Kinet.*, **5**, 385 (1973). The value reported here is corrected for new $S°$ of tBu· and Et·.
   k. K. Y. Choo, et al., *Int. J. Chem. Kinet.*, **8**, 45 (1976).
   l. D. A. Parkes and C. P. Quinn, *Chem. Phys. Lett.*, **33**, 483 (1975).
   m. Values calculated from A factor for forward reaction [L. Batt and S. W. Benson, *J. Chem. Phys.*, **36**, 895 (1962)] and thermochemistry of the reaction.
   n. N. Basco and F. G. M. Hathorn, *Chem. Phys. Lett.*, **8**, 291 (1971).
   o. M. L. White and R. R. Kuntz, *Int. J. Chem. Kinet.*, **3**, 127 (1971).

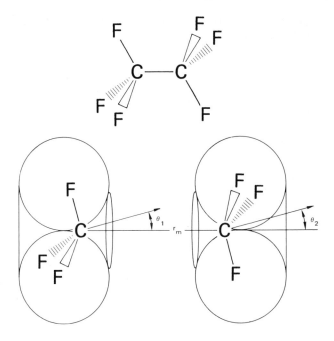

Figure 4.1   Restricted rotation model for the dissociation of C_2F_6 into $2CF_3$ radicals.

faces, they will repel each other and there is no chance of recombination. The same is true of a pyramidal approach in which their carbon atom faces approach.

This dipole interaction introduces a significant angle-dependent, repulsive force into the potential function for two dipolar radicals, which varies as r^{-3}. It favors configurations in which one dipole is at an angle to the other. However let us ignore this additional term and calculate the A factor for the decomposition of C_2F_6 by the methods we have already discussed. Then we can calculate the recombination rate from the overall equilibrium constant with the usual assumption that there is no activation energy for the back reaction. The diagram for the restricted rotation model is shown in Figure 4.1.

p. H. E. van den Bergh and A. B. Callear, *Trans. Faraday Soc.*, **66,** 2681 (1970).

q. D. M. Golden and S. W. Benson, *J. Amer. Chem. Soc.*, **91,** 2136 (1969).

r. W. C. Sleppy and J. G. Calvert, *J. Amer. Chem. Soc.*, **81,** 769 (1959).

s. D. P. Dingledy and J. G. Calvert, *J. Amer. Chem. Soc.*, **85,** 856 (1963).

t. N. Basco, D. G. L. James, and R. D. Suart, *Int. J. Chem. Kinet.*, **2,** 215 (1970).

u. H. E. Van den Bergh and A. B. Callear, *Trans. Faraday Soc.*, **67,** 2017 (1971).

v. F. W. Dalby, *J. Chem. Phys.*, **41,** 2297 (1964).

Table 4.8. ΔC_p° and Entropy of Activation for the Fission of C_2F_6 into 2CF$_3$ Groups

Degree of Freedom	$\Delta S_{300}^\ddagger$	Contributions to $\Delta C_{pT}^\ddagger$ (°K)		
		300	600	1000
Translation (no change)	0	0	0	0
Symmetry (no change)	0	0	0	0
Spin (no change)	0	0	0	0
External rotation:				
Two moments of inertia each increased by factor of 3.5	2.5	0	0	0
Barrier to torsion, 4.4 kcal $\rightarrow$ 0	1.9	−1.2	−1.2	−0.8
Four rocking modes at 250 cm^{-1} $\rightarrow$ 4 internal restricted rotations of CF$_3$	17.6	−3.0	−3.7	−4.0
One C—C stretch at 1000 cm^{-1} $\rightarrow$ reaction coordinate $\nu^\ddagger(C_p^\ddagger=2)$	−0.1	1.6	0.7	0.3
Totals	21.9	−2.6	−4.2	−4.5

$$\langle \Delta C_p^\ddagger \rangle_{1000} = -4.0$$

$$A_{300} = 10^{13.25\,+\Delta S^\ddagger/4.6} = 10^{18.1}\ \text{s}^{-1}$$
$$A_{1000} = 10^{17.0}\ \text{s}^{-1}$$

We shall make a crude estimate of the free solid angle available to each rotor as before, by considering one CF$_3$ as stationary and rotating the other about the symmetry point in the FFF plane until the F atom "hits" the C atom of the other group.

If we do this on the basis of our usual $\rho_{max} = 2.8$ (p. 89), the two CF$_3$ groups just barely interfere with each others rotation. However at this distance between C atoms, the extra dipole–dipole repulsion term makes the total interaction repulsive, not attractive, between the two groups. At $\rho_{max} = 2.6$, the CF$_3$—CF$_3$ interaction is still negative (binding) and we avoid this problem. At this distance we can make a crude estimate that each CF$_3$ considered as a sphere of 2.7 Å radius centered at the center of the FFF plane overlaps about one third the perimeter of the other, permitting a free zenith angle of about 66° and a value of $(\Delta\Omega/4\pi) \approx 0.30$. This leads to $A_{300} = 10^{18.1}$ s^{-1}. ΔS_{300}° for the dissociation is 47.4 eu, and this then leads to a value of $10^{9.5}$ l./mole-s at 300°K for the recombination of CF$_3$ radicals, in excellent agreement with the data (Table 4.7).

Especially noteworthy is the strong decrease of $\Delta S^\ddagger$ with increasing temperature arising from the negative $\Delta C_p^\ddagger$. The further contribution of decreasing ρ_{max} (p. 100) would make this decrease even larger. In the case of C_2Cl_6 the smaller dipole moment avoids the ρ_{max} problem at 2.8 and we end up calculating about the same value of $k_{recombination}$.

4.9   ASSOCIATION REACTIONS—ADDITION

The recombination of free radicals appears to have no activation energy. The process is described by a loose transition state with relatively long-range (4–6 Å) interaction between the recombining species. The same is true as we shall see later of ion–molecule reactions and, in general, of what are probably best described as acid–base interactions. The other extreme, the tight transition state is best exemplified by atom-metathesis reactions. Where might we expect to find those addition reactions which comprise the addition of an atom or radical to an unsaturated (pi) bond? Some examples are:

ΔH_r° (kcal)

$$H + C_2H_4 \rightarrow CH_3\text{—}\dot{C}H_2 \qquad -39$$
$$CH_3 + HC\equiv CH \rightarrow CH_3CH\text{=}\dot{C}H \qquad -27$$

$$HO + benzene \rightarrow \underset{}{\overset{H}{\bigcirc}}\text{—}OH \qquad -21$$

$$CH_3 + CO \rightarrow CH_3\dot{C}O \qquad -13$$

Despite the fact that these are all exothermic processes, they all seem to have some, albeit very small, activation energy. Note that not all additions to pi bonds are exothermic:

ΔH_r° (kcal)

$$I + C_2H_4 \rightarrow ICH_2\text{—}\dot{C}H_2 \qquad 6$$
$$CH_3 + CO_2 \rightarrow CH_3CO_2\cdot \qquad 13$$
$$C_2H_5 + N_2 \rightarrow CH_3CH_2N\text{=}N\cdot \qquad 23$$

For the exothermic reactions the activation energies for addition vary from about 0–8 kcal/mole. In this sense they are much like the atom-metathesis reactions which have a similarly low range of activation energies. The relation is more than superficial since both reaction types can be described as 3-electron processes from a valence-bond point of view.

$$A^{\circ} + B:C \rightleftharpoons \left[\overset{\circ}{A} \circ B \circ \overset{\circ}{C}\right]^{\ddagger} \longrightarrow A:B + \overset{\circ}{C}$$

$$A^{\circ} + {}^{\prime\prime\prime}C\text{—}C^{\prime\prime\prime\prime} \rightleftharpoons \left[{}^{\prime\prime\prime}C\text{—}C^{\prime\prime\prime\prime}\right]^{\ddagger} \longrightarrow {}^{\prime\prime\prime}C\text{—}C^{\prime\prime\prime\prime}$$

In each case an odd-electron species (A°), radical or atom, attacks an

Table 4.9. Arrhenius Parameters for Some Addition Reactions

Reference[1]	Reaction	T_m (°K)	$\log_{10} A$ (l./mole-s)	E (kcal/mole)
a	$H + C_2H_4 \rightarrow \dot{C}_2H_5$	(300–540)	10.4	2.0
b	$\dot{C}H_3 + C_2H_4 \rightarrow n\text{-}\dot{C}_3H_7$	400	8.5	7.7
a	$Cl + C_2H_4 \rightarrow CH_2Cl\dot{C}H_2$	310	10.6	0
c	$\dot{C}H_3 + \text{butadiene} \rightarrow CH_3CH_2\overline{CHCHCH_2}$	400	6.9 (8.3)	2.9 (5.4)
b	$\dot{C}_2H_5 + C_2H_4 \rightarrow n\text{-}\text{butyl}$	400	8.3	7.5
a	$O + C_2H_4 \rightarrow \text{products}$	(223–613)	9.5	1.1
d	$OH + C_2H_4 \rightarrow \text{products}$	(210–460)	9.6	0.2

1. Bibliographic sources:
   a. V. N. Kondratiev, *Rate Constants of Gas-Phase Reactions*, L. J. Holtschlag (trans.), R. M. Fristrom (ed.), NSRDS, U.S. Govt. Printing Office, Washington, D.C. (1972).
   b. J. A. Kerr and M. J. Parsonage, *Evaluated Kinetic Data on Gas Phase Addition Reactions*, Butterworth, London, 1972.
   c. L. Mandlecorn and E. W. R. Steacie, *Can. J. Chem.*, **32**, 79, 474 (1954). Values in parenthesis preferred. Calculated from minimum A factors.
   d. I. W. M. Smith and R. Zellner, *J. Chem. Soc., Faraday Trans. II*, **69**, 1617 (1973).

Table 4.10. Activation Parameters for the Addition of CH_3 to C_2H_4 using $n\text{-}CH_3CH_2\dot{C}H_2$ as a Starting Model for Transition State

Degrees of Freedom	$S^{\ddagger}_{300}$ (eu)	Contribution to $C^{\ddagger}_{vT}$ (eu)	
		300 °K	600 °K
Model reaction	−30.3	0.6	4.2
Symmetry ($\sigma = 6$; $n = 1 \rightarrow \sigma^{\ddagger} = 3$; $n^{\ddagger} = 1$)	1.4	0	0
Translation (no change)	0	0	0
Rotation (neglect)	0	0	0
CH_3 barrier (3 kcal $\rightarrow$ 0)	1.4	−1.3	−0.9
2CH_3 rocks at 1000 cm^{-1} $\rightarrow$ 700 cm^{-1}	0.4	1.0	0.6
C—$\dot{C}H_2$ rotor $V_0 = 2$ kcal $\rightarrow$ CH$_3$ $\diagdown$ C$\dot{-}$CH$_2$ torsion (400 cm^{-1})	−3.5	−0.7	0.2
C—C stretch (1000 cm^{-1}) $\rightarrow$ reaction coordinate (kT/h)	−0.1	1.6	0.7
C—C—C bend (420 cm^{-1}) $\rightarrow$ C·C—C bend (300 cm^{-1})	0.5	0.2	0.1
Totals	−30.2	1.4	4.9
		$\langle \Delta C^{\ddagger}_v \rangle = 3.2$ eu	

$A_{300} = 10^{8.6}$ l./mole-s

$A_{600} = 10^{9.1}$ l./mole-s

atom in a closed-shell system, and three electrons are severely perturbed in the overall process. In the valence-bond description shown above, we have two 1-electron bonds (the making and breaking bonds) and the third electron, which is antibonding with respect to these two bonds, is shared between the terminal atoms as shown. In principle, it is the extent of antibonding character of this third electron that determines the magnitude of the activation energy of the overall reaction. Table 4.9 shows some typical examples of Arrhenius parameters for addition reactions. We note that for the alkyl radical additions the A factors fall within the range $10^{8.5 \pm 0.5}$ l./mole-s, while for the smaller atoms and radicals $10^{9.5 \pm 0.5}$ seems more typical. The values for CH_3 addition to butadiene shown in Table 4.9 are almost certainly in error, since the method of minimum A factor gives $A_{300} \geq 10^{8.3}$ l./mole-s.

The A factor for $H + C_2H_4$ is in excellent agreement with the minimum A factor calculated as $10^{10.4}$ l./mole-s. In similar fashion, the method of minimum A factors for $CH_3 + C_2H_4$, using n-propyl radical as a model for the transition state, yields $A_{300} \geq 10^{8.2}$ l./mole-s, in good agreement with experiment. Table 4.10 shows a more precise calculation, again using the $S°(n\text{-propyl})$ as the model for the transition state. The result is not appreciably different. Note, however, the large value of $\Delta C_v^{\ddagger}$, indicating a relatively strong variation in A factor with temperature.

4.10 DISPLACEMENT REACTIONS

A related type of association reaction is one in which the intermediate state is not a stable species. These occur in the displacement reactions, of which atom transfer (metathesis) can be considered a special case. Some examples are

$$D + HOH \rightleftharpoons \left[H \overset{\overset{\textstyle D}{\textstyle |}}{\underset{\textstyle O}{\diagup}} H \right]^{\ddagger} \longrightarrow H{-}O + H;$$

$$CH_3 + H{-}O{-}O{-}H \rightleftharpoons \left[H \overset{\overset{\textstyle CH_3}{\textstyle |}}{\underset{\textstyle O{-}O}{\diagup}} H \right]^{\ddagger} \longrightarrow CH_3{-}O{-}H + \dot{O}H:$$

$$I + CH_2{\!-\!\!-\!}CH_2 \rightleftharpoons [I{-}CH_2{\!-\!\!-\!}CH_2]^{\ddagger} \longrightarrow I{-}CH_2{-}CH_2{-}\dot{C}H_2.$$

There are very few quantitative data for the reactions described above. For attachment to elements of the first row (C to F) we expect such reactions to have appreciable activation energies because they would require expansion of the valence shells (e.g., pentavalent C, trivalent O), which for these elements is energetically unfavorable. With this reasoning, we should expect boron compounds to have low activation energies for such reactions, and that is indeed the case.

For the I-atom reaction above, which is thermoneutral, E is about 18 kcal, which is·about the best guide we have to activation energies in these systems. We may expect atoms of the later rows (P, S, Si, Pb, etc.) to extend their valence shells more readily and have lower activation energies, but no good data exist for comparison. From work on translationally hot T atoms, it has been possible[10] to estimate a minimum activation energy of about 35 kcal for the displacement reaction:

$$T + CD_4 \rightleftharpoons \left[\begin{array}{c} D \quad D \\ \diagdown \quad \diagup \\ T \cdot C \cdot \dot{D} \\ | \\ D \end{array} \right]^{\ddagger} \rightleftharpoons TCD_3 + D$$

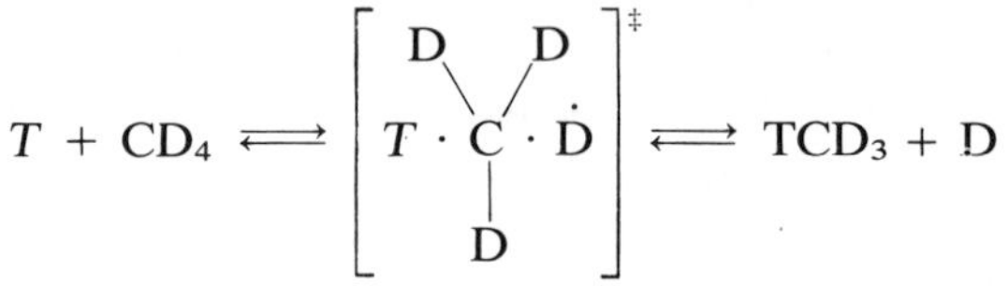

Since the overall reaction is almost thermoneutral, this suggests a very high activation energy for the formation of pentavalent carbon.

Another known example is the epoxidation, which occurs in the HCl catalyzed pyrolysis of $(t\text{-BuO})_2$:

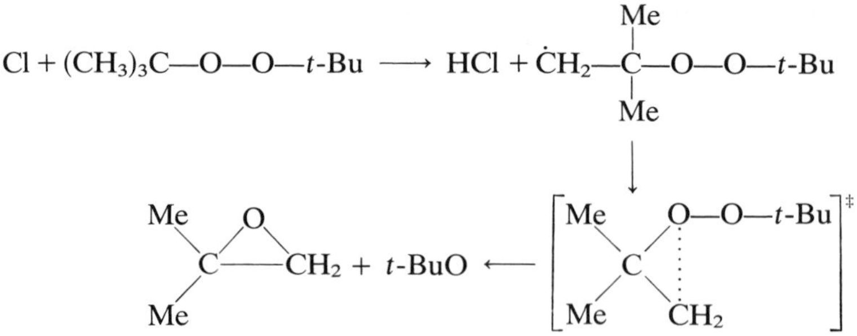

$$Cl + (CH_3)_3C\!-\!O\!-\!O\!-\!t\text{-Bu} \longrightarrow HCl + \dot{C}H_2\!-\!\overset{\displaystyle Me}{\underset{\displaystyle Me}{\overset{|}{\underset{|}{C}}}}\!-\!O\!-\!O\!-\!t\text{-Bu}$$

$$\underset{\displaystyle Me}{\overset{\displaystyle Me}{C}}\!\!\diagdown\!\!\overset{\displaystyle O}{\diagup}\!\!-\!\!CH_2 + t\text{-BuO} \longleftarrow \left[\begin{array}{cc} Me & O\!-\!O\!-\!t\text{-Bu} \\ \diagdown & \diagup \\ & C \\ \diagup & \diagdown \\ Me & CH_2 \end{array} \right]^{\ddagger}$$

It is estimated that the activation energy for this very rapid exothermic step ≤ 17 kcal/mole. This is not surprising, for in the cyclopropane system it takes 10 kcal of activation energy just to close the ring from trimethylene biradical.

Displacement reactions by radicals on saturated C, O, or N have not been observed, aside from the examples above. Such reactions have been reported for S and P and Si, but no quantitative details are available.

[10] C. C. Chou and F. S. Rowland, *J. Chem. Phys.*, **50,** 2763 (1969).

4.11 ENERGY-TRANSFER PROCESSES—DIAGNOSIS

As we have noted, association reactions produce an energized product, which, if left to itself, must ultimately redissociate. The alternative is to make a collision with some chemically inert species in which the product loses sufficient energy to become stable. When the total energy of the adduct is large and there are many internal degrees of freedom into which to share this energy, the lifetime of the adduct is long and collisions may be sufficiently frequent to deactivate the species. On the contrary, if these conditions are not satisfied, the rate of reaction can be determined by the rate of collisional deactivation, and the process is said to be controlled by the rate of energy transfer.

Schematically, we can write

$$A + B \underset{-1}{\overset{1}{\rightleftharpoons}} A - B^*;$$

$$A - B^* + M \overset{2}{\longrightarrow} A - B + M$$

where we neglect all the intermediate energy states of AB between AB^* and ground state AB. Using the steady state method for the intermediate AB^*, we find for the overall rate

$$\frac{d(AB)}{dt} = \frac{k_2 k_1 (A)(B)(M)}{k_{-1} + k_2(M)}$$

$$\overset{k_{-1} \gg k_2(M)}{\longrightarrow} \quad k_2 K_1 (A)(B)(M) \qquad (4.18)$$

$$\overset{k_{-1} \ll k_2(M)}{\longrightarrow} \quad k_1 (A)(B) \qquad (4.19)$$

Equations (4.18) and (4.19) are the reverse of the Lindemann scheme, which has been used to describe the collisional mechanism for the unimolecular decomposition of $A - B$. At high pressures, $[k_2(M) \gg k_{-1}]$, the reaction is second order, whereas at low enough pressures, $k_2(M) \ll k_{-1}$, it becomes third order. Conversely, from the point of view of the inverse reaction, the fission of $A - B$, we can see that the rate is first order at high pressures and second order at low pressures.

When the internal energy of AB^* is sufficiently large that more than one mode of chemical decomposition is available to it, we have a process that has been referred to as "chemical activation" and whose study has done much to elucidate unimolecular rate theory.[11-13] A typical example would be:

$$D + CH_3CH{=}CH_2 \underset{-a}{\overset{a}{\rightleftharpoons}} CH_3CHD{-}\dot{C}H_2^* \quad (37 \text{ kcal})$$

$$CH_3CD{=}CH_2 + H \qquad\qquad CH_3 + CHD{=}CH_2$$

The excited n-propyl radical formed by this central addition has enough internal energy to undergo any of the three reactions shown. Steps $-b$ and $-c$ have lower activation energy requirements than step $-a$ and so are preferred.

In principle, any association or inverse decomposition can go from one extreme to another with sufficiently large change in pressure. However, as of the present time, no single reaction has been followed from one extreme order to the other.

The Rice–Ramsperger–Kassel (RRK) model for these reactions gives a reasonably precise picture from which we can draw some diagnostic conclusions. Let us first see when we need to concern ourselves about pressure effects.

The mean rate constant at which an activated species $A - B^*$ will decompose is given by the RRK theory (classical form) as

$$k_{-1}(E) = A_{-1}\left(\frac{E - E^*}{E}\right)^{s-1} \tag{4.20}$$

where A_{-1} is the usual Arrhenius A factor for the unimolecular decomposition of $A - B$; E is the total energy of the complex; E^* is the activation energy for decomposition of $A - B$; and s is the number of "effective" internal degrees of freedom that can store energy in the $A - B^*$. In principle, $s \leq 3N - 6$, where $N =$ number of atoms in $A + B$. In reactions where A and B are both large groups, and the complex is loose, there is no reason that two or three rotational degrees of freedom should not participate.[14] We have already discussed examples such as $2CH_3 \rightarrow$

---

[11] B. S. Rabinovitch and M. C. Flowers, *Quart. Rev.*, **18**, 122 (1964).

[12] P. J. Robinson and K. A. Holbrook, *Unimolecular Reactions*, Wiley, New York, 1971.

[13] W. Forst, *Theory of Unimolecular Reactions*, Academic, New York, 1973.

[14] That is, rotational energy may contribute to bond breaking. However energy cannot transfer from rotational degrees of freedom to vibration because of the need to conserve angular momentum. The net effect is that the total energy required for bond breaking may be divided into two parts, $E_{rot} + E_{vib} = E$ and only E_{vib} is to be used in (4.20), whereas E^* should then be reduced by E_{rot} if most of the energy in rotation is available for bond breaking. Studies of the photolysis of NO_2 by Pitts, Sharp, and Chan have shown that about 90% of the energy in the three rotational modes can be used in bond breaking. See J. N. Pitts, Jr., J. H. Sharp, and S. I. Chan, *J. Chem. Phys.*, **40**, 3655 (1964).

C_2H_6 (p. 161) in which large changes in moments of inertia lead to large energy storage in rotation.

As an example, we can estimate the lifetime of NO_2^* formed from the attachment of O atoms to NO in the gas phase at 1000°K as follows:

EXAMPLE

We first estimate an A factor for the NO_2 "high pressure" first-order pyrolysis using our restricted rotor model and a loose transition state. From (3.23) we estimate $\rho_{max}(1000°K) = 2.47$ with $D_0(NO_2) \approx 75$ kcal (including about 2 kcal zero-point energy relative to NO). This gives $r_{\text{O} \cdots \text{N}}^\dagger = 2.20 \times 1.19 = 2.61$ Å. Using a Van der Waal radius for O of 1.6 Å, we find that the O—N—O bend has become a 1-dimensional rotation in the TS with a free cone angle of about 74° corresponding to 0.41 of a completely free rotor. The reduced moment for the rotation is about 8 amu-Å^2. Table 4.11 summarizes the contributions to $\Delta S^\dagger$ and $\Delta C_p^\dagger$ at 1000°K for NO_2 fission.

While there is no direct experimental check on this value, it does lead to a value of $10^{10.1}$ l./mole-s at 1000°K for the reverse second-order addition of $O + NO \rightarrow NO_2$, which is in the reasonable range of values for radical–atom recombination.[15]

When $O + NO \rightarrow NO_2^*$, the energy of three translations $(\frac{3}{2}RT)$ plus the two rotations of NO (RT), go into the internal modes of NO_2^*. Actually,

Table 4.11. Contributions to $\Delta S^\dagger$ for NO—O Fission at 1000°K

Degrees of Freedom	$\Delta S_{300}^\dagger$	300°K	600°K	1000°K
Symmetry $(\sigma = 2, \sigma^\dagger = 1)$	1.4	—	—	—
External rotation (two large moments change by factor of $\sim 3.3^a$	2.4			
Two stretches at 1100 and 1800 cm^{-1} → NO stretch (2200 cm^{-1}) and reaction coordinate (kT/h)	0	1.7	0.7	0.1
One bend at 600 cm^{-1} → restricted free rotation, 1-dimension	4.8	0	−0.7	−0.9
Totals	8.6	1.7	0	−0.8

$$\langle \Delta C_{p\,1000} \rangle = 0.3 \text{ (neglect)}$$
$$A_{1000} = 10^{13.25 + 8.6/4.6} \sim 10^{15.1} \text{ s}^{-1}$$

^{a} This is the result for $r^\dagger$ at 1000°K not 300°K. We use it here to avoid additional calculation.

[15] See studies at high pressures by J. Troe, *Ber. Bunsen, Geo. Phys. Chem.*, **73**, 144, 906 (1969). Troe and his colleagues have measured $NO_2 \rightleftarrows NO + O$ reactions over an extraordinary range of temperatures and pressures, and the photolysis as well. M. Quack and J. Troe, *Ber. Bunsen. Geo. Phys. Chem.*, **79**, 170 (1975), discuss this and related 3-atom systems in microscopic detail.

only a fraction $I_{NO_2}/(I_{NO}+I_{AB(NO_2)}) \sim \frac{3}{4}$ of the rotational energy of NO goes into NO_2 due to conservation of angular momentum. Here I_{NO} is the moment of inertia of NO and $I_{AB(NO_2)}$ is the product of the two large principal moments of inertia of NO_2. Furthermore, of the $\frac{3}{2}RT$ of translation, only RT goes into vibration. The other $\frac{1}{2}RT$, again because of conservation of angular momentum, will go into rotation. Thus only $\frac{3}{4}RT+RT=1.75RT$ goes into internal modes. At 1000°K this is 3.5 kcal. Moreover, not all of the 72 kcal of the newly formed bond will be found in internal modes. Because of the large change (factor of ~3.3) in two principal moments of inertia in going from transition state to ground state, the $\frac{1}{2}RT$ of rotational energy at the transition state becomes $3.5 \times \frac{1}{2}RT = 3.5$ kcal of rotational energy in ground-state geometry. Thus the difference of 3.5 kcal $-\frac{1}{2}RT = 2.5$ kcal of bond energy winds up in rotation.

This implies that the nascent molecules formed from "loose" transition complexes have large amounts of excess rotational energy, in this case, $\frac{3}{4}RT+3.5 \times \frac{1}{2}RT = 2RT = 4.0$ kcal. This also introduces a statistical weight of $(E_R^*/RT)^{1/2} \approx 1.5$, favoring these rotationally hot species.

Substituting all these numbers, we find

$$k_{1000}(NO_2) = 10^{15.1}\left(\frac{3.5}{72.5}\right)^2 = 10^{12.5} \text{ s}^{-1};$$

$$\tau_{NO_2} \sim \frac{1}{k(NO_2)} = 10^{-12.5} \text{ s}.$$

This number can be compared with the collision rate constant of about $10^{9.3} \text{ s}^{-1}$ at 1 atmosphere and 1000°K. We see that on the average NO_2^* makes only one collision per 10^3 decompositions and therefore is in the low-pressure limit for the association.

If $O + NO$ go into an upper electronic state the bond energy of which is estimated at only 40 kcal, its lifetime would have been about fourfold smaller. It is likely that collisional transfer between these states is fast.

At 300°K because $\Delta C_v^{\ddagger}$ is small we find the same result. The recombination at 1 atm is still in the third-order region because collisional deactivation is still much slower than decomposition. At low temperatures (for example, 300°K) one must use caution in estimating lifetimes by reason of the low-energy content of the associating species. The classical formula (4.20) can be replaced by one due to Marcus and Rice:

$$k(E) = \frac{W^{\ddagger}(E-E^*)}{h\rho(E)}\left(\frac{I_{ABC}^{\ddagger}}{I_{ABC}}\right)^{1/2} \tag{4.21}$$

where $W^{\ddagger}(E)$ is the number of ways of distributing energy E among the ground-state molecules, and $\rho^{\ddagger}(E-E^*)$ is the density of states for the transition state.[16] We have less energy to distribute, because E^*, the activation energy, must be localized in the potential energy of the molecule to reach the transition state. This result assumes that energy is redistributed rapidly over the internal degrees of freedom of the molecule, compared to the rate of passage to the transition state. The ratio $(I^{\ddagger}_{ABC}/I_{ABC})$ is the ratio of the principal products of inertia in transition and ground states and differs from unity significantly only for "loose" complexes, such as $NO_2^{\ddagger}$.

For "s" harmonic oscillators with a geometric mean frequency ν_m, ρ is given by $(\nu_m{}^s = \nu_1 \cdot \nu_2 \cdots \nu_s)$:

$$W(E) \approx \frac{1}{(s-1)!}\left(\frac{E}{h\nu_m}\right)^{s-1}. \tag{4.22}$$

Inserting this, (4.21) leads to the form of (4.20).

When the total excess energy in the molecule $(E-E^*)$ is of the order of that needed to excite one quantum, (4.22) can be badly in error, and it is preferable to make a discrete counting of the number of different distributions.

For $NO_2^{\ddagger}$ at 300°K there is only about $2.0RT = 1.2$ kcal of excess internal energy at the transition state, which would not be enough to excite even one quantum if the internal modes were similar to the ground state NO_2. However we have already seen that because of the loose transition state, one stretching mode (the asymmetric) has become the reaction coordinate $\nu^{\ddagger}$, while the 600-cm^{-1} bending mode has become a restricted free rotation. The quantity $W^{\ddagger}(E-E^*)$ must then reflect the ways of distributing this 1.2 kcal of excess energy into the degrees of freedom of the TS excluding the reaction coordinate. Where there has been a radical change in structure, such as a vibration becoming a virtually free rotation as for loose complexes, it leads to a multiplicative factor which is basically the partition function for the rotation:

$$W^{\ddagger}(E-E^*) \cong \frac{(E-E^*)^{s-2}}{(s-2)! (h\nu_m^{\ddagger(s-2)})}Q_r^{\ddagger} \tag{4.23}$$

with

$$Q_r^{\ddagger} = \pi^{1/2}\left(\frac{8\pi^2 I^{\ddagger}RT}{h^2}\right)^{1/2}$$

Since $s-2 = 1$ and $E-E^*$ for TS is 1.2 kcal, which is less than enough to excite the NO vibration (1900 cm$^{-1} \approx 6$ kcal), $W^{\ddagger}(E-E^*)$ becomes essentially $Q_r^{\ddagger} \approx 6$ at 300°K.

[16] In general, $\rho(E) \equiv \dfrac{dW(E)}{dE}$ is the number of energy states per unit energy range.

The frequencies of NO_2 are 1600, 1300, and 750 cm^{-1} with a geometric mean of about 1100 cm^{-1}. $(E/h\nu_m)$ at $E = 73.5$ kcal is about 24, so that $W(E) \sim 280$.

Introducing a symmetry factor of 2 for the reaction-path degeneracy, we have:

$$k_{300} = 10^{12.8} \times \tfrac{1}{280} \times 6 \times 3.5 \times 2$$

$$\approx 10^{12.0} \, s^{-1}$$

a result not too different from our earlier result at 1000°K.

EXAMPLE

Calculate the lifetime at 800°K of $C_2H_6^* \rightarrow 2CH_3$, if the $C_2H_6^*$ has been formed by $H + C_2H_5 \rightarrow C_2H_6^*$.

Using an average A factor (Table 4.6) of $10^{16.5}$ s^{-1}, we assume two thirds of the modes to be active. Thus $S = \frac{2}{3}(3N-6) = 12$. Note that if we count all the modes and exclude the internal rotation, and all C—H stretches, we find $S = 11$. The internal energy of C_2H_5 can be estimated by summing $C_V(\text{int.})$ ΔT from 0°K to 800°K. From our table of radical groups, we find $C_p^\circ(300) = 11.1$; $C_p(500) = 16.5$; and $C_p(800) = 23$. Subtracting $\frac{5}{2}R$ for translation plus external work, $\frac{1}{2}R$ for the internal rotation, and $\frac{3}{2}R$ for the external rotations, we find $C_V(300) = 2.1$; $C_V(500) = 7.5$; and $C_V(800) = 14$ gibbs/mole. This leads to $E_V(\text{int.}) = 4.6$ kcal/mole. Adding $\frac{3}{2}RT = 2.4$ kcal/mole for the translational energy of H gives $E_V(C_2H_6)^* = 7.0$ kcal/mole. The C—H bond energy is 98 kcal, whereas the C—C is 88 kcal. Hence $E - E^* = 10.0 + 7.0 = 17.0$, and $E = 98 + 7.0 = 105.0$. Thus, $\tau(C_2H_6^*) \sim 10^{-16.5}$ $(105.0/17.0)^{11} \sim 10^{7.7}$ s, which is to be compared to $10^{-9.3}$ s between collisions at 1 atm pressure. We expect less than 3% of such complexes to decompose before their next collision.

The other kind of lifetime we need to calculate is that for an ordinary thermally equilibrated molecule. In this case, we estimate the average energy content from the assumption that the transition state is at thermal equilibrium. The number of effective oscillators s can be estimated by our $\frac{2}{3}$ rule, or by estimating $C_V = C_p - 8$ at the temperature in question, and dividing by R (i.e., $s \sim C_V/R$).

EXAMPLE

Calculate the lifetime of an average energized molecule of C_2H_6 at 1000°K.

ANSWER

$$s = \frac{C_V}{R}(C_2H_6) \quad \text{at } 1000°K = \frac{29.5 - 9}{2} = 10.3;$$

$$E_V = \int_0^{1000} C_V \, dT \sim 9.4 \text{ kcal};$$

$$\tau \sim 10^{-16.5} \left(\frac{97.4}{9.4}\right)^{9.3} \sim 10^{-6.9} \text{ s}.$$

Here we have treated the internal rotations as not contributing to the energy transfer. This result is slightly high because of the low excess energy involved.

If we estimate a collisional deactivation efficiency of about $\frac{1}{3}$ per collision, then because $k_z \sim 10^{9.0}\,s^{-1}$ for C_2H_6 (1000°K, 1 atmosphere), we see that the pyrolysis of C_2H_6 will be showing very slight fall-off from unimolecular behavior at 1000°K and 1 atmosphere pressure.

If we want to find a more accurate measure of the pressure effects, the more complex method of Marcus[17] (RRKM) must be employed. This requires assignment of all frequencies in the ground state and transition state and, for complex molecules, a computer program.

For very rapid estimates, the RRK method with E_v between RT and sRT at very high temperatures is usually accurate to better than a power of 10 when E_v exceeds two quanta. When it is less, it is best to count the distributions of energy among the oscillators and use RRK for the ground state.

4.12 ATOM RECOMBINATION

The recombination of two atoms X in the presence of simple third bodies M (Ar, Ne, H_2, etc.) can be considered to occur by a sequence of bimolecular steps:

$$X + X \rightleftarrows X_2^*;$$
$$X_2^* + M \rightleftarrows X_2^v + M;$$
$$X_2^v + M \rightleftarrows X_2^{v-1} + M$$

$$\cdot$$
$$\cdot$$
$$\cdot$$

$$X_2^{1} + M \rightleftarrows X_2^{0} + M$$

(4.24)

where X_2^* is an unstable atom pair[18] and X_2^v is among the top vibrational states of X_2. It has been shown[19] that such a scheme leads to a very good agreement with experimental results over a large temperature range.

Steady state treatment of the excited vibrational states leads to an apparent third-order rate constant:

$$\frac{d(X_2)}{dt} = \text{rate} = k_r(X)^2(M);$$

$$k_r = K^* k_z^* \lambda G(T)$$

(4.25)

[17] See discussions in B. S. Rabinovitch and D. W. Setser, *Adv. Photochem.*, Interscience, New York, 1964, Vol. 3, p. 1.
[18] X_2^* is a pair of atoms whose internuclear separation $r \leq r_m$ [see (4.16)].
[19] S. W. Benson and T. Fueno, *J. Chem. Phys.*, **36**, 1597 (1962). H. O. Pritchard, *Reaction Kinetics*, I, Specialist Periodical Report, Chem. Soc., London (1975); this gives a current, detailed review.

with K^* the equilibrium constant between X and X_2^*; k_z^* the rate constant for collision of X_2^* with M; and λ the probability that such a collision leads to deactivation of X_2^* to one of the bound levels. $G(T) =$ $1 -$ (the probability that $X_2^{(v)}$ will be reactivated to X_2^*) and is approximately equal to $(N+1)^{-1}$ where N is the number of vibrational levels of S_2 within energy range RT of the dissociation threshold. It can be shown that $K^* \sim (\frac{4}{3}\pi r_m^3 Q^*/Q_x^2)$, where (Q^*/Q_x^2) is the ratio of electronic partition functions for the particular ground state chosen for X_2 and for X. Thus, for $H + H \rightleftarrows H_2^*$, $(Q^*/Q_H^2) = \frac{1}{4}$, whereas for $2I \rightleftarrows I_2^*$; $(Q^*/Q_I^2) = \frac{1}{16}$, it being assumed that only $^1\Sigma$ ground-state species are formed as X_2^*.

The efficiency of deactivation λ^* is usually close to unity but can decrease if X and M differ greatly in their masses.[20] The quantity r_m is calculated by the same methods as those employed in discussing radical recombinations [see (4.8)] and has a negative temperature coefficient if we use the Boltzmann distribution to assign the rotational energy of the colliding pair at RT.

The value of k_z^* is different for symmetrical and unsymmetrical molecules. Excited unsymmetrical molecules with atoms of appreciably different masses, such as HCl^* or SO^*, have most of the kinetic energy of the vibration present in the lighter atom. Hence collision of M with the molecule can only be effective in deactivation if the lighter atom is struck by M. In cases such as HCl or HF, it means that less than half of the collisions are effective.

Although the measurements on atom recombination are not very precise, it is well established that the rate constant has a negative temperature coefficient. This arises in the present theory from the negative temperature contribution of K^* and $G(T)$. The contribution from K^* just about cancels the $T^{1/2}$ of k_z^*, whereas $G(T) \propto 1/T$ for most molecules except H_2 and $H—X$ at low temperatures. Experimentally, T^{-1} seems to be about what is observed for the recombination rate constant.

A negative activation energy of the order of RT implies that in the reverse reaction (dissociation of X_2) the activation energy is less than the energy of dissociation at the mean reaction temperature by RT. At $3500°K$, this amounts to 7 kcal, and most workers using shock tubes to follow dissociation reactions find activation energies appreciably below the ΔE_T° for the reaction. Many workers choose to refer their observed dissociation energies to ΔE_0°, the dissociation energy at absolute zero. However $\Delta E_T^\circ = \Delta E_0^\circ + \langle \Delta C_v \rangle T$, and $\langle \Delta C_v \rangle = +1.0$ gibbs/mole at low T and changes over to -1.0 at high T. It changes sign at about $T = (h\nu/3k)$, where $\nu =$ vibration frequency. For I_2, this is $70°K$; for Cl_2, $140°K$, and

[20] S. W. Benson and G. Berend, *J. Chem. Phys.*, **40**, 1289 (1964); **44**, 470 (1966).

for H_2, 2000°K. For practically all molecules except H_2 and HX, $\langle\Delta C_v\rangle \sim$ -1 above 500°K, and $\Delta E < \Delta E_0^\circ$. For H_2 above 3000°K, this is also true, and if the recombination rate constant at high temperatures varies at T^{-1}, the activation energy for dissociation should be $\Delta E_0^\circ - \frac{3}{2}RT$.

Table 4.11 lists some typical atom recombination rate constants. Note that they all seem to lie close to $10^{9.5 \pm 0.5}$ liter^2/mole^2 sec at 300°K, and about a power of 10 smaller at 3000°K. The experimental uncertainties are usually about a factor of 2.

An alternative mechanism exists for recombination if the third-body M is a polyatomic molecule capable of forming a long-lived complex with one of the recombining atoms X. Thus, it is observed that n-pentane (Table 4.12) is ten times more effective than Ar in recombining I atoms. In these cases, we write the mechanism as:

$$M + X + M \underset{-1}{\overset{1}{\rightleftharpoons}} MX + M;$$

$$MX + X \overset{2}{\longrightarrow} X_2 + M. \tag{4.26}$$

Note that the formation of the MX complex is itself termolecular. The overall steady state rate is now ($K_1 = k_1/k_{-1}$);

$$\frac{d(X_2)}{dt} = \frac{k_1 k_2 (X)^2 (M)^2}{k_{-1}(M) + k_2(X)} \approx K_1 k_2 (X)^2 (M) \tag{4.27}$$

with an apparent termolecular rate constant k_r:

$$k_r = K_1 k_2. \tag{4.28}$$

If MX is a reasonably tight complex, we can expect $\Delta S_1^\circ \approx$ -29 gibbs/mole. If k_2 has a typical atom abstraction $A = $ factor of $10^{10.5}$,

$$\log(k_r) \sim 6 - \frac{E_2 + \Delta H_1}{\theta} (\text{l.}^2/\text{mole}^2\text{-s}) \tag{4.29}$$

with an abnormally small preexponential factor.

If E_2 is small or nearly zero, there will be a relatively large negative activation energy because ΔH_1 is negative for an exothermic reaction. Such a case occurs in the O_2 or NO catalyzed recombinations of O atoms at low pressures where the steps $O + NO_2 \rightarrow O_2 + NO$ and $O + O_3 \rightarrow \sim O_2$ are both very fast (Table 4.1). NO acts as a catalyst for I atom recombination in the same way.[21]

For the more weakly bound van der Waal or charge-transfer complexes, such as I atom forms with large molecules or unsaturates, the

[21] H. Van den Bergh and J. Troe, *Chem. Phys. Lett.*, **31**, 351 (1975).

Table 4.12. Termolecular Rate Constants for Some Atom Recombinations

Reference[1]	Reaction	Temperature (°K)	$\log k^2$ (M)
a	$H+H+M \rightarrow H_2+M$	1072	9.5 (H_2)
			9.3 (N_2, H_2O)
b		4000	9.3 (H_2), 8.3 (Ar)
c		4500	8.6 (Ar)
d		300	10.0 (H_2)
e		300	9.2 (H_2)
h	$D+D+M \rightarrow D_2+M$	3500	8.5 (D_2), 9.3 (D)
f	$O+O+M \rightarrow O_2+M$	300	8.9 (O_2)
i			9.0 (N_2), 8.5 (Ar, He),
			9.2 (N_2O)
j		2200	8.85 (O_2)
k		2000	7.4 (Ar)
l		4000	7.3 (Ar), 7.6 (Kr), 8.1 (Xe)
d	$N+N+M \rightarrow N_2+M$	300	8.9 (N_2)
g		300	9.5 (N_2)
m		300	9.3 (N_2)
n	$I+I+M \rightarrow I_2+M$	300	9.3 (Ne)
o			9.50 (He), 9.64 (Ar)
p			10.82 (n-C_5H_{12}), 9.36 (n-C_4H_{10})
c	$H+F+M \rightarrow HF+M$	4500	8.7 (Ar)
q	$H+Cl+M \rightarrow HCl+M$	3500	8.4 (Ar)
d	$N+O+M \rightarrow NO+M$	300	9.5 (N_2)
r	$F+F+Ar \rightarrow F_2+M$	295	7.5 ± 0.3

1. Bibliographic sources:
    a. G. Dixon-Lewis, M. M. Sutton, and A. Williams, *Disc. Faraday Soc.*, **33**, 205 (1962).
    b. R. W. Patch, *J. Chem. Phys.*, **36**, 1919 (1962).
    c. T. A. Jacobs, R. R. Giedt, and N. Cohen, *J. Chem. Phys.*, **43**, 3688 (1965).
    d. C. B. Kretschmer and H. L. Petersen, *J. Chem. Phys.*, **39**, 1772 (1963).
    e. I. R. Hurle, A. Jones, and J. L. Rosenfeld, *Proc. Roy. Soc.* (London), **A310**, 253 (1969).
    f. I. M. Campbell and B. A. Thrush, *Trans. Faraday Soc.*, **62**, 3366 (1966).
    g. J. A. Golden and A. L. Meyerson, *J. Chem. Phys.*, **28**, 978 (1958).
    h. J. P. Rink, *J. Chem. Phys.*, **36**, 1398 (1962).
    i. J. E. Morgan and H. I. Schiff, *J. Chem. Phys.*, **38**, 1495 (1963).
    j. J. H. Kieffer and R. W. Lutz, *J. Chem. Phys.*, **42**, 1709 (1965).
    k. K. L. Wray, *J. Chem. Phys.*, **38**, 1518 (1963).
    l. J. P. Rink, *J. Chem. Phys.*, **36**, 572 (1962).
    m. L. I. Avramenko and V. M. Krasnen'kov, *Izv. Akad. Nauk USSR, Otd. Khim. Nauk*, 1196 (1963).
    n. G. Porter, Z. G. Szabo, and M. G. Townsend, *Proc. Roy. Soc.* (London), **A270**, 493 (1962).
    o. M. I. Christie et al., *Proc. Roy. Soc.* (London), **A216**, 152 (1953).
    p. R. Marshall and N. Davidson, *J. Chem. Phys.*, **21**, 659 (1953).
    q. T. A. Jacobs, N. Cohen, and R. R. Giedt, *J. Chem. Phys.*, **46**, 1958 (1967).
    r. P. S. Ganguli and M. Kaufman, *Chem. Phys. Lett.*, **25**, 221 (1974).
2. Units are in $l.^2$ mole^2-s.

complex is extraordinarily loose; $\Delta S_1^\circ \sim -15$ gibbs/mole,[22] and the A factor appears normal, that is $\sim 10^{9.8}$ l.^{2}/mole^2-s. It is only the magnitude of the negative temperature coefficient of k_r, -1 to -3 kcal, rather than $-RT$, which is a clue to the behavior.

4.13   ENERGY-TRANSFER PROCESSES—LOW-PRESSURE LIMIT

When the process of energy transfer becomes rate limiting, decomposition reactions become second order and association processes become third order. From the analysis given in Section 4.9, we should expect this to be true (at 1 atmosphere) of most three- or four-atom molecules above 300°K and for five- to seven-atom molecules above 1000°K. The rate constants for such processes can be examined from the point of view of the association. The dissociation reaction can then be deduced by use of the dissociation equilibrium constant.

Let us consider the association of $NO + O$:

$$O + NO \underset{-1}{\overset{1}{\rightleftharpoons}} NO_2^*;$$

$$NO_2^* + M \overset{2}{\longrightarrow} NO_2 + M;$$

$$[O + NO_2 \overset{3}{\longrightarrow} O_2 + NO] \qquad \text{(fast)}.$$

The steady state rate of disappearance of O atoms at low pressure is given by

$$\frac{-d(O)}{dt} = \frac{2k_1 k_2 (M)(O)(NO)}{k_{-1} + k_2(M)} \qquad (4.30)$$

$$\xrightarrow[k_{-1} \gg k_2(M)]{\text{low pressure}} 1 K_1^* k_2(O)(NO)(M) = 2k_r(O)(NO)(M) \qquad (4.31)$$

where $K_1^* = k_1/k_{-1}$, and k_r is the termolecular recombination-rate constant.

Step 2, a deactivation step, can be readily estimated by assuming the usual van der Waal radii for NO_2^* and M, and further assuming that every collision of M with NO_2^* will be effective in deactivating NO_2^*. A typical collision rate for unlike species at 300°K is $k_2 \sim 10^{11.4}$ l./mole-s with a spread of about a factor of 2 to allow for different masses and cross

---

[22] O. K. Rice and D. Atack, *J. Phys. Chem.*, **58**, 1017 (1954).

sections. Our problem then reduces to making estimates of the thermochemistry of the nascent, energetic species NO_2^*.

Two quite different models for NO_2^* can be considered. In the first we assume that the newly acquired internal energy from the bond formation is partitioned among all the degrees of freedom of the molecule. For NO_2^* this would suggest a geometry not very different from that of the ground-state molecule with about 24 kcal in each of three internal degrees of freedom. This corresponds to a tight structure.[23] Since reaction (1) involves no energy change, being adiabatic,[24] the constant $K_1^* = \exp(\Delta S_1^*/R)$ reflects only the entropy change in the reaction. The entropy of NO_2^* is then given by the entropy of ordinary thermal NO_2 plus the additional entropy arising from the excess internal energy.

The equilibrium constant for NO_2^* is given by

$$K_1^* = \exp\left(\frac{\Delta S_1^\circ}{R}\right) \cdot W(E) \cdot \left(\frac{I_{NO_2}^\ddagger}{I_{NO_2}}\right)^{1/2} \tag{4.32}$$

where the last term in parentheses corrects for the excess rotational entropy of NO_2^* and is ~ 1.8, ΔS_1° is the normal entropy change for the association to ground state, NO_2, and $W(E)$ is the number of ways in which the internal energy of NO_2^* can be distributed among the three internal vibrational modes the mean frequency of which is taken at $1100\ cm^{-1}$. The total energy E can be taken as the bond dissociation energy E° (72 kcal) plus the thermal energy of the recombination RT. From the formula for $W(E)$ [(4.22)] $\Delta S_{300}^\circ = -31.3$ gibbs/mole, $(\Delta C_v) \cong 0$:

$$K_1^* = 10^{-5.0} \times \frac{1.8}{2}\left(\frac{72.5}{3.2}\right)^2 = 10^{-2.3}\ \text{liter/mole}.$$

Combining this with our value of $k_2 = 10^{11.4}$, we obtain $k_r = 10^{9.1}\ l.^2/mole^2$-s, which is about a factor of 20 smaller than the reported value of $10^{10.3}$ (Table 4.13). Thus the tight complex cannot represent the recombination path.

The alternative model is the loose complex, which is appropriate for bimolecular radical recombination. In applying it to the termolecular process we are assuming that the collision complex does not resemble an

---

[23] Since each degree of freedom has about 8 quanta, which is very much lower in energy than the dissociation limit, the structure is not very far from the ground state geometry.

[24] Actually, the collision process favors formation of NO_2^* with $\frac{1}{2}RT$ more energy (from translational approach) than would correspond to thermodynamic equilibrium. Thus $\Delta H_1 = -\frac{1}{2}RT$ and $\Delta E = \frac{1}{2}RT$. However these quantities are only of importance in connection with the entropy, not the pseudoequilibrium constant of the complex.

Table 4.13. Arrhenius Parameters for Some Third-Order Addition Reactions Involving Molecules

Reference	Reaction	$T_m(°K)$	$\log k_1(l.^2/mole^2\text{-s})$
a	$O+O_2+M \rightarrow O_3+M$	380	8.1 (O_2), 8.2 (N_2), 8.5 (O_3)
b		300	8.1 (O_2)
c	$O+NO+M \rightarrow NO_2+M$	300–500	$9.16+1.9/\theta$ (O_2)
d		200–300	$8.9+1.8/\theta$ (O_2)
e		300	10.46 (O_2)
f	$H+NO+M \rightarrow HNO+M$	300	10.17 (H_2)
g	$H+O_2+M \rightarrow HO_2+M$	300 and up	$8.67+1.6/\theta$ (Ar)
h	$Cl+NO+M \rightarrow NOCl+M$	245 ± 50	$9.7+1.1/\theta$ (Ar), 10.5 (He), 10.5 (N_2), 10.6 (O_2), 10.5 (Cl_2), 10.5 (SF_6)

1. Bibliographic sources:
    a. S. W. Benson and A. E. Axworthy, Jr., *J. Chem. Phys.* **42**, 2614 (1965).
    b. N. Basco, *Proc. Roy. Soc.* (London), **A283**, 302 (1965).
    c. F. S. Klein and J. T. Herron, *J. Chem. Phys.*, **41**, 1285 (1964).
    d. M. A. A. Clyne and B. A. Thrush, *Proc. Roy. Soc.* (London), **A269**, 404 (1962).
    e. A. A. Westenberg and N. DeHass, *J. Chem. Phys.*, **40**, 3087 (1964).
    f. M. A. A. Clyne and B. A. Thrush, *Disc. Faraday Soc.*, **33**, 139 (1962).
    g. M. A. A. Clyne and B. A. Thrush, *Proc. Roy. Soc.* (London), **A275**, 559 (1963).
    h. T. R. Clark, M. A. A. Clyne, and D. H. Stedman, *Trans. Faraday Soc.*, **62**, 3354 (1966).

equilibrated, energized NO_2^*, but on the contrary spends most of its time in the transition-state region looking like $NO_2^{\ddagger}$. This is the model employed to account for termolecular atom recombination. The equilibrium constant K_1^* then represents the probability of finding an O atom in the spherical shell surrounding NO of radius $r^{\ddagger}$ and thickness Δ.[25] Then:

$$K_1^* = 4\pi(r^{\ddagger})^2 \Delta \tag{4.33}$$

The value of $r^{\ddagger}$ is calculated from the potential energy curve [see (4.8)]

---

[25] See footnote 26.

[26] A simple dynamical argument yields the same value. The rate constant for formation of this loose complex, k, is given by $\pi(r^{\ddagger})^2 \bar{v}$, where $\bar{v}$ is the mean relative thermal velocity of approach of O and NO and $r^{\ddagger}$ is now the impact parameter for the collision. During such a collision $r(O \cdots NO)$ will decrease from $r^{\ddagger}$ to r_{min}, which could be $r_0(O—NO)$ and then if no other forces intervene, back out to $r^{\ddagger}$ and dissociation. The time τ_c for the collision which is the inverse of k_{-1} is then given by the total radial distance traversed, $\Delta \approx 2(r^{\ddagger} - r_0)$ divided by the mean radical velocity v_r, which is $\bar{v}$ at $r^{\ddagger}$ and zero at r_0 or $v_r \approx \frac{1}{2}\bar{v}$. Hence $\tau_c \approx 4(r^{\ddagger} - r_0)/\bar{v}$ and $K_1 = k_1 \tau_c \approx 4\pi(r^{\ddagger})^2(r^{\ddagger} - \sigma)$.

and at 300°K is given by:

$$r^{\ddagger} = r_0 \rho_{max} = 3.6 \text{ Å}$$

The shell thickness Δ can be set equal to $r^{\ddagger} - r_0 = 2.4$ Å with some steric correction for $O \cdot \cdot \cdot O$ van der Waal interference, or can be arbitrarily set equal to 1 Å with not much loss in accuracy. This former yields $K_1^* \leq 0.23$ while the latter yields $K_1^* = 0.10$ l./mole. This last value, together with the collision deactivation rate constant $k_2 = 10^{11.4}$ l./mole-s gives $k_r = 10^{10.4}$ l.^{2}/mole^2-s, in excellent agreement with the observed data. Throughout, we have assumed close to unit efficiency for collisional deactivation which is reasonable for these small systems with high energy.

If we apply these same methods to the termolecular recombination $O + O_2 \rightarrow O_3$, we find that the loose complex gives a value about 50-fold too high while the tight complex fits the data very closely. The reactions $Cl + NO + M \rightarrow ClNO + M$ and $H + NO + M$ also fit the loose model, while $H + O_2 + M$ fits the tight model if the collisional deactivation is considered inefficient for Ar by a factor of 3.

While we can only speculate on the differences in these reactions, they do suggest that the answer may lie in the potential energy surface. They suggest that our universal application of the Lennard–Jones or Morse functions does not work for reactions with O_2, or alternatively that O_2 is not really a radical (or biradical), but more like a stable, closed-shell pi-bonded system. The relatively high ionization potential of O_2 (12.0 eV) compared to most free radicals (8–10 eV) suggests a more tightly bound electron system. NO is a normal radical in this respect in having an ionization potential of 9.25 eV. We shall return to this question in our discussion of activation energies of chemical reactions.

4.14 REACTIONS OF IONS

In the past decade, interest in gas-phase ion reactions and studies of the reactions of ions with molecules have increased exponentially. At the same time information about the thermochemistry of ions, both positive and negative, has increased in both quantity and quality. It is possible to estimate the entropies and heat capacities of ions by our standard techniques and with comparable precision. In each case we need to know the electronic state and the structure. The sensitive feature in the latter has to do with overall symmetry and with internal barriers to torsion in larger molecules. The simple octet rule of valence is a valuable guide to guessing structure where direct information is not available. Thus we would guess that the very stable CH_5^+ has the structure (trigonal bipyramid):[27]

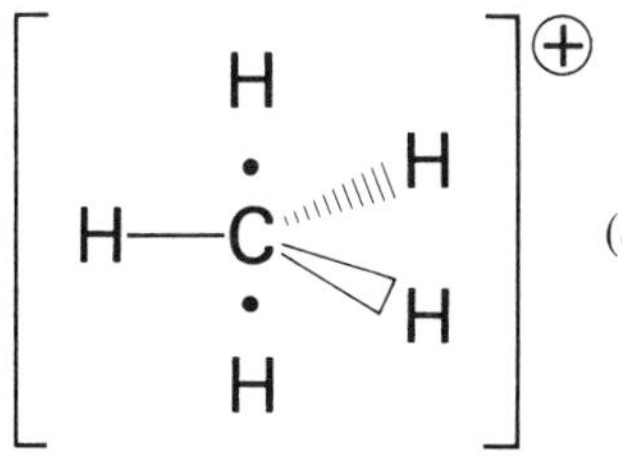

$$\left[\begin{array}{c} H \\ \vdots \\ H-C\cdots\cdots H \\ \vdots \\ H \end{array} \right]^{\oplus} \qquad (\sigma = 6)$$

while $C_2H_5^+$ is expected to be somewhat close to diborane in structure:

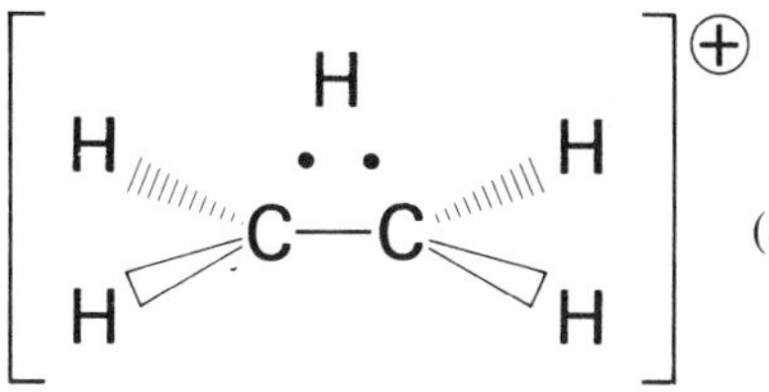

$$\left[\begin{array}{c} H \\ H_{\cdots}\quad \vdots\vdots \quad{}_{\cdots}H \\ {}^{}C-C \\ H \qquad\qquad H \end{array} \right]^{\oplus} \qquad (\sigma = 2)$$

The anion $C_2H_5^-$ by way of contrast, is expected to look like C_2H_5 with an internal torsion barrier like that in the isoelectronic methyl amine:

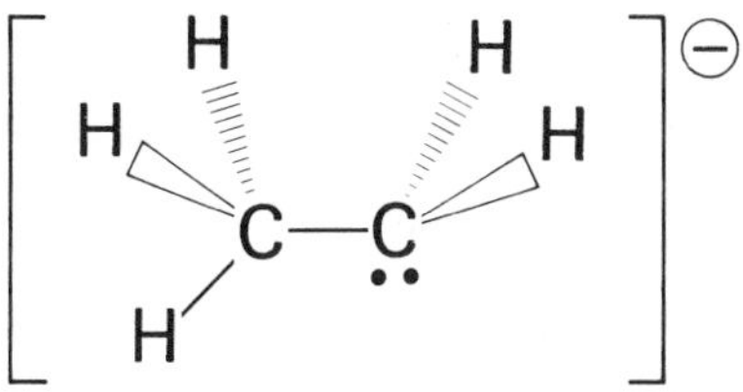

$$\left[\begin{array}{c} H \qquad\quad H \\ H\quad\quad\quad H \\ C-C \\ H \qquad :: \end{array} \right]^{\ominus} \qquad (\sigma = 3)$$

The enthalpy of formation, however, is apt to be much more sensitive to charge than is the case for uncharged molecules, and we expect to see systematic deviations from additivity behavior even for the case of group additivity. The reason for this is that the presence of an atomic charge produces an intense nearby field ($\sim 10^9$ V/cm) and correspondingly large induced dipoles in surrounding atoms, which tends to augment the field within the structure even at points distant from the charge center. The electrostatic polarization energy V_{pol} produced by a unit charge e, acting on a polarizable atom at a distance r, is given by:

$$V_{pol} = -\frac{1}{2}\frac{\alpha e^2}{r^4} = -\frac{166\alpha}{r^4} \text{ kcal/mole} \qquad (4.34)$$

[27] Some molecular orbital calculations suggest a structure looking more like a pyramidal CH_3^+, coupled to an H_2 molecule, the axis of which is perpendicular to the symmetry axis of the CH_3^+.

where the polarizability of the atom α is expressed in Å^3 and r is in Å. Thus the interaction of a C atom which has $\alpha_c \sim 1.0 \, \text{Å}^3$ with a unit charge 2.5 Å away is 4.3 kcal/mole. The polarizability of atoms in molecules follows with reasonable accuracy a simple atom additivity law and interactions can be estimated by such methods[28] to about ±20%.

One further consequence of these large polarization interactions is that there will always be a strong attractive interaction between ions and neutral molecules at distances shorter than 4 Å. This is in contrast to the very small van der Waal interactions between two molecules or a molecule and a radical at such distances. Radical–radical interactions or valence forces as we have already noted have ranges comparable to ion–molecule interactions. If the molecule has a dipole moment, then the ion–dipole interaction is superposed on the polarization attraction. The former is given by:

$$V_{\text{(ion–dipole)}} = \left(-\frac{e\mu}{r^2} \right) \cos \theta$$

$$= \left(-\frac{69\mu}{r^2} \right) \cos \theta \ (\text{kcal/mole}) \qquad (4.35)$$

where μ is in Debye units. At 2.5 Å, for example, the interaction of the 1.85 Debye H_2O dipole with a unit electronic charge is $(\theta = 0°)$ 20.4 kcal/mole.

From an electronic bonding point of view the class of ion–molecule reactions should be related to radical–molecule reactions. They both correspond to an open-shell species reacting with a closed-shell species. The intrinsic activation energy of a reaction (defined as the activation energy of the reaction in its exothermic direction) takes on values in the range of 0–15 kcal/mole for radical–molecule reactions. We might expect a similar range of intrinsic activation energies for ion–molecule reactions. However this is probably less than the various electrostatic interactions involved in ion–molecule collisions and so we may expect to find that ion–molecule reactions are determined by their collision frequency alone. This latter can be determined very simply by consideration of the attractive polarization forces and the centrifugal forces in an ion–molecule encounter.

If we consider the collision trajectory of an atom B with an ion A^+ having an initial relative velocity v_r^0 and an impact parameter b (Figure 4.2), conservation of energy permits us to write the equation of motion

[28] R. J. W. LeFevre, *Adv. Phys. Org. Chem.* **3**, 1 (1965); G. G. LeFevre and R. J. W. LeFevre, in *Physical Methods of Organic Chemistry*, Vol. 1, Part III-C, A. Weissberger and B. W. Possiter, (eds.), Wiley, New York, 1972.

which takes place in a plane, in the polar form:

$$E_r^\circ = \tfrac{1}{2}\mu v_r^{\circ 2} = \tfrac{1}{2}\mu(\dot{r})^2 + \frac{p_\theta^2}{2\mu_{AB}r^2} - \frac{e^2\alpha}{2r^4} \qquad (4.36)$$

where the terms are defined in Figure 4.2, and we have omitted repulsive energy and any other attractive interaction except polarization. In Figure 4.2, it should be noted that $\dot{\theta}$ = constant of motion. There are no forces acting at right angles to $\vec{r}$. This gives Kepler's law of planetary motion: $\theta = ct +$ constant; equal angles are swept out in equal intervals of time.

Rearranging and substituting for the constant angular momentum p_θ, we find for the relation of the distance of closest approach r_c to the impact parameter b ($\dot{r} = 0$ at $r = r_c$):

$$\left(\frac{r_c}{b}\right)^4 - \left(\frac{r_c}{b}\right)^2 + \frac{e^2\alpha}{2b^4 E_r^\circ} = 0 \qquad (4.37)$$

This can be solved to determine the impact parameter b, which will produce a distance of closest approach r_c at a given initial kinetic energy E_r°. Note $r_c < b$. To a first approximation we find:

$$\left(\frac{r_c}{b}\right)^2 \approx \frac{e^2\alpha}{2b^4 E_r^\circ} \qquad (4.38)$$

or solving for b:

$$b \approx \left(\frac{e^2\alpha}{2r_c^2 E_r^\circ}\right)^{1/2} \qquad (4.39)$$

The value of r_c is determined by the structure of the transition-state complex and can be taken for the tight transition state. Thus in the very exothermic reaction:

$$CH_4 + H_2^+ \rightarrow CH_5^+ + H + 66 \text{ kcal}$$

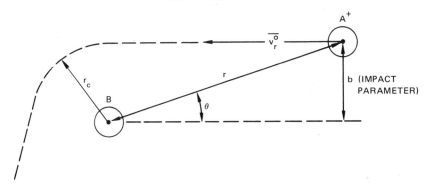

Figure 4.2   Collision trajectory of an ion A^+ with a polarizable atom B.

we might guess a transition state structure, such as:

$$
\begin{bmatrix}
\text{H} \quad \text{H} \\
\text{H} \cdot \text{C} \cdot \text{H} \cdot \text{H} \\
\text{H}
\end{bmatrix}^{\oplus}
\quad \text{OR} \quad
\begin{bmatrix}
\text{H} \qquad \text{H} \\
\text{C} \cdot \text{H} \\
\text{H} \ \text{H} \cdot \text{H}
\end{bmatrix}^{\oplus}
$$

The C·H distance would then be $1.10 + 0.4 = 1.5$ Å, and since the H·H distance is 1.1 Å, the distance between the centers of gravity of the CH_4 and H_2^+ at closest approach would be $r_c = 1 + \frac{1}{2}(1.1) \cong 2.0$ Å. Substituting this value in (4.39) together with the values of E_r° and α would give the impact parameter b for the reaction.

The reaction cross section Q or collision area in this simple collision model is given by:[29]

$$Q = \pi b^2 = \frac{\pi e^2 \alpha}{2 r_c^2 E_r^\circ} \tag{4.40}$$

which can be written as:

$$Q = \pi r_c^2 \left(\frac{e^2 \alpha / 2 r_c^4}{E_r^\circ} \right) \tag{4.41}$$

$$= \pi r_c^2 \left(\frac{E_{\text{pol}(r_c)}}{E_r^\circ} \right)$$

Since πr_c^2 is the collision cross section we would have used in a force-free, hard-sphere model for the reaction, we see that the polarization interaction increases this cross section by the ratio of the polarization energy at r_c to the initial translational energy E_r°. If we are dealing with thermal ions, E_r° can be taken as RT.

The problem we have just chosen can be used to illustrate an important aspect of ion–molecule reactions, namely the facility of electron transfer. If we were to substitute in (4.41) the value of α, for CH_4, we would have calculated too large a cross section. The reason is that the ionization potential of H_2 (15.4 eV) is much greater than that of CH_4 (12.6 eV). At distances of the order of 5–6 Å, where the polarization interaction becomes of the order of RT, the electron can jump from the CH_4 to H_2, yielding the charge-exchange process:

$$CH_4 + H_2^+ \rightarrow CH_4^+ + H_2 + 64 \text{ kcal}$$

[29] Note that the "chemical" reaction rate constant k has the simple relation to cross section given by: $k = Q\bar{v}$ [see (p. 162)], where $\bar{v}$ is the mean thermal velocity of collision in the system.

Note that such charge transfer requires two conditions. The first is that the rate at which the electron can tunnel from the CH_4 to the H_2^+ be rapid compared to the relative translational motion of the two ions. Such tunneling rates can be calculated from simple 1-dimensional quantum-mechanical models for the process and will depend exponentially on the ratio of the de Broglie wavelength λ of the electron to the separation distance r, $e^{-r/\lambda}$. It is not a significant condition at distances of the order of $r \leq 10$ Å.

The second condition for exothermic processes is that once the electron has transferred there must be a mechanism available that can absorb the kinetic energy of the reaction; otherwise the electron will return.[30] In the above example, such a relaxation process requires that there be some vibrational excitation of the newly formed H_2 molecule or CH_4^+. The rate at which such relaxation occurs is actually the rate-limiting step in the charge transfer process. In the present instance, the H—H distance is appreciably different in H_2 and in H_2^+ so that the moment of inertia will change appreciably and both rotational, as well as vibrational, modes may absorb the excess energy of the transfer. In consequence, the actual chemical encounter is between a CH_4^+ ion and an excited H_2. Thus the polarizability to use in solving for the impact parameter should be that appropriate to H_2.

$$Q(CH_4 + H_2^+) = \frac{\pi e^2 \alpha_{H_2}}{2 r_c^2 E_r^\circ} \tag{4.42}$$

For thermal ion–molecule reactions at 300°K, the polarization energies are generally appreciable relative to RT at distances of the order of about 5 Å, so that we may expect such reactions to have cross sections of the order of about 80 Å^2 or chemical rate constants in the range of $10^{-9.5 \pm 0.3}$ cc/particle-s $= 10^{11.3 \pm 0.3} M^{-1} s^{-1}$. This is generally the case as can be seen from some sample reactions shown in Table 4.14.

One final point is worthy of mention and that has to do with reactions which have activation energies. If this activation energy E_a is small, in the range of 0–15 kcal/mole, and it is not available from charge-transfer processes or internal excitation, we can then require that it be present in the form of radial kinetic energy at the distance of closest approach. That is, in (4.36), we set $\frac{1}{2}\mu \dot{r}_c^2 = E_a$ instead of zero. Equation (4.32) then becomes:

$$\left(\frac{r_c}{b}\right)^4 - \left(\frac{r_c}{b}\right)^2 + \frac{e^2 \alpha}{2 b^4 (E_r^\circ - E_a)} = 0 \tag{4.43}$$

---

[30] Thus the exothermic charge transfer process, $Na^+ + Cs \rightarrow Na + Cs^+ + 29$ kcal cannot occur at large distances since there is no way in which the excess energy of 29 kcal can be absorbed in the system. It can occur in a "hard" collision since then the 29 kcal can be transferred into repulsive energy of translation of the separating pair, $Na + Cs^+$.

Table 4.14.   Rate Constants for Some Ion–Molecule Reactions at 300°K

Reference[1]	Reaction	$\alpha(\text{Å}^3)$	r_c	log k (l./mole-s)
a	$H_2^+ + H_2 \rightarrow H_3^+ + H$	0.8	2.2	11.3
a	$CH_4^+ + CH_4 \rightarrow CH_5^+ + CH_3$	2.6	3.0	10.9
a	$Kr^+ + H_2 \rightarrow KrH^+ + H$	0.8	2.5	10.7
a	$CH_3^+ + CH_4 \rightarrow C_2H_5^+ + H_2$	2.6	2.0	11.6
b	$H^- + O_2 \rightarrow HO_2 + e^-$	1.4	2.0	11.8
b	$O^- + NO \rightarrow NO_2 + e^-$	1.6	1.6	11.1
b	$H^- + HOH \rightarrow OH^- + H_2$	1.6	2.1	12.3
b	$O^- + CH_4 \rightarrow OH^- + CH_3$	2.6	2.9	10.8
b	$CO_3^- + NO \rightarrow CO_2 + NO_2^-$	1.6	3.8	10.1

1. Bibliographic sources:
    a. J. O. Hirschfelder, C. F. Curtiss, and R. B. Bird, *Molecular Theory of Gases and Liquids*, Wiley, New York, 1959, p. 950.
    b. J. L. Franklin and P. W. Harland, *Ann. Rev. Phys. Chem.*, **25**, 485 (1974).

When the collisional energy is sufficient to provide E_a, we find a solution similar to the previous one except that $E_r^\circ - E_a$ becomes the effective energy in the encounter. For thermal encounters, however, where $E_r^\circ < E_a$, the activation energy must be provided by the polarization interaction and only "hard" or small impact parameter collisions can be effective. That is $b < r_c$ and the approximate solution is given by:

$$b \leq r_c \sim \left[\frac{e^2\alpha}{2(E_a - E_r^\circ)} \right]^{1/4} \tag{4.44}$$

4.15   ACTIVATION ENERGY OF CHEMICAL REACTIONS—BIMOLECULAR REACTIONS

The Arrhenius parameter that turns out to be most sensitive to chemical structure and least accessible to theoretical calculation is the activation energy for a chemical reaction. In consequence, we will discuss activation energies from a more qualitative and empirical point of view. The fact that exothermic chemical reactions have any activation energy requirements at all tells us that the electronic clouds in molecules are not "soft" but "hard." The intrinsic activation energy of an elementary reaction (for the exothermic direction) is a quantitative measure of the amount by which the electron clouds of the reactants must be deformed to permit the reaction to proceed. When two uncharged, nonpolar molecules, such as H_2 and D_2, come together, the predominant force at close distances

(<2.5 Å) is a very "hard" repulsion. The molecules act as if they were impenetrable. This general "impenetrability of matter" is itself a reflection of the Pauli exclusion principle—the fact that no more than two electrons (of opposite spin) can occupy the same spatial region.[31]

For purposes of discussion it is convenient to categorize bimolecular reactions into three classes:

Reaction Type	Electronic Structures	Range of Intrinsic E_{act} (in kcal/mole)
Molecule + Molecule	(two closed shells)	20–50
Radical + Molecule	(one closed shell/one open shell)	0–15
Radical + Radical	(two open shells)	0

From a purely empirical point of view we observe that the intrinsic activation energies of these classes decrease monotonically as the number of open-shell species[32] increases until finally, radical–radical reactions have zero activation energy. To understand such behavior, let us consider a simple atom-metathesis reaction from a valence-bond point of view.

In the reaction $D\cdot + H_2 \rightarrow H—D + H\cdot$, which is only slightly exothermic, the activation energy is about 8 kcal. The reaction proceeds over a potential hill. In the course of a reactive collision, we must form a bond between D and H and break one between H and H. If this were to happen gradually, by the H_2 molecule losing some fraction of its bonding electrons to the D atoms, we might imagine the entire process to be energetically downhill. However it takes much more energy to remove the bonding electron from H_2 than we gain back from attaching it to the D atom. One quantitative measure of this is the difference between the ionization potential (IP) of H_2, and the electron affinity (EA) of D less the coulombic interaction between the two, ϵ^2/r. Since $IP(H_2) = 13.6$ eV; $EA(D) = 0.77$ eV, and $\epsilon^2/r = 8.73$ eV (at $r = 1.6$ Å), we see that the ionic interaction V_{ionic} between H_2 and D lacks about 4.1 eV of being thermoneutral. Neglecting van der Waal and repulsive forces:

$$V_{ionic}^{(H_2^+ D^-)} = IP(H_2) - EA(D) - \frac{\epsilon^2}{r} \approx 4.1 \text{ eV} \qquad (r = 1.6 \text{ Å})$$

This suggests that a gradual shift of charge from one species to the other would also be endothermic. Alternatively, since this particular reaction is so symmetrical, we can imagine the transition state as looking

---

[31] What occupies "space" in molecules is not the nuclei but the light electrons.

[32] By open shell we mean those molecules, atoms, or radicals that have at least one orbital vacancy in their valence shells. Thus all atoms except the rare gases are open-shell, as are all radicals and positive ions.

like the average of the canonical structures:

$$D\cdot + H:H \rightleftarrows ([\dot{D}\cdot H\cdot H]^{\ddagger} \rightleftharpoons [D\cdot H\cdot \dot{H}]^{\ddagger}) \rightleftarrows D:H + \dot{H}$$

Here we have, in the three-center, three-electron system, two bonding electrons, and one nonbonding electron. The third electron, which is shared between the terminal atoms, has a node at the central atom. We thus have the same number of bonding electrons in this hypothetical transition state that we had in the reactants and to a first approximation the total energies of the system are about the same, and basically this is why atom metathesis reactions have small intrinsic activation energies.

If it is not favorable to shift charge gradually from one species to the other, we must picture the collision occurring with the activation energy being used to deform the electron clouds sufficiently to permit at some point, the sudden transformation of the electron structure to something like that shown above in the *TS* whence the reaction continues to completion. Johnston and Parr[33] have constructed a model in which they permit the energy of a bond to vary nonlinearly, $E_B \propto n^p$, with the electron density in it (bond order). By further assigning a repulsive energy to the nonbonding electron which depends only on the distance between the end atoms, and assuming the reaction proceeds with conservation of total bond order, they can assign empirical parameters to various bonds, which permits them to estimate metathesis activation energies to about ± 2 kcal.

However Alfassi and Benson have noted that there is a very strong correlation between activation energies and the electron affinities of the terminal atoms in a typical metathesis reaction $A + BC \rightarrow AB + C$. That such a relation exists is suggested by the following series of metathesis reactions which are all nearly thermoneutral:

Reaction $[A + BC \rightarrow AB + C]$	ΔH_r° kcal/mole	Intrinsic E_{act} kcal/mole	Electron Affinity of A(eV)
$^*CH_3 + CH_4 \rightarrow {}^*CH_4 + CH_3$	0	14.6	~0.3
$H + CH_4 \rightarrow H_2 + CH_3$	0	12.2	0.8
$CH_3O + CH_4 \rightarrow CH_3OH + CH_3$	0	11.0	1.4
$O + CH_4 \rightarrow HO + CH_3$	2	7.0	1.5
$Cl + CH_4 \rightarrow HCl + CH_3$	1	2.8	3.6

It is found that a good empirical relation is:

$$E_{act}(\text{intrinsic}) = 13.0 - 3.30I \tag{4.45}$$

where I is the sum of the electron affinities in eV of A and C and E_{act} is in units of kcal/mole. Negative values of E_{act} are arbitrarily set equal to zero.

[33] H. S. Johnston and C. Parr, *J. Amer. Chem. Soc.*, **85**, 2544 (1963).

This expression fits some 30 reactions to about ± 1.7 kcal/mole. A correction for the exothermicity of the reaction ΔH_r gives an equation of the form:

$$E_{act}(\text{intrinsic}) = \frac{14.8 - 3.64I}{1 + \Delta H/40} \tag{4.46}$$

which has an average error of ±1.4 kcal/mole.

An even simpler relation with better empirical fit is given by a bond-additivity relation:

$$E_{act} = X_A + X_C \tag{4.47}$$

where X_A and X_C are additive contributions of the groups to E_{act} and are listed in Table 4.15. Finally, the relation:

$$E_{act} = (F_A)(F_C) \tag{4.48}$$

gives perhaps the best fit without the need to reassign negative values of E_{act}. For 50 reactions the last two expressions give agreement to within ±1 kcal/mole with a maximum deviation of 2.6 kcal/mole. Most of the latter can be accounted for by dipole–dipole repulsion. Thus the reactions of CF_3 with RH generally have E_{act} about 2 kcal less than the corresponding metathesis $CH_3 + RH$. However when R is polar, such that the bond is polarized $R^{-\delta'}H^{+\delta'}$, then the approach of the polar radical $F_3^{-\delta}C^{+\delta}$ to the H atom gives rise to a dipole–dipole repulsion before the transition state is reached, of the order of 2–4 kcal/mole. This is observed

Table 4.15. The Contributions of the Different Endgroups to the Activation Energies of Metathesis Reactions*

Atom or Group	X (in kcal mol^{-1})	F (in kcal $mol^{-1})^{1/2}$
H	4.7	3.00
F	−3.8	0.35
Cl	−2.3	0.57
Br	−4.8	0.32
I	−5.7	0.15
O	2.0	2.15
Na	2.6	2.30
OH	−0.2	1.30
SH	−1.7	0.84
NH_2	−1.6	1.30
HO_2	1.1	1.70
CHO	0.3	1.55
CH_3	6.2	3.50
CF_3	4.1	2.95
C_2H_5	4.2	2.85

* See Equations 4.47, 4.48.

in the case of $RH = HCl$, H_2S, NH_3, and HBr.

$$\begin{array}{c} F \\ \diagdown_{+\delta} \\ F^{\prime\prime\prime\prime\prime}C \cdots\cdots \overset{+\delta'}{H} \underset{}{\overset{-\delta'}{\text{---}R}} \\ \diagup \\ F \end{array}$$

with $-\delta$ on the left F group.

An extreme form of the charge-transfer mechanism occurs in the case of metal-atom metathesis with halogens. Thus $K + Cl_2 \rightarrow KCl + Cl$ has zero activation energy and a rate constant of about $10^{11.5}$ l./mole-s. Here, because the electron transfer is favorable at large distances,

$$K + Cl_2 \rightarrow K^+ + Cl_2^- \rightarrow [K^+Cl^-Cl]^{\ddagger} \rightarrow K^+Cl^- + Cl.$$

The distance at which the electron jumps is given approximately by the ionic potential:

$$V_{ionic} = IP(K) - EA(Cl_2) - \frac{e^2}{r} + E_{pol} \qquad (4.49)$$

The closest distance r_c at which transfer can take place is given by setting $V_{ionic} = 0$ and, neglecting the polarization interaction:

$$r_c = \frac{e^2}{IP - EA} = \frac{14.4}{IP - EA} \text{ Å} \qquad (4.50)$$

Using the experimental values $IP(K) = 4.3$ eV and $EA(Cl_2) = 2.4$ eV, we find $r_c = 7.6$ Å, in good agreement with the data on the reaction cross section.

As we have already noted, addition reactions of radicals to pi bonds have the same electronic characteristics as metathesis reactions. They are three-center, three-electron transition states with E_{act} in the same ranges as those for metatheses. We again find the same polar effects at work. If we take the sequence of addition of atoms or radicals to olefins, we find that anything that decreases the ionization potential of the olefin or that increases the electron affinity of the radical or atom will act to decrease the overall activation energy. The olefin acts as a donor of electrons (Lewis base) and the radical as an acceptor (Lewis acid).

The following exothermic series of additions to C_2H_4 illustrates these effects:

Reaction	E_{act} kcal/mole	EA(eV) of Radical	ΔH_r° kcal/mole
$CH_3 + C_2H_4 \rightarrow n\text{-}\dot{C}_3H_7$	7.7	~0.3	−25
$H + C_2H_4 \rightarrow \dot{C}_2H_5$	2.0	0.8	−39
$O + C_2H_4 \rightarrow \dot{O}CH_2\dot{C}H_2$ (triplet)	1.6	1.5	−30
$HO + C_2H_4 \rightarrow HOCH_2\dot{C}H_2$	0.5	1.8	−32
$Cl + C_2H_4 \rightarrow ClCH_2\dot{C}H_2$	0	3.6	−22

As we substitute CH_3 for H in the olefin, thus lowering the ionization potential of the olefin, the activation energy for each of those reactions decreases until they are all very small for addition to tetramethyl ethylene.

In the case of the $Cl + C_2H_4$ addition, we calculate that at $r_{C\cdots Cl} = 2.1$ Å the ion pair has zero energy. Although this is barely larger than the C—Cl covalent radius (1.7 Å), it neglects a polarization interaction and a van der Waal attraction at much larger distances, all of which are in accord with the experimental finding of a very loose transition state for this exothermic addition.

4.16 ACTIVATION ENERGIES OF MOLECULE–MOLECULE REACTIONS

Perhaps the most complex to analyze of the classes of activation energies are molecule–molecule. These include the four, five, or six-center addition reactions of HX to olefins, as well as the six-center Diels–Alder reactions and the inverse 1,5-H transfer-addition reactions (ENE reactions) illustrated by the reverse of ester pyrolysis. E_{act} for all of these reactions vary from 20–50 kcal/mole and cluster in the range of 35 kcal/mole.

The four-center HX addition reactions have been the 'most tractable and have been quantitatively interpreted in terms of what has been called the semi-ion pair model.[34,35] It is assumed that H—X and the olefins both must be activated by promoting a bonding electron to a localized, ionic site:

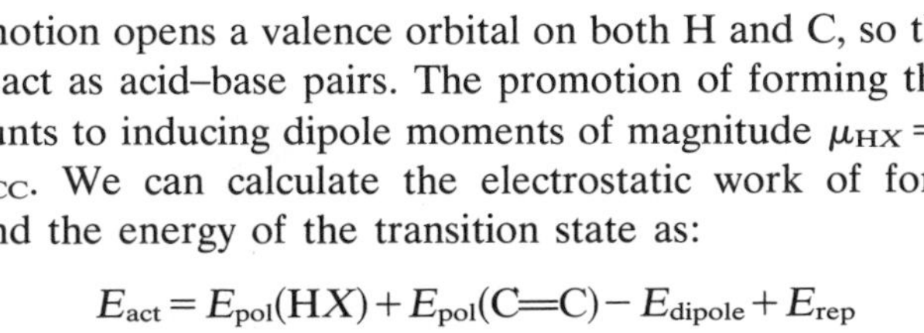

This promotion opens a valence orbital on both H and C, so that the two species react as acid–base pairs. The promotion of forming the semi-ion pair amounts to inducing dipole moments of magnitude $\mu_{HX} = \frac{1}{2}er_{HX}$ and $\mu_{CC} = \frac{1}{2}er_{CC}$. We can calculate the electrostatic work of forming such dipoles and the energy of the transition state as:

$$E_{act} = E_{pol}(HX) + E_{pol}(C\!\!=\!\!C) - E_{dipole} + E_{rep}$$

where E_{dipole} is the dipolar attraction of the two semi-ion pairs and E_{rep} is the Born repulsion of the two species. On substituting the various terms and differentiating at the TS $(\partial E_{act}/\partial r)_{TS} = 0$ to obtain the maximum, r_m,

---

[34] S. W. Benson and G. R. Haugen, *J. Amer. Chem. Soc.*, **87**, 4036 (1965).

[35] S. W. Benson and G. R. Haugen, *Int. J. Chem. Kinet.*, **2**, 235 (1970).

we find:

$$E_{act} = \frac{e^2}{8}\left[\frac{r_{HX}^2}{\alpha_{HX}} + \frac{r_{C=C}^2}{\alpha_{C=C}} - \frac{3}{2}\frac{r_{HX}r_{C=C}}{r_m^3}\right]$$

$$= 41.5\left\{\frac{r_{HX}^2}{\alpha_{HX}} + \frac{r_{C=C}^2}{\alpha_{C=C}} - \frac{3}{2}\frac{r_{HX}r_{C=C}}{r_m^3}\right\} \text{ kcal/mole} \qquad (4.51)$$

when r are in Å and α in Å^3.

These expressions yield E_{act} to within about ± 1 kcal/mole for over 60 reactions of HX addition to variously substituted olefins and dienes. X can be any of the halogens, OH, NH$_2$, HS, or PH$_2$. The model gives a particularly satisfying explanation of the Markownikoff effect, which states that on addition of HX to a substituted olefin, X will go to the most heavily substituted C atom. Relative to C$_2$H$_4$, a CH$_3$-group substitution lowers the activation energy by about 6 kcal/mole, the effect being nearly additive for two groups as in (CH$_3$)$_2$C=CH$_2$. Substitution of a third and fourth CH$_3$ has relatively little effect, about 1.5 kcal each in lowering E_{act}. Explanation for such a strong effect has been given in terms of the CH$_3$ polarization energy by the adjacent positively charged C$^{+1/2}$ atom. Negatively charged C atoms will not polarize as effectively because the negative charge is associated with the extra electron and is farther from the CH$_3$ group than the C$^{+1/2}$ nucleus.[36]

On examining the inverse ENE reactions (six-center addition of olefins to unsaturates), we find very similar effects of CH$_3$ substitution on E_{act}, making it likely that these also involve semi-ion pairs:[37]

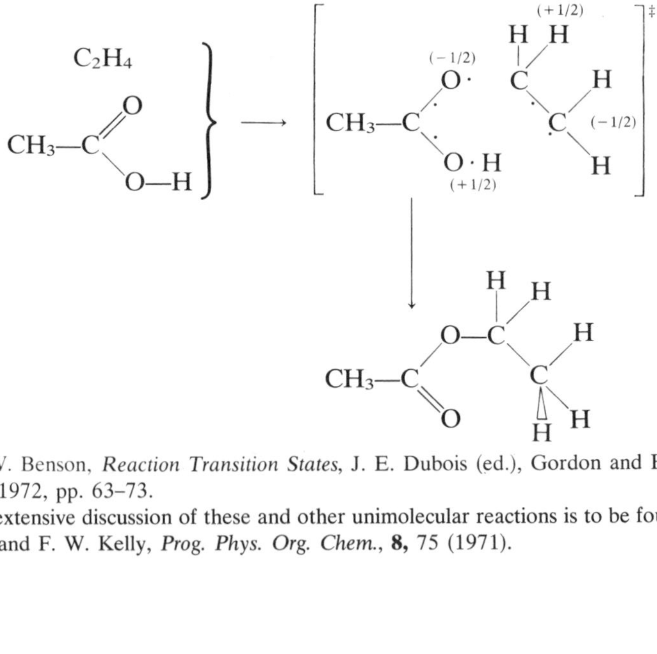

[36] S. W. Benson, *Reaction Transition States*, J. E. Dubois (ed.), Gordon and Breach, New York, 1972, pp. 63–73.

[37] An extensive discussion of these and other unimolecular reactions is to be found in G. G. Smith and F. W. Kelly, *Prog. Phys. Org. Chem.*, **8**, 75 (1971).

E_{act} (intrinsic) for these reactions range about 36 kcal and are also decreased about 5 kcal for each α-CH_3 substitution in the ester.

The Markownikoff effect can be reversed by placing substituents on the double bond of polarity opposite to that required so stabilize positive carbon. Thus CF_3, CN, and $C=O$ groups substituted on an olefin will cause abnormal anti-Markownikoff addition:

$$CF_3-CH=CH_2 + HBr \rightarrow CF_3CH_2-CH_2Br.$$

Such effects have not been documented quantitatively. The model seems to break down when halogen groups are directly substituted on carbon, as in $CF_2=CH_2$ and $CCl_2=CH_2$.[38]

Because of the very polar nature of the bonds, interactions calculated in the usual way tend to become very large. If, as a first approximation, we treat such polar groups as though they were CH_3 groups, we arrive at a calculated E_{act} not too different from the observed value.

4.17 DYNAMICAL EFFECTS IN CHEMICAL REACTIONS

Transition-state theory is an equilibrium theory of chemical reactions. It makes the assumption that the species at the transition state is in effective thermodynamic equilibrium with reactants. In consequence, the detailed dynamical path followed by reactants in arriving at the transition state is of no importance. Note that this doesn't require that any single species crossing the transition state be in equilibrium with reactants or have passed through all possible partitions of its internal energy. It just requires that the ensemble of transition state species satisfy the equilibrium requirement.

For simple fission reactions or more complex unimolecular reactions in which the ΔH_r and intrinsic activation energy are both small, TS theory presents a relatively simple picture and one that poses no peculiar dynamical problems. However as soon as we consider reactions for which E intrinsic is large or bimolecular metathesis reactions which are very exothermic, dynamical questions do arise. Thus in the very exothermic, metathesis reaction:

$$H + F_2 \rightarrow HF + F + 98 \text{ kcal}; \qquad E_{act} = 2 \text{ kcal}$$

we may consider how the energy of the reaction $98 + 2 = 100$ kcal is distributed among the degrees of freedom of the product molecules. Such questions have taken on great importance with the discovery and

---

[38] E. Tschuikow-Roux and K. R. Maltman, *Int. J. Chem. Kinet.*, **7**, 363 (1975). These authors have attempted to solve this problem by a modified bond energy—bond order approach.

widespread use of lasers which depend for their action on the production of chemical species in highly excited electronic, vibronic, or rotational states in concentrations exceeding that in the less excited states. If there should be a preferred, that is, nonstatistical distribution of the reaction energy in the degrees of freedom of the product molecule, then detailed balancing tells us that the product molecules in the reverse reaction will have higher rate constants when the activation energy is localized in the same preferred modes.

In the above example of $H + F_2$, the major pathway produces $HF^{(v=4)}$ in the fourth vibrational level and less than 5% of the reaction goes to produce nonexcited $HF^{(v=0)}$. Detailed balancing then tells us that in the reverse reaction, $HF^{(v=0)}$ will react with F more than ten times slower than $HF^{(v=4)}$, even though the total activation energies in both systems is the same.

The four-center elimination reactions, $CH_3—CH_2X \rightarrow C_2H_4 + HX$, provide another interesting example. Although the endothermicities of these reactions are quite small, in the range 10–20 kcal, they have activation energies some 20–50 kcal in excess of this. Here there are many internal degrees of freedom in the center-of-mass system ($3n - 6 = 18$ for C_2H_5X), only one of which corresponds to relative translational energy of the products. Hence one might expect on purely statistical grounds to find very little of the intrinsic activation energy in relative translation and the bulk of it in the internal modes.[39] Again, the reverse reaction in such cases will occur with highest rates from reactants with excess energy in these same internal modes. This is of some practical interest since it suggests the possibility of performing specific and selective chemical syntheses by exciting molecules vibrationally or electronically at otherwise low ambient temperatures.

The precise partitioning of reaction energy will be determined by the dynamics with which the energized particle crosses the transition state barrier and the dynamical relations between the internal motions. Thus in the reaction of $F + H_2$, if we assume that the *TS* is nonlinear and tight, then we may picture it as shown in Figure 4.3. The asymmetric stretch of the *TS*, indicated by vector arrows in the diagram, becomes the reaction coordinate. Conservation of momentum and angular momentum requires that any relative motion of separation of the two H atoms

---

[39] However, the intrinsic activation energy is not released until the transition state is passed. Intuitively, we do not expect to see energy released in those degrees of freedom which are about the same in *TS* and reactants. We do expect to see the energy develop in modes which change appreciably. These would be relative repulsion of olefin and HX and vibration of HX.

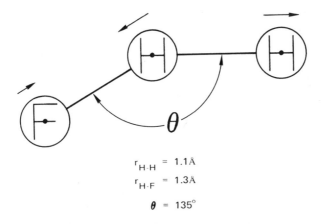

$$r_{H\cdot H} = 1.1\text{Å}$$
$$r_{H\cdot F} = 1.3\text{Å}$$
$$\theta = 135°$$

Figure 4.3 Tight, nonlinear transition state for $F+H_2$ metathesis. (Arrows represent directions of atom motions in the asymmetric stretch which is the reaction coordinate.)

impart a component along the HF bond and one at right angles to it. Because of its large mass, the F atom is virtually stationary during the trajectory.

Such dynamical considerations then require that about half the reaction energy appear in relative translation and the remainder appear in the newly formed HF, partitioned between rotation and vibration in a ratio determined by the angle θ. Thus when $\theta = 0$, no energy will be in rotation of HF except for that amount present in the angular momentum of the initial collision[40] $F+H_2$. The *TS* entropy is smallest for this linear configuration and a maximum for $\theta \sim 135°$. The observed results[41] indicating that $HF^{(v=2)}$ is most favored and with high rotational excitation are in rough agreement with such a picture. About 60% of the total reaction energy appears as vibrational energy, the remainder being partitioned between rotation and translation. What is perhaps at first surprising, is that so little $HF^{(v=0)}$ is produced. However, when it is considered that the H—F distance changes by some 0.4 Å in passing from the *TS* to product and that the $H_2\cdot\cdot\cdot F$ collision has about $\frac{1}{2}$ of the activation energy present as motion of the H atom relative to F, this begins to appear quite reasonable.

---

[40] Note that the collision is in three dimensions, not the two suggested by Figure 4.3. The two-dimensional representation is, however, a useful approximation.

[41] J. C. Polanyi and K. B. Woodall, *J. Chem. Phys.*, **57**, 1574 (1972); T. P. Schafer, et al., *ibid.*, **53**, 3385 (1970).

Similar considerations suggest that in the four-center elimination reactions about 50% of the energy of the products may appear as vibrational energy in the departing HX and momentum conservation súggests that the rest may be partitioned between relative translation and rotation of the olefins. Rotational energy of the HX is expected to be small because of conservation of momentum restrictions.

Perhaps the most famous example of a dynamically controlled reaction is the bimolecular collision between two HI to form $H_2 + 2I$.[42] The activation energy for this process is about 50 kcal and experiments of the molecular beam type have shown that collisions between two HI molecules[43] even with as much as 200 kcal of relative translational energy have less than 0.1 Å^2 cross section. This very low effectiveness of translational energy in a chemical reaction is not as surprising as it appears. Because of the large difference in masses between I (127 amu) and H (1 amu), an HI molecule with, for example, 50 kcal of translational energy has only $\frac{1}{127}$ of E_{tran} in the kinetic energy of the H atom.[44] The overwhelming bulk of it is associated with the I atom.

The only way in which this kinetic energy of the HI molecules can be communicated to the H atoms is in a nearly colinear collision in which the two H atoms are caught between the two I atoms:

$$[I \cdots H \cdots H \cdots I]^{\ddagger}$$

If we treat such a linear TS by our tight TS methods,[45] we find $\Delta S^{\ddagger}_{300} = -36.2$ eu, giving $A_{300} = 10^{7.2}$ l./mole-s, corresponding to $10^{-13.6}$ cc/particle-s and a reaction cross section aside from activation energy of about 0.02 Å^2. In contrast, the nonlinear TS with one extra external rotation and one internal rotation has a cross section about 200-fold higher, in excellent agreement with the observed rate data. However, such transition-state geometry does not allow for the translational energy to be used in work on the HI bonds.

$$\begin{bmatrix} I\cdot & & & \\ & \cdot & & \\ & \cdot H & \cdots & \cdot H\cdot \\ & & & \cdot \\ & & & \cdot I \end{bmatrix}^{\ddagger}$$

[42] J. H. Sullivan, *J. Chem. Phys.*, **46**, 73 (1967).
[43] S. B. Jaffe and J. B. Anderson, *J. Chem. Phys.*, **49**, 2859 (1968).
[44] Since $E_{tran} = \frac{1}{2}mv^2$ and both atoms have the same velocity, $E_H/E_I = m_H/m_I = \frac{1}{127}$.
[45] We can use I_2 as a model for TS and make corrections for increased moment of inertia. The high-frequency internal motions contribute very little.

4.18 ENERGY TRANSFER IN COLLISIONS

A chemical change consists of a breaking of bonds in the reactants and the formation of new bonds to create the products. Our consideration of elementary process has been pretty much confined, up to this point, to the consideration of such changes, that is, chemical reactions. However there is an even more elementary interaction between molecular species necessary for the occurrence of chemical reactions, and that is the exchange of energy between molecules.

If we divide molecular energy into the four conveniently separable categories, translational (T), rotational (R), vibrational (V), and electronic (E), the simple, hard-sphere collision model of gas-kinetic theory gives us an adequate picture of the exchange of translational energy between two particles. This is a simple process and is governed in a fairly straight-forward manner by the laws of conservation of momentum and energy. These latter operate in nearly equivalent fashion for both classical and quantized systems and as a consequence we find that translational energy is exchanged $(T-T)$ most readily between species of the same total mass. Light and heavy particles exchange translational energy very slowly. In a collision of two "smooth," "hard" spheres, the two particles will exchange components of velocity along their line-of-centers when their masses are equal. When the masses m_1 and m_2 are unequal, the fraction of line-of-center energy exchanged varies as m_1/m_2 $(m_1 < m_2)$. Because of such relations, very light particles such as electrons come into translational equilibrium with heavy ions and atoms only very slowly.

As a crude estimate we can consider that translational energies equilibrate between two populations of comparable mass in about three collision times.[46] Because rotational velocities and energies tend to be nearly "classical" and comparable to translational energies for all except diatomics containing H atoms $(H-X)$, very similar considerations apply to exchanges in energy between rotational and translational degrees of freedom $(R-T)$. They are very fast and relaxation times are also of the order of a few collision times.

The energy exchange between vibrational degrees of freedom and translation $(V-T)$ is on the contrary, much slower. The basic reason has to do with the higher velocities of vibrational motions (arising from quantization) compared to translational motion. Thus H atoms in the first vibrational level of H_2 $(\bar{\nu} = 4400 \text{ cm}^{-1})$ have a total vibrational energy of

---

[46] Because most collisions are grazing or sideway collisions, rather than "head-on" collisions, molecules moving very rapidly in a gas lose their excess translational energy only over several collisions. The components of velocity perpendicular to the line-of-centers at collision tend to be conserved.

$3h\nu/2$ and an average kinetic energy one half of this or 3300 cm^{-1}. This is about 16 times larger than RT at $300°K$, which is 210 cm^{-1}. In the average room-temperature collision between H_2 and a free H atom, the latter is moving about four times more slowly than either of the bound atoms. Similar considerations apply to most other diatomic molecules. It is not until we get to diatomics with frequencies in the region of 200 cm^{-1} that we expect to find vibrational motion behaving classically in energy-transfer processes. Alternatively, we may consider that $(V–T)$ energy transfer is rapid at temperatures such that $T \sim h\bar{\nu}/k$. For H_2 this would be at about $6000°K$ and for N_2 $(\bar{\nu} \sim 2300 \text{ cm}^{-1})$, it would be at about $3000°K$.

An alternative way of regarding the process is to say that $V–T$ exchange will be fast only when the times of collision, τ_c, is comparable to, or shorter than, the time of a vibration $(1/\nu)$. Because the interaction of two atoms during a collision is not infinitely short, but extends over some finite distance a, when this time of encounter is nearly resonant with the vibration time, we have our greatest probability of exchanging a quantum.

Landau and Teller[47] showed that one might expect an exponential relation between the probability of $V–T$ transfer, P_{V-T} and the ratio of $(\tau_c/\tau_{\text{vib}})$:

$$\ln P_{V-T} \propto \frac{\tau_c}{\tau_{\text{vib}}} \cong \frac{4a/v_r}{1/\nu} = \frac{4a\nu}{v_r} \tag{4.52}$$

Here a, the range of the repulsive forces acting between the colliding atoms, can be taken from potential functions, such as the Lennard–Jones or Morse potential.[48] It is of the order of 0.2 Å for most atoms. v_r is the initial relative velocity of the collision, and the factor of 4 takes into account the total duration of the collision. Since $P_{V-T} \leq 1$, we can write:

$$P_{V-T} \propto P_0 \exp \left(\frac{-4a\nu}{v_r} \right) \tag{4.53}$$

Since the probability of a collision with relative velocity v_r is itself from simple kinetic theory given by:

$$P(v_r) \propto \exp \left(\frac{-v_r^2}{2kT} \right) \tag{4.54}$$

the average of P_{V-T} over a Maxwell–Boltzmann distribution of thermal velocities will have its major contribution from regions where the product $(P_{V-T}) \cdot P(v_r)$ has a maximum value. This leads to the result, confirmed

[47] L. Landau and E. Teller, *Physik. Zeit. Sov.*, **11**, 18 (1937).
[48] See p. 87.

experimentally[49,50] that the temperature dependence of $V–T$ exchange is given by the Landau–Teller relation:

$$\langle P_{V-T} \rangle \sim A \exp\{-BT^{-1/3}\} \tag{4.55}$$

Typical $V–T$ transfer rates are about 10^{-6} per collision at $1000°K$ for nonpolar diatomic molecules without H atoms and about 10^{-2} to 10^{-4} for very polar diatomics or H-atom containing molecules. It is found that $V–V$ transfer rates between two diatomics tend to be fast ($\sim 10^{-2}$ per collision) when the exchanging quanta are nearly the same.[51] The rates decrease exponentially as the difference in energies of the quanta increases. Any appreciable intermolecular forces of attraction, such as dipole–dipole (in HF and HCN) or dipole–quadrupole (CO_2–CO) also make $V–V$ transfer more efficient.

Very large polyatomic molecules possessing many internal degrees of freedom tend to exchange energy both $V–V$ and $V–T$ more efficiently ($\sim 10^{-2}$ to 10^{-3} per collision at $25°C$). What limited evidence exists at the moment, also suggests that very highly excited molecules,[52] that is, molecules containing many quanta of vibrational energy may exchange $V–T$ almost as efficiently as translational energy. This has been the basis for expecting that in unimolecular reactions, very highly excited molecules with energies in excess of the activation energy for reactions are deexcited (below the barrier) at almost every collision (the "strong collision hypothesis").[53]

Electronic energy transfer ($E–E$) appears to be very efficient between molecules even when the two energy levels are not resonant. $V–E$ transfer seems also to be very efficient. Thus excited Na atoms can transfer their energy to N_2 or CO at almost every collision. About half the electronic energy so transferred ends up as vibrational energy in CO and N_2. Such efficiency in $E–E$ and $V–E$ transfer seems very reasonable since it involves interactions through the electrons in the colliding partners. Thus the interaction of excited Na* with N_2 can be thought of as forming a metastable complex ($Na^+N–N^-$) that permits a very close bond between Na^+ and one of the two N atoms. Even though the electron

[49] *Transfer and Storage of Energy by Molecules*, Vol. 2, *Vibrational Energy*; Vol. 3, *Rotational Energy*; G. M. Burnett and A. M. North (eds.), Wiley, Chichester, U.K., 1970.

[50] K. F. Herzfeld and T. A. Lotovitz, *Absorption and Dispersion of Ultrasonic Waves*, Academic, New York, 1959.

[51] Atoms such as O, S, or molecules such as NO having orbital degeneracy are very effective in vibrational relaxation because of a resonant interaction produced by the collision.

[52] B. S. Rabinovitch, et al., *J. Phys. Chem.*, **75**, 1366, 2177, 3037 (1971) and earlier papers.

[53] Evidence is now, however, very strong that the collisional deactivation efficiency usually lies in the range 0.1–1.0.

affinity of N_2 is nearly zero, the dipole–quadrupole interaction of the excited Na* (dipole) with the very polarizable N_2 bond guarantees strong interaction. By way of contrast, closed-shell systems of low polarizability such as He, Ne, or Ar are very inefficient in quenching Na*. That is, V–E processes are generally inefficient. There is an exception to this lattter statement and that involves very highly excited atoms that are close to their ionization limits. The ions of such species can form bonds even with rare gas atoms and a chemical ionization can occur:[54]

$$He^* + H \rightarrow HeH^+ + e^-$$
$$Ar^* + Na \rightarrow ArNa^+ + e^-$$
$$He^* + Ar \rightarrow HeAr^+ + e^-$$
$$Hg^* + K \rightarrow HgK^+ + e^-$$

The bonds in these molecule–ions are generally very weak, of the order of 0.5 to 2.0 eV, that is, about 10–50 kcal/mole. The reverse process, dissociative recombination occurs very rapidly and has a bimolecular rate constant of about $10^{14.3 \pm 0.5}$ l./mole-s. While at first glance such a rate constant seems abnormally high, it must be realized that it corresponds to a product of a thermal velocity (which for an electron is about $10^{2.3}$ faster than for most atoms) and a potential interaction that extends out to much larger distances since it is ionic and not short-range. These large values are also compatible with the overall entropy change in these reactions which, as written, are about -14.0 eu. This would then yield for the rate constant for the forward reaction values in the neighborhood of $10^{11.2 \pm 0.5}$ l./mole-s, corresponding to reaction at every gas-kinetic collision.

4.19 REACTION DYNAMICS—CONSERVATION RULES

While we intuitively think of a collision as a violent event in which everything gets shaken up, for molecular and atomic systems this is not necessarily true. We have already seen some examples of this in the preferred forms of activation energy (p. 200) for the HI reaction and in the difficulty of transferring energy between disparate masses on collision (p. 201). In each case, and in most examples, the reasons are basically the same; namely, that the interacting masses are so different that energy and momentum conservation that governs such interactions precludes strong coupling and efficient transfer of energy. When to this constraint is added the restriction of quantization which makes available energy levels that

---

[54] A summary of recent work is given by A. Fontijn, *J. Pure Appl. Chem.*, **39**, 287 (1970).

are discrete and with large separations for light atoms, we find that the efficiency is further reduced.

The result of such weak coupling is that certain degrees of freedom tend to be preserved with little change in collisions. They are said to be adiabatic. One of the best conserved properties of a collision is its total electron spin. The spatial orientation of the magnetic field of an electron can only be changed by a torque acting on it. In molecular systems, such a torque can only arise from a magnetic field. The only strong magnetic fields in collisions are those due to the spin of other electrons and the dynamic field created when electrons move rapidly across the electric field lines of another charged particle; this is essentially a case of magnetic induction. Since electrons strongly repel each other, their mutual magnetic interactions, which depend on the sixth power of the distance between them, tend to be very small. However when they come close to the positive nucleus, they have high velocity and feel, consequently, a very large induced magnetic field which can alter their spin direction.

A consequence of this is known as the "rule of spin conservation." In elementary reactions, total electron spin is usually conserved. Thus we find that ground-state O atoms (^3P) will not recombine with N_2 ($^1\Sigma$) to form the stable $N_2O(^1\Sigma)$. The same limitation applies to $O(^3P) + CO(^1\Sigma)$. In both cases an appreciable activation energy is required to modify the spin of the system from its triplet state to the singlet state of the product. The reaction also has an A-factor characteristic of a tight transition state. In contrast, the excited state of $O(^1D)$ can combine readily with either N_2 or CO to form products which if quenched before they dissociate, will remain as stable species. The usual way to describe these interactions is to say that triplet CO_2 and N_2O are not stable while the singlet species are. Figure 4.4 shows two potential energy curves for N_2O, the stable lower one (singlet) and the repulsive upper one (triplet). A similar diagram can be drawn for CO_2. The crossing point for the two curves occurs at a distance slightly larger than the stable r_{N_2-O} and at an energy above the level of $N_2 + O(^3P)$. Thus there will be an activation energy for the so-called "crossing" (~ 20 kcal). This amounts to an energy sufficiently great so that on collision, one of the valence electrons of the O atom will come close enough to one of the nuclei to change spin to the singlet.

Most symmetry rules are of this nature. They require an excess activation energy to break down the conservation of the required property so that the conservation-breaking reaction can occur.

The same type of spin restrictions operate in ionic reactions as well. Thus the ion analog to the above reactions is:

$$O^+ + N_2 \rightarrow NO^+ + N + 26 \text{ kcal}$$

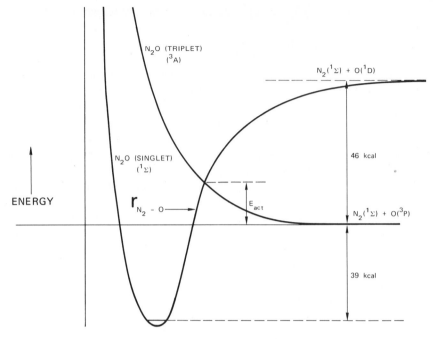

Figure 4.4 Crossing of singlet and triplet curves for N_2O.

Despite its exothermicity, this reaction has about a 7-kcal activation energy, very analogous to the isoelectronic neutral exchange reaction:

$$^{15}N + N_2 \rightleftarrows {^{15}NN} + N,$$

which appears to have about a 6 kcal activation energy. In both cases the total spin of the reactants is $\frac{3}{2}$ (quartet), so that the TS would have to be a quartet. However, the ground states of either N_2O^+ or N_3 can both be shown to be spin $\frac{1}{2}$ (doublet), so that the quartet is an excited, probably repulsive, state.

For heavier atoms beyond the second row in the periodic table (beyond argon), valence electrons have deeply penetrating orbits, and spin-orbit coupling tends to be large. Hence spin rules do not hold so well for these systems.

The other conservation rules which appear to hold well in unimolecular reactions of organic molecules have to do with the orbital symmetry conservation rules proposed by Woodward and Hoffman (*loc. cit.*). By constructing molecular orbital diagrams for reactants and products, these authors were able to show that symmetry properties of the ground-state reactants correlated with excited states of products and hence could not proceed directly without large activation energies. In this class were direct concerted reactions, such as:

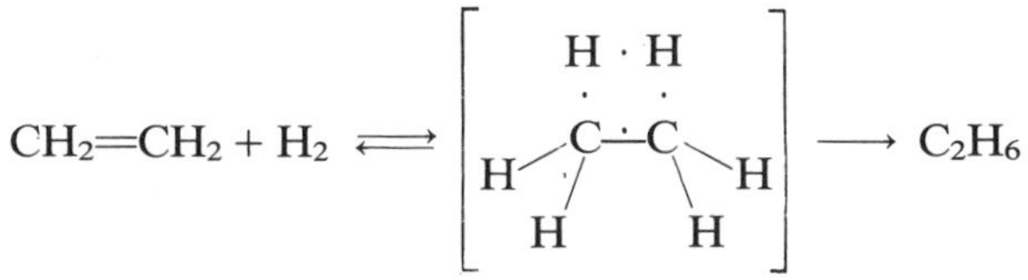

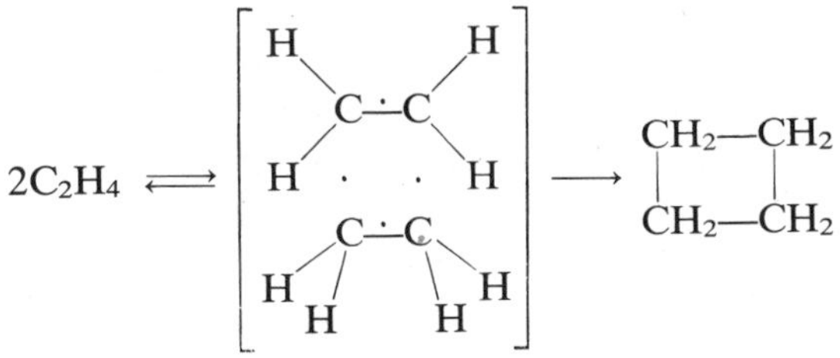

$$CH_2{=}CH_2 + H_2 \rightleftharpoons \begin{bmatrix} H \cdot H \\ \overset{\cdot\;\;\cdot}{\underset{H}{C}{\cdot\cdot}{C}}\diagdown_{H} \\ H \quad H \end{bmatrix} \longrightarrow C_2H_6$$

$$2C_2H_4 \rightleftharpoons \begin{bmatrix} H \diagdown \quad \diagup H \\ C{\cdot\cdot}C \\ H \quad \cdot \quad \cdot \quad H \\ C{\cdot\cdot}C \\ H \diagup \quad \diagdown H \\ H \quad\quad H \end{bmatrix} \longrightarrow \begin{matrix} CH_2{-}CH_2 \\ | \qquad | \\ CH_2{-}CH_2 \end{matrix}$$

In contrast, the excited states (triplet) of the olefin do have the correct symmetry properties to go directly to products and proceed without activation energy.

4.20 ISOTOPE EFFECTS

The substitution of an isotopic atom in a molecule has an effect on its thermochemical and kinetic properties; mainly through its effect on zero-point energies. The simplest case is that found in the ionization potentials of the H and D atoms. Because the Rydberg constant contains the reduced mass of the nucleus–electron system, the electron in deuterium has a lower energy than the electron in hydrogen by about $1/1836$ of the IP(H) or about 0.01 eV $= 0.23$ kcal/mole.

We find a much larger effect on vibrational energy levels. The zero point energy of any vibration is $\frac{1}{2}h\nu$, and the frequencies in two isotopically substituted diatomic molecules will be in the ratio:

$$\frac{\nu_1}{\nu_2} = \left(\frac{\mu_2}{\mu_1}\right)^{1/2} \tag{4.56}$$

where μ_2 and μ_1 are the reduced masses for the vibration. These differences in frequencies and reduced masses are only appreciable for H and D atoms where we find large effects on the activation energies. Thus the dissociation energies of H_2 and D_2 with vibration frequencies of about 4400 cm^{-1} and 3100 cm^{-1}, respectively, differ by $h(\nu_1 - \nu_2)/2 = 650$ cm$^{-1} \simeq 1.9$ kcal. In polyatomic molecules the substitution of each isotopic atom changes at least three frequencies so that polysubstituted molecules like CD_4 can have large differences in total zero-point energies from their lighter analogues.

For very accurate estimates of the changes in thermochemistry, we are obliged to make an assignment of frequencies and then calculate the changes due to isotopic substitution. This can be done, generally, by making a total vibrational analysis of the molecule and estimating the frequency changes. For chemical reactions, this would have to be done for the *TS* state, as well. Standard methods now exist for making vibrational assignments and estimating the effects of isotopic substitution. Where the molecule is relatively simple and the frequencies of the unsubstituted molecule are known, it is not too difficult to make quick estimates.

Thus in CO_2 the reduced masses of the different frequencies are given by:

$$\text{Asymmetric stretch:} \quad \mu_a = \frac{2m_O m_C}{m_C + 2m_O}$$

$$\text{Symmetric stretch:} \quad \mu_s = m_O/2$$

$$\text{Degenerate bends:} \quad I_b \propto \frac{m_C m_O}{m_C + m_O}$$

The changes in the corresponding frequencies are then calculated from (4.56). Normally, it is considered that only primary isotope effects involving the breaking of H or D bonds will be large. However secondary effects can be equally large. Thus in the C_2H_6 molecule, the breaking of the C—C bond involves no primary isotope effects for H atoms. However, in breaking the C—C bond we change, in going to the loose transition state, $4CH_3$ rocking modes at about 950 cm^{-1} into free restricted rotations with no zero-point energy. The corresponding frequencies for C_2D_6 would be about 670 cm^{-1} so that $\Delta E_0^\circ \approx 560 \text{ cm}^{-1} = 1.6 \text{ kcal/mole}$, and we should expect a stronger bond by this amount in the heavier molecule.[55] Since isotopic effects are used as a diagnostic tool in determining the mechanism of unimolecular and bimolecular reactions, it is important to pay attention to these secondary isotope effects which can be quite large.[56]

---

[55] There is a smaller effect on the reaction coordinate, the C—C stretch. Its reduced mass changes in the ratio $(m_{CH_3}/m_{CD_3}) = \frac{5}{6}$. This gives rise to a 45 cm^{-1} difference in E_0° or ~0.15 kcal.

[56] A comprehensive review is *Isotope Effects in Chemical Reactions*, C. J. Collins and N. S. Bowman, Eds., Am. Chem. Soc. Monographs, New York (1970).

5

Analysis of Complex Reaction Systems

5.1 CHAIN REACTIONS

In Chapters 3 and 4 we have been primarily concerned with the kinetics of what may be described as elementary chemical reactions. These are reactions that proceed from reactants to products by passage over a single energy barrier or well. Very few chemically reacting systems are so simple. Most can be shown to be complex in that usually two or more elementary reaction steps can be demonstrated to coexist. The most common type of complexity found is that in which highly reactive intermediates, atoms, or free radicals are formed and act as catalysts for the reaction. The mechanism of this catalysis consists of a cycle of elementary steps (propagations), each involving the reaction of a free radical and, finally, reforming the original chain carrier. The net effect of the chain cycle is to produce a molecular reaction without changing the concentrations of chain carriers. The reaction of $H_2 + Br_2$ provides a classical example:

$$\text{Chain} \quad \begin{cases} H + Br_2 \rightleftarrows HBr + Br \\ Br + H_2 \rightleftarrows HBr + H. \end{cases} \quad (5.1)$$

Net reaction: $H_2 + Br_2 \rightleftarrows 2HBr$

At each propagation step in the chain, one chain carrier is converted into another ($H \rightleftarrows Br$), but the sum of the total carrier concentrations $[(H) + (Br)]$ is left unchanged.

Reactions in which the total carrier concentration is increased by the chain are called branching chains, the $H_2 + O_2$ reaction providing an important example.

$$H + O_2 \rightleftarrows HO + O;$$

$$O + H_2 \rightleftarrows OH + H; \tag{5.2}$$

$$HO + H_2 \rightleftarrows HOH + H.$$

There is no way of adding these equations to yield a net molecular reaction. The first two reactions in the scheme give two chain carriers for one ($H \rightarrow HO + O$, or $O \rightarrow OH + H$), and the result is an exponential growth of radicals in the system once the chain has been initiated. The onset of lasing in a cavity is also an autocatalytic chain branching reaction. We shall restrict our discussion here to the nonbranching chains.

5.2 SOME CRITERIA FOR SIGNIFICANCE OF CHAINS

In principle, there is no unique path from some particular set of reactants to products. The possible number of alternative paths is limited only by the imagination of the kineticist analyzing the reaction. However in the laboratory system, "the race is to the swiftest," and, of the infinity of possible paths, only those leading most rapidly to products under the experimental conditions will be observed. In trying to make an a priori assessment of the preferred path for a chemical reaction, we should compare all possible paths. One of these may be the "concerted" process involving a single transition state. Thus, in the reaction $H_2 + Br_2 \rightarrow 2HBr$, the simplest path would be the four-center path involving $H_2 + Br_2$, for example

$$H{-}H + Br{-}Br \rightleftarrows \begin{bmatrix} H & \cdot & H \\ \dot{Br} & \cdot & \dot{Br} \end{bmatrix}^{\ddagger} \rightleftarrows 2HBr \tag{5.3}$$

Such concerted paths involving stable molecules generally have high activation energies, and if they are bimolecular, low A factors. Radical chains can compete favorably with them only if the propagation steps are sufficiently rapid, meaning usually that they have low activation energies. In our analyses of complex reaction systems, we always have to consider the competition of concerted processes with radical processes. Pyrolysis of large molecules, particularly unsaturates, may show the coexistence of both paths.

Not all molecules are capable of decomposing by fast, simple chains. For example, the pyrolysis of C_2H_6 to $H_2 + C_2H_4$ admits of a simple chain:

$$\text{Chain} \quad \begin{cases} H + C_2H_6 \rightarrow H_2 + C_2H_5 \\ C_2H_5 \rightarrow C_2H_4 + H. \end{cases} \tag{5.4}$$

Net reaction: $\qquad C_2H_6 \rightarrow H_2 + C_2H_4$

However, there is no simple chain decomposition of CH_4 to form either $C_2H_6+H_2$ or $C_2H_2+2H_2$. Let us see why.

A simple method of testing to see if a molecule X can undergo chain decomposition is to start with an unspecified radical R that can attack the molecule X in any of the standard bimolecular reactions to propagate a new radical, which will ultimately give back R. We shall consider the reactions of addition, atom abstraction, and unimolecular fission among our possible propagation steps. In order to be fast bimolecular steps must have low activation energies ($E/\theta \leq 5.5$), whereas, for unimolecular steps, $E/\theta \leq 14$.[1] These restrictions come about from the requirement that the rate of the chain cycle be reasonably rapid (ten to 100-fold), compared to the rate of formation of chain centers. If this is not the case, the initiation reactions will begin to dominate the system, and the chain is not important.[2] Such a case is provided by the chain decomposition of hydrazine: $2N_2H_4 \rightarrow 2NH_3+N_2+H_2$.

The chain steps are:

$$H+N_2H_4 \xrightarrow{\ 1\ } H_2+N_2H_3$$
$$N_2H_3 \xrightarrow{\ 2\ } N_2H_2+H$$
$$\phantom{N_2H_3 \xrightarrow{\ 2\ }} \llcorner\!\rightarrow N_2+H_2$$

fast

and the second step is too slow below 1200°K to permit rapid chain cycles.

The activation energy restrictions usually rule out steps in which saturated C, N, O, F atoms, and so on increase their valence shells (pentavalent carbon or trivalent O). Thus we rule out

$$CH_3+CH_4 \longrightarrow \left[H_3C \cdots \overset{\overset{\displaystyle H}{|}}{\underset{\underset{\displaystyle H}{|}}{C}} \cdots H \right] \longrightarrow C_2H_6+H$$

as an important propagation step in the CH_4 pyrolysis.

Let us consider a pyrolytic chain (one reactant) decomposition of C_2H_5Cl, as an example. A radical R can only abstract an atom, hence

$$R+C_2H_5Cl \begin{cases} \xrightarrow{\ 1\ } \cdot CH_2CH_2Cl+RH \\ \xrightarrow{\ 1'\ } CH_3\dot{C}HCl+RH \\ \xrightarrow{\ 1''\ } CH_3\dot{C}H_2+RCl \end{cases} \qquad (5.5)$$

[1] $\theta = 2.303RT$ in kcal/mole.

[2] It must not be too rapid, that is, too long a chain length, or else the system may not reach a stationary state.

are the three possible propagation reactions. The subsequent steps would have to be

$$\dot{C}H_2CH_2Cl \xrightarrow{\ 2\ } C_2H_4 + Cl$$

$$CH_3\dot{C}HCl \xrightarrow{\ 2'\ } C_2H_3Cl + H \qquad (5.6)$$

$$CH_3\dot{C}H_2 \xrightarrow{\ 2''\ } C_2H_4 + H$$

and to constitute a chain, the radical R would have to be H or Cl.

From our bond-energy tables, we see that a Cl atom could not abstract Cl because it would be endothermic by 22 kcal and thus have too high an activation energy. Therefore only HCl and H_2 are possible products of the various combinations of the chain cycles. These are $1 + 2$ yielding $C_2H_4 + HCl$ with $R = Cl$; $1' + 2'$ yielding $C_2H_3Cl + H_2$ with $R = H$; and, $1'' + 2''$ with $R = H$ yielding $C_2H_4 + HCl$. Again, from bond dissociation energies we can show that $2'$ and $2''$ are endothermic by ~ 39 kcal, whereas 2 is only 20 kcal endothermic. Hence only the sequence $1 + 2$ can give a fast chain with $C_2H_4 + HCl$ as products and Cl· and ·C_2H_4Cl as chain carriers.

The molecule Cl_2O cannot undergo simple chain pyrolysis. Abstraction of a Cl atom leads to the tightly bound radical ClO·, which has no effective path for reacting with Cl_2O, according to our rules above. The same is true of F_2O, N_2O, and ClO_2.[3]

The molecule H_2O_2, by our rules, does not lend itself to a fast chain, for only abstraction of H can occur from it, leading to HO_2:

$$R· + H_2O_2 \rightarrow RH + H\dot{O_2} \qquad (5.7)$$

However, $H\dot{O_2}$ cannot undergo ready decomposition because the bond strengths are $DH°(H-O_2) = 47$ kcal and $DH°(HO-O) = 63$ kcal. This is in accord with the facts that show H_2O_2 decomposing by a simple nonchain, radical process, even up to 550°C.

$$H_2O_2 + M \rightarrow 2OH + M;$$

$$HO + H_2O_2 \rightarrow HOH + HO_2; \qquad (5.8)$$

$$2HO_2 \rightarrow H_2O_2 + O_2.$$

5.3 RATES OF CHAIN CYCLES

The rate of a cycle consisting of elementary steps is measured by the reciprocal of the total time required to pass through each step in the

---

[3] However a chain propagation involving two radicals is possible:

$$2ClO· \rightarrow Cl-O-O· + Cl·;$$

$$Cl-O-O· + M \rightleftarrows Cl· + O_2 + M;$$

$$Cl· + Cl_2O \rightleftarrows Cl_2 + ClO·;$$

cycle. Thus, in the H_2—Br_2 chain [(5.1)] the mean time required for a H atom to react, τ_H is given by

$$\tau_H \equiv \frac{(H)}{d(H)/dt} = \left[\frac{d[\ln (H)]}{dt}\right]^{-1} = \frac{1}{k_a(Br_2)} \qquad (5.9)$$

where we have neglected the reverse of this reaction ($H + Br_2 \rightarrow HBr + Br$). Similarly, for Br atoms ($Br + H_2 \rightarrow HBr + H$),

$$\tau_{Br} = \frac{1}{k_b(H_2)}. \qquad (5.10)$$

If we wish to correct for the back reaction, we would multiply this by $1/(1-\alpha)$ where α is the fraction of Br atoms returned by the back reaction.

The total chain cycle requires a time $\tau = \tau_{Br} + \tau_H$. For a chain of m steps, the chain cycle time is given by a similar sum, and we can see that, if one of the steps k is significantly slower than the rest, the chain time τ can be approximated by τ_k, and this becomes the rate determining step in the chain. Not all chains admit of such simplification, but many do.

Although the total concentration of all radicals is unaffected by the propagation steps, the ratio of radical concentrations is given uniquely by the propagation steps under the conditions of the usual quasisteady state analysis. Thus in the H_2–Br_2 reaction,

$$H + Br_2 \underset{-a}{\overset{a}{\rightleftharpoons}} HBr + Br,$$
$$Br + H_2 \underset{-b}{\overset{b}{\rightleftharpoons}} HBr + H, \qquad (5.11)$$

the condition of long chain lengths ($\lambda \geq 10$)[4] implies that the slowest chain step is at least ten times faster than any step involving initiation or destruction of radicals. Hence the quasisteady state hypothesis applied either to (H) or (Br) leads to the same equation:

$$\frac{d(H)}{dt} = \frac{d(Br)}{dt} \approx 0, \qquad (5.12)$$
$$k_a(H)(Br_2) + k_{-b}(HBr)(H) = k_{-a}(Br)(HBr) + k_b(Br)(H_2),$$

or, neglecting the slow, very endothermic, reaction step $-a$,

$$\frac{(H)}{(Br)} = \frac{k_b(H_2)}{k_a(Br_2) + k_{-b}(HBr)}. \qquad (5.13)$$

---

[4] The kinetic chain length λ is defined as the rate at which radicals propagate the chain to produce product molecules relative to their rate of termination, which by hypothesis is also equal to their rate of initiation. Hence:

$$\lambda = \frac{\text{rate of reaction}}{\text{rate of initiation}}.$$

The rate of reaction $R(HBr)$ is given by

$$R(HBr) = R_a - R_{-a} + R_b - R_{-b}, \qquad (5.14)$$

which, on substituting from (5.12), becomes

$$R(HBr) = 2(R_a - R_{-a}) \approx 2R_a$$

$$= 2k_a(H)(Br_2), \qquad (5.15)$$

or, in terms of (Br) atoms from (5.13),

$$R(HBr) = \frac{2k_a k_b (H_2)(Br_2)(Br)}{k_a(Br_2) + k_{-b}(HBr)}. \qquad (5.16)$$

In similar fashion, it will always be possible to write a steady-state rate expression for long-chain reactions in terms of the concentration of any one of the chain carriers. As we shall see later, the absolute concentrations of the chain carriers will be determined by the initiation and termination reactions.

As an additional example, consider the chain pyrolysis of neopentane (NpH) to form isobutene + CH_4. Applying our usual test, we write:

$$R + C(CH_4)_4 \xrightarrow{\ a\ } RH + \dot{C}H_2{-}C(CH_3)_3;$$

$$\qquad (5.17)$$

$$\dot{C}H_2{-}C(CH_3)_3 \xrightarrow{\ b\ } \dot{C}H_3 + CH_2{=}C(CH_3)_2.$$

If R is CH_3, we have a chain with two carriers, the neopentyl radical ($\dot{N}p$) and ($\dot{C}H_3$).

The steady-state concentration ratio for long chains is given by

$$\frac{(\dot{N}p)}{(\dot{C}H_3)} = \frac{k_a(NpH)}{k_b}, \qquad (5.18)$$

whereas the steady-state rate is

$$\frac{-d(NpH)}{dt} = R_a = R_b = k_a(CH_3)(NpH) = k_b(\dot{N}p). \qquad (5.19)$$

5.4 PRODUCTS OF COMPETING CHAINS

For more complex molecules, more than one chain cycle is possible, as we noted for C_2H_5Cl. If the chain lengths of any one of these is long, the relative product distribution is determined uniquely by the propagation steps. Thus in the pyrolysis of n-butane we find by our usual method the following chain cycles:

$$R \cdot + CH_3CH_2CH_2CH_3 \begin{array}{c} \xrightarrow{a} \dot{C}H_2CH_2CH_2CH_3 + RH; \\ \\ \xrightarrow{b} CH_3\dot{C}HCH_2CH_3 + RH \end{array}$$

$$\dot{C}H_2CH_2CH_2CH_3 \xrightarrow{c} C_2H_4 + \dot{C}_2H_5;$$ (5.20)

$$CH_3\dot{C}HCH_2CH_3 \xrightarrow{d} C_3H_6 + \dot{C}H_3.$$

Hence the two cycles produce $CH_4 + C_3H_6$ if abstraction occurs at the secondary H atom and $C_2H_6 + C_2H_4$ if it occurs at the primary H atom. However, either carrier $\dot{C}H_3$ or $\dot{C}_2H_5$ can abstract at either position, so steps a' and b' must be added to steps a and b to represent this.

The ratio of C_2H_4 to C_3H_6 is given by

$$\frac{R(C_2H_4)}{R(C_3H_6)} = \frac{k_c(n\text{-Bu}^{\cdot})}{k_d(sec\text{-Bu}^{\cdot})} = \frac{R_a + R_a'}{R_b + R_b'}$$ (5.21)

where primes refer to $\dot{C}_2H_5$ abstraction, and unprimed R to $\dot{C}H_3$.

The steady-state equations are

$$R_c = R_a + R_a' = R_a' + R_b';$$
$$R_d = R_b + R_b' = R_a + R_b;$$ (5.22)

so that

$$\frac{(CH_3)}{(C_2H_5)} = \frac{k_b'}{k_a}$$

and

$$\frac{R(C_2H_4)}{R(C_3H_6)} = \frac{k_b' + k_a'}{k_b + k_a}.$$ (5.23)

We have, of course, neglected any secondary reactions, so that the simple expressions above apply only to the initial stages of the reaction.

The overall logarithmic rate of decomposition of the n-butane is

$$\frac{R(C_4H_{10})}{(C_4H_{10})} \equiv \frac{-d(C_4H_{10})}{(C_4H_{10})\,dt} = \frac{(R_a + R_a' + R_b + R_b')}{(C_4H_{10})}$$

$$= [(k_a + k_b)(\dot{C}H_3) + (k_a' + k_b')(\dot{C}_2H_5)]$$

$$= \left[(k_a + k_b)\frac{k_b'}{k_a} + (k_a' + k_b') \right](\dot{C}_2H_5).$$ (5.24)

As a final example, let us consider the chain pyrolysis of n-PrCl. The chain test leads to

$$R + CH_3CH_2CH_2Cl \rightarrow \begin{cases} \xrightarrow{1} \dot{C}H_2CH_2CH_2Cl + RH \\ \xrightarrow{2} CH_3\dot{C}HCH_2Cl + RH \\ \xrightarrow{3} CH_3CH_2\dot{C}HCl + RH \\ \xrightarrow{4} CH_3CH_2C\dot{H}_2 + RCl \end{cases} \qquad (5.25)$$

followed by

$$\dot{C}H_2CH_2CH_2Cl \xrightarrow{5} C_2H_4 + \dot{C}H_2Cl$$
$$CH_3\dot{C}HCH_2Cl \xrightarrow{6} C_3H_6 + Cl$$
$$CH_3CH_2\dot{C}HCl \xrightarrow{7} CH_2{=}CHCl + \dot{C}H_3$$
$$CH_3CH_2\dot{C}H_2 \xrightarrow{8} C_2H_4 + \dot{C}H_3$$

$$(5.26)$$

We have neglected back reactions, abstraction reactions by large radicals, secondary reactions with products, and the more endothermic competing reactions, which eliminate H atoms ($\dot{C}H_2CH_2CH_2Cl \rightarrow CH_2{=}CHCH_2Cl + H$). We note three competing paths, leading to $C_3H_6 + HCl$ ($R = Cl$); $C_2H_4 + CH_3Cl$ ($R = \dot{C}H_2Cl$ or CH_3); and $CH_2{=}CHCl + CH_4$ ($R = CH_3$).

The steady-state relations yield

$$\begin{aligned} R_1' + R_2' + R_3' + R_4' &= R_1 + R_1' + R_1'' = R_5; \\ R_1 + R_2 + R_3 &= R_2 + R_2' + R_2'' = R_6; \\ R_3 + R_3' + R_3'' &= R_7; \\ R_4' + R_4'' &= R_8; \\ R_1'' + R_2'' + R_3'' + R_4'' &= R_7 + R_8 \end{aligned} \qquad (5.27)$$

where unprimed rates are for Cl, primed for $\dot{C}H_2Cl$, and R'' for $\dot{C}H_3$.

Solution of the relations given above can be accomplished, and they lead to unique solutions for all radical ratios and rates. Inspection of these equations will show that the products are given by

$$\frac{R(C_2H_4)}{R(C_3H_6)} = \frac{R_1 + R_1' + R_1'' + R_4' + R_4''}{R_1 + R_2 + R_3};$$
$$\frac{R(CH_2CHCl)}{R(C_3H_6)} = \frac{R_3 + R_3' + R_3''}{R_1 + R_2 + R_3}.$$

$$(5.28)$$

More detailed analysis will show that not all of the paths described above are followed appreciably because the rate of splitting of C—C bonds is sufficiently slow, compared to the reverse reaction of the radicals with HCl, that even a few percent of reaction is sufficient to suppress all paths except $C_3H_6 + HCl$.

5.5 SOME SIMPLE CHAIN CATEGORIES

The nonbranching chain reactions fall into three basic categories, corresponding to metathesis, addition, or pyrolysis reactions. The latter gives overall reactions from a single reactant, whereas the former involve the exchange or addition of species between two reactants.

Some examples of pyrolysis reactions that have been studied are:

1. $C_2H_6 \rightarrow C_2H_4 + H_2$
2. Neopentane $\rightarrow$ i-butene $+ CH_4$
3. $(CH_3)_2O \rightarrow CH_2O + CH_4$
    $ \llcorner \rightarrow CO + H_2$
4. $C_2H_5Br \rightarrow C_2H_4 + HBr$
5. $CH_3COCH_3 \rightarrow CH_2CO + CH_4$
    $ \llcorner \rightarrow \tfrac{1}{2}[CH_2\!=\!C\!=\!CH_2 + CO_2]$

The reactions listed above all have in common the fact that they are endothermic and exentropic, so that, in principle, they run to a measurable equilibrium state and should be characterized by a "ceiling" temperature, below which the equilibrium constant becomes rapidly unfavorable for reaction. The overall driving force is the entropy increase for the reaction.

The reverse reactions should be observable as addition reactions below the ceiling temperature, proceeding by the same chain mechanism. In practice, however, this is rarely the case. Usually the addition rates are too slow to observe below the ceiling temperature, whereas at slightly higher temperatures and concentrations, secondary reactions complicate the system.

During the addition of H_2 to C_2H_4, for example, polymerization of C_2H_4 competes with C_2H_6 formation.[5] The addition of CH_4 to CH_2O to form $(CH_3)_2O$, or of H_2 to CO to form CH_2O, both of which are nearly thermoneutral, are too far above their ceiling temperatures even at

---

[5] This probably would not be the case for sufficiently high $(H_2)/(C_2H_4)$ ratios (i.e., >20), but the reaction has not been studied under these conditions).

1000-atm pressure to ever yield significant products at any temperature where the reverse homogeneous reactions are measurably rapid.

Some examples of metathesis chains that have been studied are

$$H_2 + Br_2 \rightleftarrows 2HBr$$

$$H_2 + D_2 \rightleftarrows 2HD$$

$$H_2 + CH_2{=}CHCH_3 \rightarrow C_2H_4 + CH_4$$

$$D_2 + C_2H_2 \rightleftarrows C_2HD + HD$$

$$Cl_2 + CH_4 \rightarrow CH_3Cl + HCl$$

$$Cl_2 + C_6H_5Br \rightarrow C_6H_5Cl + BrCl$$

These reactions are generally exothermic, sometimes thermoneutral, and involve very little entropy change. Thus they can frequently be observed to come to equilibrium, and the reverse reaction can be measured independently.

The addition reactions are related to the pyrolysis reactions and are generally exothermic and endentropic. They have ceiling temperatures T_c, above which they may be expected to show decreasing overall rates. In addition, the reactions can show abnormally small activation energies due to a reversible, exothermic step. Telomerization is a special case of an addition reaction, and the simplest case of a polymerization. Thus CCl_4, CF_3CN, or CCl_3Br can add to a double bond, as follows:

$$CCl_4 + \quad \underset{/}{\overset{\backslash}{C}}{=}\underset{\backslash}{\overset{/}{C}} \quad \rightarrow CCl_3{-}\underset{/}{\overset{\backslash}{C}}{-}\underset{\backslash}{\overset{/}{C}}{-}Cl$$

$$CF_3CN + \quad \underset{/}{\overset{\backslash}{C}}{=}\underset{\backslash}{\overset{/}{C}} \quad \rightarrow CF_3{-}\underset{/}{\overset{\backslash}{C}}{-}\underset{\backslash}{\overset{/}{C}}{-}CN$$

$$CCl_3Br + \quad \underset{/}{\overset{\backslash}{C}}{=}\underset{\backslash}{\overset{/}{C}} \quad \rightarrow CCl_3{-}\underset{/}{\overset{\backslash}{C}}{-}\underset{\backslash}{\overset{/}{C}}{-}Br$$

or generally,

$$A{-}X + \quad \underset{/}{\overset{\backslash}{C}}{=}\underset{\backslash}{\overset{/}{C}} \quad \rightarrow A{-}\underset{/}{\overset{\backslash}{C}}{-}\underset{\backslash}{\overset{/}{C}}{-}X$$

If the olefin concentration is sufficiently high, polymeric species of various chain lengths can be formed, such as $CCl_3{-}({>}C{-}C{<})_n{-}Cl$ with $n = 1$, 2, and so on. These complexities generally restrict the study of addition reactions to very low olefin concentrations if quantitative data are sought.

Common to each of the chain categories outlined above is a characteristic propagation step. In the pyrolytic chains it is the unimolecular fission of a radical. Thus in the C_2H_6 pyrolysis, the fission step is

$$\dot{C}_2H_5 \rightleftharpoons C_2H_4 + H.$$

These steps are generally the principle source of the endothermicity of the overall reaction and have relatively high activation energies.[6] Because of this endothermicity, the back reactions are usually fast and appreciable, even when small amounts of product are formed.

As we note in Chapter 3, very little direct experimental data are available on the rate constants for such propagation steps, and the estimated rate constants, by our usual methods, are probably as reliable as some of the quoted data. The largest uncertainty associated with these fissions is their pressure dependence. They generally tend to be in an intermediate pressure regime, so that the high-pressure Arrhenius parameters are not applicable.

In a similar fashion the reverse addition step is characteristic of addition chains, and there is the same lack of knowledge regarding the pressure dependence of the addition rate. As an example, we can consider the addition of Br_2 or HBr to olefins:

$$Br + \quad \overset{\diagup}{\underset{\diagdown}{C}} = \overset{\diagup}{\underset{\diagdown}{C}} \quad \underset{-1}{\overset{1}{\rightleftarrows}} \quad Br - \overset{\diagup}{\underset{\diagdown}{C}} - \dot{\overset{\diagup}{\underset{\diagdown}{C}}} \quad ,$$

$$Br - \overset{\diagup}{\underset{\diagdown}{C}} - \dot{\overset{\diagup}{\underset{\diagdown}{C}}} + Br_2 \overset{2}{\longrightarrow} Br - \overset{\diagup}{\underset{\diagdown}{C}} - \overset{\diagup}{\underset{\diagdown}{C}} - Br + \dot{Br}$$

$$+ HBr \overset{2'}{\longrightarrow} Br - \overset{\diagup}{\underset{\diagdown}{C}} - \overset{\diagup}{\underset{\diagdown}{C}} - H + \dot{Br}.$$

Here the second step (2) is irreversible, being about 23 kcal exothermic (2′ is about 10 kcal/mole exothermic). The steady state rate of the disappearance of Br_2 is given by

$$\frac{-d(Br_2)}{dt} = \frac{k_1 k_2 (Br_2)(\text{olefin})(Br)}{k_{-1} + k_2(Br_2)}. \tag{5.29}$$

[6] For the decomposition $\Delta H° = 39$ kcal, and because the activation energy for the reverse addition reaction is about 3.5 kcal, $E = 42.5$ kcal. At 900°K, which is a typical pyrolysis temperature for C_2H_6, the lifetime of a $\dot{C}_2H_5$ radical would be about 10^{-4} s, if the reaction were at its high pressure limit. However, $E/RT \sim 23$, a low value, corresponding to relatively few vibrational quanta, and this reaction is almost certain to be in an intermediate pressure region, even at 1 atm of total pressure.

This has two limiting cases:

$$\frac{-d(Br_2)}{dt} \xrightarrow{k_{-1} \ll k_2(Br_2)} k_1(Br)(olefin) \tag{5.30}$$

$$\xrightarrow{k_{-1} \gg k_2(Br_2)} K_1 k_2(Br_2)(olefin)(Br) \tag{5.31}$$

where $K_1 = k_1/k_{-1}$. (Note that both 1 and -1 may be in pressure-dependent regions.)

In the second, or equilibrium case, the activation energy for the propagation scheme is $\Delta E_1 + E_2$. Because $\Delta E_1 \sim -10$ kcal (see bond energies), whereas $E_2 \sim 1.5$ kcal, we see that the propagation scheme can have an overall negative activation energy of about -8.5 kcal/mole. In point of fact the photochemical bromination of C_2H_4, in which the $(Br)_{ss}$ concentration has no temperature dependence, does decrease markedly with increasing temperature.[7]

For the metathesis chains that can reach equilibrium, and that are nearly thermoneutral, all of the chain steps are, in principle, metathetical, and reversible. The H_2–D_2 exchange is a simple example:

$$H + D_2 \underset{-1}{\overset{1}{\rightleftharpoons}} HD + D;$$

$$D + H_2 \underset{-2}{\overset{2}{\rightleftharpoons}} HD + H.$$

The rate of exchange at steady state is

$$\frac{d(HD)}{dt} = (R_1 - R_{-1}) + (R_2 - R_{-2})$$

$$= 2(R_1 - R_{-1})$$

$$= 2k_1(H)(D_2) \left\{ 1 - K_{-1} \frac{(HD)}{(D_2)} \frac{(D)}{(H)} \right\}$$

where

$$(D)/(H) = \frac{k_1(D_2) + k_{-2}(HD)}{k_2(H_2) + k_{-1}(HD)},$$

so that

$$\frac{d(HD)}{dt} = \frac{2k_1(H)(D_2)}{1 + \left(\dfrac{k_{-1}}{k_2}\right)\left(\dfrac{HD}{H_2}\right)} \left\{ 1 - \frac{(HD)^2}{K_{eq}(H_2)(D_2)} \right\} \tag{5.32}$$

where K_{eq} is the equilibrium constant for the net reaction, $H_2 + D_2 \rightleftharpoons 2HD$. Because k_{-1} and k_2 are comparable to each other in absolute

[7] G. B. Kistiakowsky and J. C. Sternberg, *J. Chem. Phys.*, **21**, 2218 (1953).

magnitude, we note that the effect of (HD) inhibition appears long before the overall reverse reaction is significant.[8]

5.6   INITIATION STEPS

For long chain reactions, we have seen that the rates can be derived from the propagation steps and expressed in terms of the concentration of any one of the radical carriers. The absolute concentrations of these carriers is fixed by the rates at which initiation and termination of radicals occur. If we restrict ourselves for the moment to homogeneous systems, stable molecules can only give rise to mono radicals in pairs, and conversely, mono radicals can only disappear two at a time.

All possible reactions must be considered as radical sources, the rule being that the most rapid will be the one observed. The same is true for the termination reactions. Our criteria for selecting initiation steps is based on the lowness of the activation energy and the order of the reaction. Because unimolecular fission reactions have generally high A factors, $10^{16\pm1}$ s^{-1}, they will be favored over bimolecular processes, which have A factors of about $10^{9\pm1}$ l./mole-s for radical disproportionations.

Consider the hypothetical example of radical production by competing unimolecular and bimolecular paths from a reactant A:

$$A \xrightarrow{\ 1\ } 2 \text{ radicals}$$

$$2A \xrightarrow{\ 2\ } 2 \text{ radicals}$$

At a pressure of 100 torr ($\sim10^{-2.4}$ M), the ratio of rates is

$$\frac{R_1}{R_2} = \frac{k_1(A)}{k_2(A)^2} \approx \frac{10^{16\pm1-E_1/\theta}}{10^{9\pm1-E_2/\theta}(A)}$$

$$\approx 10^{9.4\pm2-(E_1-E_2)/\theta}.$$

The rates R_1 and R_2 are competitive when $E_1 - E_2 = (9.4 \pm 2)\theta$. At 2000°K, a typical temperature for shock tube studies, $\theta = 9.2$ kcal/mole, which corresponds to about 88 ± 18 kcal. This is so high that we can reasonably conclude that bimolecular initiations will almost never compete with unimolecular at 2000°K. At 1000°K, a typical pyrolysis temperature for hydrocarbons, $\theta = 4.6$ kcal/mole and $E_1 - E_2 = 44 \pm 9$ kcal. This is still sufficiently large that unimolecular steps are still by far the

---

[8] Thus when $(HD)/(D_2) = 0.20$ ($\sim9\%$ reaction), the denominator in (5.32) is 1.2, whereas the bracketed term is negligibly different from unity [0.997 for $(H_2) = (D_2)$].

favored ones at 1000°K. That is, it will be exceptional for a bimolecular initiation step to have an activation energy lower than that for a unimolecular step by 44 kcal.

Such exceptions may occur in the pyrolysis of the olefins. Thus for C_3H_6 the two competing paths will be

$$C_3H_6 \xrightarrow{\ 1\ } \dot{C}_3H_5 + H, \qquad \Delta H_1^\circ = 87 \text{ kcal/mole};$$

$$2C_3H_6 \xrightarrow{\ 2\ } \dot{C}_3H_5 + \dot{C}_3H_7, \qquad \Delta H_2^\circ = 48 \text{ kcal/mole}.$$

At 900°K, $\Delta E_1^\circ = \Delta H_1^\circ - RT \simeq 85$ kcal/mole, and the difference in activation energies is expected to be ~37 kcal/mole, which would account for a factor of $10^{9.0}$ in the ratio of rates. This is close enough to the expected differences in A factors to make it likely that both paths will play a role at various pressures and temperatures in this range. Unfortunately, the pyrolysis itself is sufficiently complex to obscure the details of the precise initiation.

In a number of instances, positive, quantitative evidence has been adduced for bimolecular initiation processes. Stable radicals, such as NO and NO_2, are capable of abstracting atoms to form molecules and initiate chains. Thus it is found that NO_2 is a very powerful chain initiator for hydrocarbon reactions. The reaction is

$$NO_2 + RH \rightarrow HONO + \dot{R}.$$

Because the H—ONO bond strength is about 78 kcal/mole, and $DH°(R\text{—}H)$ is in the range 85–100 kcal/mole, such initiation reactions will have extremely low activation energies, about 2 to 5 kcal in excess of the endothermicity of the reactions, that is from 10 to 25 kcal/mole. In similar fashion, $DH°(H\text{—}NO) \sim 49$ kcal/mole, so that NO is expected to act as an initiator of chain reactions at sufficiently high concentrations. This is in fact well documented.[9]

It has been often proposed that termolecular processes of initiation are possible. An example would be olefins reacting with diatomic molecules:

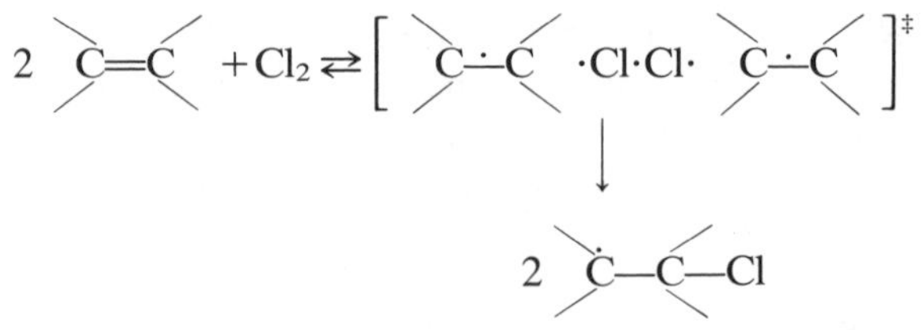

[9] P. Goldfinger et al., *Trans. Faraday Soc.*, **57**, 2197, 2210, 2220 (1961); L. V. Karmilova, N. S. Enikolopyan, and A. B. Nalbandjan, *Zhur. Fiz. Khim.*, **30**, 748 (1956); B. W. Wojciechowski and K. J. Laidler, *Can. J. Chem.*, **38**, 1027 (1960).

Such a reaction would be endothermic by only about 20 kcal/mole. However, because the intermediate is not a stable species, there would be an appreciable intrinsic activation energy in addition to this endothermicity, which would make the overall step prohibitive.[10] Even more important, however, is the unfavorable A factor. The entropy change for three molecules forming a single transition state can be estimated at $\sim -60 \pm 10$ gibbs/mole. At 1 mole/liter standard state, this would become -47 gibbs/mole, yielding an A factor of about $10^{4\pm2}$ liter^2/mole^2-sec with an accompanying increase in activation energy of $2RT$.[11]

Another example of a bimolecular initiation is afforded by the HBr catalyzed isomerization of olefins. Butene-1 undergoes a chain catalyzed isomerization[12] initiated by

$$CH_3CH_2—CH\!=\!CH_2 + HBr \rightleftarrows CH_3CH_2—\dot{C}H—CH_3 + Br$$

for which $\Delta H° = +48$ kcal/mole. The most rapid competing unimolecular reaction would be the scission of the C—C bond to form CH_3 + allyl at a cost of 74 kcal/mole. At 550°K, $\theta = 2.5$ kcal/mole, so that the difference of 26 kcal/mole amounts to a difference of $10^{10.4}$ in relative rates, making the bimolecular act the favored one. The rest of the chain is

$$Br + CH_3—CH_2—CH\!=\!CH_2 \rightleftarrows CH_3—\dot{C}H—CH\!=\!CH_2 + HBr;$$
$$HBr + CH_3\dot{C}H—CH\!=\!CH_2 \rightleftarrows CH_3—CH\!=\!CH—CH_3 + Br.$$

Each step in the chain is very fast with an activation energy in the range of 3 to 6 kcal/mole.

The competing reaction of HBr addition is considerably slower because of the unfavorable precursor equilibrium step[13]:

$$Br + CH_3—CH_2—CH\!=\!CH_2 \rightleftarrows CH_3—CH_2—\dot{C}H—CH_2Br.$$

A very unexpected initiation step turns out to be important in many reactions. This is the case in which initiation involves products, so that the reaction becomes autocatalytic. An excellent example occurs in the

[10] What is implied here is that the exothermic back reaction of two chloroalkyl radicals to give two olefins + Cl_2 is expected to have an appreciable activation energy. This is not the case for the analogous observed reaction, $2ROO\cdot \rightleftarrows [ROOOOR] \rightarrow 2RO\cdot + O_2$, which proceeds by way of a stable tetroxide.

[11] Such a case does occur in the decomposition reaction:

$$2NOCl \rightarrow 2NO + Cl_2$$

[12] A. Maccoll and R. A. Ross, *J. Am. Chem. Soc.*, **87**, 4997 (1965). The authors have offered a molecular mechanism for the reaction, but the radical chain gives an excellent quantitative fit.

[13] $K \approx 10^{-6.3+10/\theta}$ atm$^{-1} \sim 10^{-3.3}$ at 0.1 atm olefin (550°K), from our tables.

metathesis reactions of organic iodides with HI:

$$RI + HI \rightleftharpoons RH + I_2.$$

This reaction can be recognized as the reverse of the usual halogenation reaction, $RH + X_2 \rightarrow RX + HX$. For iodine the equilibrium lies well over in favor of I_2, in contrast to the reverse case for F_2, Cl_2, and Br_2. In the HI reactions, very rapid initiation is provided by I_2[14]:

$$I_2 \rightleftharpoons 2I \text{ (rapid, equilibrium } K_{I_2}\text{)};$$

$$\left. \begin{array}{l} I + RI \overset{1}{\rightleftharpoons} R\cdot + I_2 \\[2ex] R\cdot + HI \overset{2}{\rightleftharpoons} RH + I \end{array} \right\}$$

The overall rate of reaction is given by

$$\frac{d(I_2)}{dt} = -\frac{d(RI)}{dt} = \frac{k_2 k_1 (I)(RI)(HI)}{k_{-1}(I_2) + k_2(HI)}$$

$$= \frac{k_1 K_{I_2}^{1/2}(I_2)^{1/2}(RI)}{1 + k_{-1}(I_2)/k_2(HI)} \tag{5.33}$$

$$\xrightarrow{k_{-1}(I_2) < k_2(HI)} k_1 K_{I_2}^{1/2}(I_2)^{1/2}(RI). \tag{5.34}$$

The last equation predicts a rate approaching zero in the early stages of the reaction when $I_2 \rightarrow 0$. This in principle would imply an induction period early in the reaction, during which some less favorable initiation, such as $RI \rightarrow R + I$, would occur. These latter reactions are so much slower than I_2 initiation that even at 0.01% reaction, with $(I_2)/(RI) = 10^{-4}$, the I_2 initiation is competing favorably. In practice, such early regimes are never seen because spurious, though slow, wall reactions make it almost impossible to avoid small initial traces of I_2 in these systems.

An equally interesting example is provided by the hydrogenation of C_2H_4 at high temperatures (800 to 1200°K). In the early stages of the reaction, initiation is provided by the bimolecular process:

$$C_2H_4 + H_2 \overset{1}{\rightleftharpoons} H + \dot{C}_2H_5.$$

However depending on the precise conditions, the competing unimolecular initiation by the product can become of comparable importance:

$$C_2H_6 \overset{1'}{\rightleftharpoons} 2\dot{C}H_3.$$

[14] It should be pointed out that in these systems the chain length $\lambda \ll 1$. Under the usual experimental conditions, it is in the range of from 10^{-2} to 10^{-3}. This does not quite fit the usual concept of chain reaction, but there is no reason to exclude it.

The relative rates of the two steps above are given by

$$\frac{R_1'}{R_1} = \frac{k_1'(C_2H_6)}{k_1(C_2H_4)(H_2)} \approx \frac{10^{16.5-87/\theta}[(C_2H_6)/(C_2H_4)]}{10^{10.1-64/\theta}(H_2)} \tag{5.35}$$

and at $(H_2) = 10^{-2}$ M and 750°K ($\theta = 3.4$ kcal/mole)

$$\frac{R_1'}{R_1} \approx 10^{8.0-23/\theta}\frac{(C_2H_6)}{(C_2H_4)} \tag{5.36}$$

$$= 10^{1.7}\frac{(C_2H_6)}{(C_2H_4)}$$

We see that $R_1 = R_1'$ at $(C_2H_6)/(C_2H_4) = 0.02$, whereas at higher temperatures, such as 1000°K (or lower H_2 concentrations), this is true at $(C_2H_6)/(C_2H_4) = 3 \times 10^{-4}$.

In the calculation above we estimate the rate of $H_2 + C_2H_4$ by estimating the rate constant for the back reaction (-1), (the disproportionation of $H + \dot{C}_2H_5$) as 10^{10} liters/mole-sec with no activation energy.

5.7 TERMINATION REACTIONS

We have seen in Chapter 4 that recombination reactions of large radicals do not show much variation with structure and mostly fall in the range $10^{9.5\pm1.0}$ liters/mole-s for either recombination or disproportionation.

Under these conditions, the termination reaction in a complex chain is predominantly between the radicals present in largest concentrations. Because this is determined by the propagation steps, it is possible to make a fairly direct choice of termination reaction, or reactions, from estimates of the chain steps. Difficulties intrude into this picture as soon as we are involved with small radicals or atoms. For these species, recombination may frequently be pressure dependent. Thus, in the case just considered, present experimental evidence indicates that at 1000°K, $2CH_3 \rightarrow C_2H_6$ is pressure dependent below 400 torr.[15] This means that we must make some estimate of the correction to be applied for this falloff in rate. A relatively rapid, though crude, method for doing this is to decompose the recombination into the steps

$$R + R \underset{-1}{\overset{1}{\rightleftharpoons}} R_2^*;$$

$$R_2^* + M \overset{k_z}{\longrightarrow} R_2 + M.$$

[15] M. C. Lin and M. H. Back, *Can. J. Chem.*, **44**, 2357 (1967).

The steady state rate of recombination is then given by

$$\frac{-d(R)}{dt} = \frac{2d(R_2)}{dt} = \frac{2k_1 k_z (R)^2 (M)}{k_z(M) + k_{-1}} \tag{5.35}$$

As a first approximation, we take k_z equal to 0.5 times collision frequencies, and k_{-1} would be estimated by the methods discussed in Section 4.20. The rate constant k_1 could be taken as the normal pressure-independent rate constant. Thus in the case of CH_3 recombination at 1000°K, and 100 torr, we can choose $k_1 = 10^{9.8}$ liter/mole-s, $k_z = 10^{11.0}$ liter/mole-s, and $(M) = 10^{-2.8}$ M; k_{-1} is given by:

$$k_{-1} = 10^{16.5} \left(\frac{E - E^*}{E} \right)^{s-1} s^{-1}$$

with $s = [C_{p1000}(C_2H_6) - 9]/R = 10.2$,[16]

$$E - E^* = \int_0^{1000} (C_p - 9)\, dT \approx 9.4 \text{ kcal/mole},$$

Now CH_3 recombination has a loose transition state in which the two large moments of inertia change by about a factor of 2.5 each (Sect. 3.7) at 1000°K. In consequence of the 87 kcal of energy liberated by bond formation, a significant amount remains fixed in the two external rotational degrees of freedom. If the collision of $2CH_3$ has RT of rotational energy at inception, then in the nascent $C_2H_6^*$, the moment of inertia change will raise this to $2.5^2 RT = 12.5$ kcal at 1000°K. Subtracting RT from this, which is the rotational energy of the collision, we arrive at 10.5 kcal coming from bond formation. Hence the residue of the bond energy, $E^* = 87 - 10.5 = 76.5$ kcal is the amount in the other vibrational degrees of freedom. Thus our equation becomes:

$$k_{-1} = 10^{16.5} \left(\frac{9.4}{85.9} \right)^{9.2} = 10^{7.7} \text{ s}^{-1}$$

The ratio of the pressure-dependent rate to pressure-independent rate is given by

$$\frac{k_z(M)}{k_z(M) + k_{-1}} = \frac{10^{8.2}}{10^{8.2} + 10^{7.7}} \approx 0.75$$

This is a small correction in view of the uncertainty of the various estimates. However at 10 torr it is down to $\frac{1}{4}$ and significant. Of particular

---

[16] We are subtracting internal rotation and external rotation and translation from energy-storing modes.

importance is the apparent negative activation energy in the recombination-rate constant.

The case of atoms is of special interest because the atom–atom, or atom–diatomic molecule, recombination is always in the third-order region with a recombination rate constant of about $10^{9.5\pm0.5}$ liter^2/mole^2-s. At $(M) = 10^{-2.8}$ this is sufficiently slow that these combinations are only important when the atom concentrations exceed the concentrations of complex radicals by at least 10^3. This can happen in reaction systems containing I_2 and I, but in very few others.

In the $C_2H_6 \rightleftarrows C_2H_4 + H_2$ chain, the radical species are C_2H_5, CH_3, and H. We can neglect $H + H + M$, as an important recombination under all conditions. In the system $CHCl_3 + Br_2 \rightleftarrows CBrCl_3 + HBr$, in contrast, we can neglect $\dot{C}Cl_3$ radicals in termination because their concentration is indeed less than 10^3(Br).

Once we have passed the small radicals, such as CH_3, termination need not be restricted by third body effects because disproportionation is always in competition with combination and is not pressure dependent. Thus two C_2H_5 radicals will disproportionate at about one-seventh their rate of recombination, whereas branched or polar radicals will prefer disproportionation. Two $CH_3\dot{O}$ radicals will disproportionate with a rate constant estimated at $10^{10.5}$, whereas the recombination rate constant is about $10^{9.0}$ liter/mole sec. Similarly, two t-butyl radicals disproportionate about 3.5 times faster than they recombine.

5.8 TRANSFER REACTIONS AND SENSITIZATION

Because the dominant initiation reaction is always the most rapid of the possible initiation reactions in the system, it does not necessarily give rise to the active chain carriers. The metathetical reactions relating active chain carriers to other nonpropagating radicals are referred to as transfer reactions, and they constitute an important aspect of chain systems.

The C_2H_6 pyrolysis provides a good example of a transfer reaction. The initiation produces $2CH_3$ radicals, whereas the chain carriers are H and C_2H_5. The scheme is as follows.

$$\text{Initiation:} \quad C_2H_6 \rightarrow 2\dot{C}H_3;$$

$$\text{Transfer:} \quad \dot{C}H_3 + C_2H_6 \rightleftarrows CH_4 + \dot{C}_2H_5;$$

$$\text{Chain:} \quad \left\{ \begin{array}{l} \dot{C}_2H_5 \rightleftarrows C_2H_4 + H \\ H + C_2H_6 \rightleftarrows \dot{C}_2H_5 + H_2. \end{array} \right.$$

In many cases the transfer is so rapid and irreversible that the initial radicals never appear in the scheme. The thermal decomposition of F_2O_2

provides another example. The products at -80 to $0°C$ are almost solely $F_2 + O_2$, whereas the initiation could be the cleavage of the O—O bond.

Initiation: $F_2O_2 + M \rightarrow 2FO + M;$

Transfer: $F\dot{O} + F_2O_2 \rightarrow F_2O + FO\dot{O};$

Chain: $\begin{cases} M + FO\dot{O} \rightarrow \dot{F} + O_2 + M \\ F + F_2O_2 \rightarrow F_2 + FO\dot{O} \end{cases}$

Here the chain lengths are extremely long, so that F_2O is a very minor product of the reaction. It is not clear in this system, however, what the termination is.

Many systems are induced to undergo chain reactions by initiating them with radicals introduced by either photochemical means or from thermally unstable species.

In the range 40–80°C diacyl peroxides (e.g., CH_3COO—$OCOCH_3$) are excellent thermal sensitizers. They produce acyl radicals that cleave to form alkyl or aryl radicals:

$$(RCOO)_2 \rightarrow 2RC\dot{O}_2 \rightarrow 2\dot{R} + 2CO_2.$$

These radicals can then, by transfer, initiate a chain.

In the range 100–150°C dialkyl peroxides can provide the same initiation. They form alkoxy radicals, which can further cleave into alkyl radicals plus ketones or aldehydes. Ditertiary butyl peroxide is a well-known example.

$$(t\text{-}BuO)_2 \rightarrow 2t\text{-}Bu\dot{O} \rightarrow 2CH_3 + 2 \text{ acetone}.$$

In the range 250–350°C azo compounds form radicals:

$$R—N{=}N—R \rightarrow 2\dot{R} + N_2.$$

Above 300°C a variety of substances ranging from metal alkyls, such as HgR_2, to ethylene oxide can act as radical sensitizers. The reactions are

$$HgR_2 \rightarrow Hg + 2\dot{R}$$

$$\overline{CH_2—CH_2—O} \rightarrow (CH_3CHO)^* \rightarrow \dot{C}H_3 + \dot{C}HO$$

In the latter instance the intermediate acetaldehyde is vibrationally excited and a competing side reaction is its quenching to ground state, stable CH_3CHO.

5.9   STEADY-STATE HYPOTHESIS

Most of the transfer and propagation steps which occur in ordinary nonbranching chain reactions have rates proportional to the first-order of a radical concentration. Hence the precise rate law for each radical, R_i is given by an equation of the form:

$$\frac{d(R_i)}{dt} = \sum_j k_{ij}(R_j)(\text{molec.}) - \left[\sum_j k_j(\text{molec.})\right](R_i)$$

$$+ \text{terms for initiation and termination} \qquad (5.36)$$

The terms involving initiation and termination may be zero and second order in radical concentrations, respectively. With modern computers, it is possible when the rate constants and initial conditions are known to integrate these coupled equations numerically.[17] Under certain conditions, however, it is possible to make an assumption, known as the quasisteady-state hypothesis, or steady state for short, which permits a much simpler solution. The assumption is that all radical and intermediate concentrations are negligible compared to those of reactants and products, and hence, $d(R_i)/dt \approx 0$. This converts the above equations from differential to algebraic equations that can then be solved explicitly for the steady-state radical concentrations. This, then, permits an explicit solution for the rate law.

The limits of validity of such an assumption have been examined in detail[18,19] and shown to require only that the total radical concentration be negligible compared to reactant and product concentrations. This will begin to fail for iodination reactions near 700°K and for bromination reactions near 900°K. It is a condition which must be examined in every case to determine its validity. It becomes particularly suspect in the very early stages of a reaction when product concentrations are still small and may be comparable to those of the intermediates.

When it is applicable, one particular equation can always be separated out of the total set of equations for the radical concentrations. It is equivalent to stating that the sum of rates of all initiation processes for radicals must equal the sum of all rates of radical termination; this considerably facilitates the algebra.

In the $H_2 + Br_2$ chain reaction considered earlier (p. 213), this initiation–termination condition leads to:

$$Br_2 + M \underset{-1}{\overset{1}{\rightleftharpoons}} 2Br + M$$

[17] C. W. Gear, *Numerical Initial Value Problems in Ordinary Differential Equations*, Prentice-Hall, Englewood Cliffs, N. J., 1971.

[18] S. W. Benson, *J. Chem. Phys.*, **20**, 1605 (1952); S. W. Benson, *Foundations of Chemical Kinetics*, McGraw-Hill, New York, 1960.

[19] D. L. Allara and D. Edelson, *Int. J. Chem. Kinet.*, **7**, 479 (1975).

as the only equations involved in these processes and hence:

$$(Br)_{ss} = K_1^{1/2}(Br_2)^{1/2}$$

where K_1 is the equilibrium constant for Br atom formation. Note that it does not matter whether this process is at its high pressure or low-pressure limit or is even heterogeneously catalyzed. It only matters that $(Br)_{ss}$ is small at all times under consideration compared to reactant and product concentrations.

When this initiation–termination equation is combined with the other algebraic equations for the m radical stationary state reactions, it can further be shown that this set now reduces to $(m-1)$ algebraic equations, linear and homogeneous in radical concentrations which can be solved explicitly for the ratio of concentrations of the radical intermediates.

5.10 PYROLYSIS OF CH₃CHO

One of the most thoroughly studied pyrolysis systems that undergoes a simple chain decomposition, is that of CH_3CHO. The reaction is easily studied from 450 to 550°C. It forms $CH_4 + CO$ as major products and minor amounts of H_2, CH_3COCH_3, and C_2H_4.[20] The main features of the chain are very much what we would expect from our preceding discussion.

Initiation: $CH_3{-}CHO \xrightarrow{i} \dot{C}H_3 + \dot{C}HO;$

Chain:
$$\dot{C}H_3 + CH_3CHO \xrightarrow{1} CH_4 + CH_3\dot{C}O$$
$$CH_3\dot{C}O + M \xrightarrow{2} \dot{C}H_3 + CO + M;$$

Termination: $2CH_3 \xrightarrow{t} C_2H_6;$

Transfer:
$$\dot{C}HO + M \xrightarrow{3} H + CO + M$$
$$H + CH_3CHO \xrightarrow{4} CH_3\dot{C}O + H_2.$$

The steady state radical concentrations are given by (long chains):

$$(CH_3\dot{C}O)/(\dot{C}H_3) \simeq \frac{k_1(CH_3CHO)}{k_2(M)},$$

[20] K. J. Laidler and M. T. H. Lin, *Proc. Roy. Soc.* (London), **A297**, 365 (1967); *Can. J. Chem.*, **46** (1968). See also M. B. Colket, D. W. Naegeli, and I. Glassman, *Int. J. Chem. Kinet.*, **7**, 223 (1975).

or with

$$(CH_3CHO) = (M) \sim 10^{-2.5} M$$

$$(CH_3\dot{C}O)/(\dot{C}H_3) \approx \frac{10^{8.0-8.0/\theta}}{10^{11.0-10/\theta}} \sim 10^{-3.0+2.0/\theta}$$

$$\approx 10^{-2.3} \text{ at } 700°K.$$

$$(\dot{C}HO) = \frac{k_i}{k_3} = \frac{10^{16-80/\theta}}{10^{11.5-20/\theta}} = 10^{4.5-60/\theta}$$

$$\approx 10^{-14.5} M \text{ (700°K).}$$

$$(\dot{H}) = \frac{k_i}{k_4} \approx \frac{10^{16-80/\theta}}{10^{10.3-5/\theta}} = 10^{5.7-75/\theta}$$

$$\approx 10^{-17.7} M \text{ (700°K)}$$

$$(CH_3) = \left[\frac{k_i(CH_3CHO)}{k_t} \right]^{1/2} \approx \left[\frac{10^{16-80/\theta}(10^{-2.5})}{10^{10.5}} \right]^{1/2}$$

$$\approx 10^{-11.3} M \text{ (700°K).}$$

We can see that only CH₃ radicals are of importance, and this justifies using a single termination reaction of 2CH₃. The overall rate of the chain is

$$\frac{-d(CH_3CHO)}{dt} = k_1(CH_3)(CH_3CHO) = k_1\left(\frac{k_i}{k_t}\right)^{1/2}(CH_3CHO)^{3/2}. \quad (5.36)$$

This equation is in excellent agreement with the observed data that verify the $\frac{3}{2}$ order and also the assignments of the individual rate constants and activation energy:

$$k_{obs} = 10^{10.8-48/\theta} \text{ (liter/mole)}^{1/2} \text{ s}^{-1}.$$

The rate of H₂ production should be equal to the rate of initiation, so that the kinetic chain length λ is given by[21]

$$\lambda = \frac{d(CH_4)/dt}{d(H_2)/dt} = \frac{k_1(CH_3CHO)^{1/2}}{(k_ik_t)^{1/2}}$$

$$= \frac{10^{8.0-8.0/\theta}(10^{-2.5})^{1/2}}{10^{13.3-40/\theta}} = 10^{-6.6+32/\theta}$$

$$= 10^{3.4} \text{ at } 700°K$$

$$= 10^{2.2} \text{ at } 800°K.$$

$$= 0.5 \text{ at } 1100°K$$

---

[21] See footnote page 266.

The reaction is extremely sensitive to small traces of O_2, and the walls of the vessel must be conditioned to yield reproducible rate data.[22] Ethylene oxide, $Hg(CH_3)_2$, and I_2 all will sensitize the pyrolysis at temperatures about 60°C lower than the usual pyrolysis range.

5.11 "WRONG" RADICALS

As we have already noted, the existence of more than one abstractable atom in a molecule makes for a diversity of possible products and raises, as well, the question of the importance of alternate radical species. The pyrolysis of CH_3CHO presents a simple example. In this case, the abstraction of an alkyl H atom leads to a relatively inert radical, $\cdot CH_2CHO$. The rate at which this may be expected to continue the chain by fission is limited by the endothermicity of the reaction

$$\cdot CH_2CHO \xrightarrow{\ 5\ } CH_2CO + H \quad \Delta H° = 34 \text{ kcal/mole.}$$

If we assume by analogy with other unsaturates an activation energy of about 3 kcal for the addition of H to ketene, $E_5 \approx 37$ kcal/mole. With an A factor of about $10^{13.5}$ s^{-1}, this gives a steady-state concentration of $\cdot CH_2CHO$, as follows:

$$\cdot CH_3 + CH_3CHO \xrightarrow{\ 1'\ } \cdot CH_2CHO + CH_4;$$

$$\dot{C}H_2CHO + CH_3CHO \xrightarrow{\ 6\ } CH_3CHO + CH_3 \cdot CO$$

$$\frac{(\cdot CH_2CHO)}{(\cdot CH_3)} = \frac{k_1'(CH_3CHO)}{k_6(CH_3CHO) + k_5}$$

$$\approx \frac{k_1'}{k_6} \approx 1.$$

This last result[23] is obtained by estimating k_1' and k_6 both as $10^{8.0-10/\theta}$ liter/mole-s, so that at $(CH_3CHO) = 10^{-2.5}$ M; $k_6(CH_3CHO)/k_5 \sim 3$ at 700°K, but only 0.3 at 800°K. In such a case, we should change our termination to include this inert radical. We should add the steps

$$CH_3 + \cdot CH_2CHO \xrightarrow{\ k_t'\ } CH_3CH_2CHO;$$

$$2 \cdot CH_2CHO \xrightarrow{\ k_t''\ } CHOCH_2CH_2CHO.$$

---

[22] At high temperatures (>900°K) the O_2 effect disappears. See Colket et al. (*loc. cit.*).

[23] K. J. Laidler and M. T. H. Lin arrive at about the same ratio, but for different reasons. They omitted reaction 6 and underestimated the rate of step 1'. See *Proc. Roy. Soc.* (London), **A297**, 365 (1967); *Can. J. Chem.*, **46** (1968). See also I. Bárdi et al., *Int. J. Chem. Kin.*, **8**, 285 (1976).

The final expression becomes

$$\frac{-d(CH_3CHO)}{dt} = \frac{k_1 k_i^{1/2}(CH_3CHO)^{3/2}}{[k_t + (k_1'/k_6)k_t' + (k_1'/k_6)^2 k_t'']^{1/2}}.$$

Using the generally valid geometric mean rule, $k_t' = 2(k_t k_t'')^{1/2}$, we obtain

$$\frac{-d(CH_3CHO)}{dt} = k_1 \left(\frac{k_i}{k_t}\right)^{1/2} \frac{(CH_3CHO)^{3/2}}{[1 + (k_1'/k_6)(k_t''/k_t)^{1/2}]}$$

$$\approx \frac{k_1}{2}\left(\frac{k_i}{k_t}\right)^{1/2}(CH_3CHO)^{3/2}.$$

In the last step we have assumed that $k_1' \sim k_6$ and $k_t'' \sim k_t$.

The path taken above also gives an additional contribution to H_2 production, which is almost exactly equal to the initiation rate at $10^{-2.5}\ M$ of CH_3CHO. The total H_2 production is now

$$\frac{d(H_2)}{dt} = k_i(CH_3CHO) + \left(\frac{k_i}{k_t}\right)^{1/2}\left(\frac{k_1'}{k_6}\right)k_5(CH_3CHO)^{1/2}.$$

This will modify our previous discussion of H_2 production and chain length. We note that C_2H_5CHO should be as abundant as H_2 production, and CH_2CO should be about half of H_2. Acetone production has been reported as another minor component, comparable to H_2.[24] Acetone can arise from two sources. The major one appears to be the endothermic displacement of H from CH_3CHO:

$$CH_3 + CH_3CHO \rightleftarrows (CH_3)_2C\dot{H}O \rightarrow (CH_3)_2CO + H.$$

A second is the addition of CH_3 to ketene to produce the inert radical $\cdot CH_2COCH_3$, which can abstract from CH_3CHO to form acetone:

$$\cdot CH_3 + CH_2CO \rightarrow \cdot CH_2COCH_3;$$

$$CH_3CO\dot{C}H_2 + CH_3CHO \rightarrow CH_3COCH_3 + CH_3\dot{C}O.$$

The alternative source of acetone from $CH_3\dot{C}O + CH_3$ is not significant because the $(CH_3\dot{C}O)$ concentration is so low.

We see that in the present instance the production of "inert" or "wrong" radicals does not produce a significant change in the kinetics or rate law. Although this may often be the case, it also may happen that such species will be important in modifying the rate law or quantitative Arrhenius parameters.

Another example is to be found in the chain pyrolysis of C_2H_5Br,

---

[24] K. J. Laidler and M. T. H. Lin, *Proc. Roy. Soc.* (London), **A297**, 365 (1967); *Can. J. Chem.*, **46** (1968).

where the "wrong" radical is $CH_3\dot{C}HBr$. The chain steps are

$$Br + C_2H_5Br \underset{-1}{\overset{1}{\rightleftharpoons}} HBr + \dot{C}H_2CH_2Br$$

$$\underset{-1'}{\overset{1'}{\rightleftharpoons}} HBr + CH_3\dot{C}HBr.$$

$$M + \dot{C}H_2CH_2Br \underset{-2}{\overset{2}{\rightleftharpoons}} C_2H_4 + Br + M$$

$$CH_3\dot{C}HBr \overset{3}{\rightleftharpoons} CH_2{=}CHBr + H\cdot.$$

Step 3 has an endothermicity of about 34 kcal and an expected activation energy of about 37 kcal. It is very slow compared to the exothermic, bimolecular reaction with HBr ($-1'$), which has an estimated rate constant $10^{9.5-3/\theta}$ liter/mole-s. Even at $(HBr) = 10$ torr $= 10^{-3.5}$ M, Step $-1'$ will be 10^3 to 10^4 times faster than Step 3.[25]

The steady-state concentration of the $CH_3\dot{C}HBr$ is then given by

$$\frac{(CH_3\dot{C}HBr)}{(B\dot{r})} = K_1'\left(\frac{C_2H_5Br}{HBr}\right)$$

$$\approx 10^{1.3-7/\theta}\left(\frac{C_2H_5Br}{HBr}\right)$$

$$\approx 10 \text{ (700°K and 1\% decomposition)}$$

$$\approx 1 \text{ (700°K and 10\% decomposition).}$$

Here the wrong radical is a major radical species early in the reaction and can account for much of the termination because the other radical $\dot{C}H_2CH_2Br$ is present in negligible amounts. We can estimate its steady state concentration from the near equilibrium which is established in Step 2:[26]

$$\frac{(\dot{C}H_2CH_2Br)}{(Br)} = \frac{k_{-2}(C_2H_4)(M) + k_1(EtBr)}{k_2(M) + k_{-1}(HBr)}$$

$$\approx K_{-2}(C_2H_4) + \frac{k_1(EtBr)}{k_2(M)}$$

At 700°K and 10 torr C_2H_4, $(\dot{C}H_2CH_2Br)/(Br) \approx 10^{-3.4}$. At 650°K, which is the middle of the temperature range and with 100 torr of added C_2H_4, it becomes $10^{-2.6}$. Because $Br + Br + M$ is relatively slow, due to its

---

[25] We assume $T \approx 700°K$, $A_3 = 10^{14} \text{ s}^{-1}$.

[26] In making these estimates we assign values as follows:

$$k_{-2} \sim 10^{10-2/\theta} \text{ liter}^2/\text{mole}^2\text{-s}; \quad k_1 = 10^{10.5-13/\theta} \text{ liter/mole-s};$$

$$k_{-1} = 10^{9.5-2/\theta} \text{ liter/mole-s}; \quad k_2 = K_2 k_{-2} \text{ with } K_2 \sim 10^{3-9/\theta} \text{ mole/liter}.$$

third-order rate constant, all termination is with $CH_3CHBr + Br$.

Termination: $Br + CH_3\dot{C}HBr \xrightarrow{t'} CH_3CHBr_2$.

The absolute concentration of Br atoms is then determined by the initiation, which is most probably the unimolecular fission of the weakest bond, the C—Br bond. However, bimolecular metathesis from products can become competitive:

$$C_2H_5Br \xrightarrow{i} C_2H_5 + Br;$$

$$C_2H_4 + HBr \xrightarrow{i'} C_2H_5 + Br.$$

The ratio of rates by our usual methods is

$$\frac{R_i}{R_i'} = \frac{10^{14.8-68/\theta}(EtBr)}{10^{10-46/\theta}(C_2H_4)(HBr)}$$

$$= 10^{4.8-22/\theta} \frac{(EtBr)}{(C_2H_4)(HBr)}.$$

At 650°K and 10% decomposition with $(EtBr)_0 = 10^{-2.5}$ M, this ratio has the value 10^2. However, large added quantities of either HBr or C_2H_4 at 600°K, or at 50% decomposition, could reduce it to below unity and shift the initiation mechanism. With C_2H_5Br as the source, however, we have

$$(Br)^2 = \frac{k_i(C_2H_5Br)}{k_t'K_1'\left(\dfrac{C_2H_5Br}{HBr}\right)}$$

$$= \frac{k_i(HBr)}{k_t'K_1'}$$

Inserting the first-order concerted, four-center path, k_A, the overall rate of the reaction becomes

$$\frac{-d(EtBr)}{dt} = k_A(EtBr) + k_1(Br)(EtBr)$$

$$\frac{-d(EtBr)}{(EtBr)\,dt} = k_A + k_1\left(\frac{k_i}{k_t'K_1'}\right)^{1/2}(HBr)^{1/2} \tag{5.37}$$

The apparent first-order chain contribution term can be estimated from our previous assignments (with $HBr = 10$ torr $\sim 10^{-3.6}$ M) as $10^{10.5-43.5/\theta}$ s^{-1}. This is in excellent agreement with the observations of Goldberg and Daniels[27] who reported, between 310 and 476°C, an overall first-order rate law $10^{11.8-46.4/\theta}$ s^{-1}.

---

[27] A. E. Goldberg and F. Daniels, *J. Am. Chem. Soc.*, **79**, 1314 (1957). C_2H_4 inhibition can only arise from the back reaction, not from $\dot{C}H_2CH_2Br$ recombination.

In agreement with (5.37), Goldberg and Daniels observed catalysis by HBr, inhibition by large amounts of C_2H_4, and an induction period that was extremely pronounced at the lowest temperatures and that was decreased markedly by HBr addition. They also verified the autocatalytic behavior. However they ascribed the catalysis to Br_2 production arising from the reaction of HBr with EtBr. This seems very unlikely.

The preceding analysis, on the contrary, shows that at low HBr the "wrong" radical cuts the chain lengths down to negligible proportions, and the major effect of HBr is on suppressing this species. There is, as well, a small suppression of the "right" radical, $\cdot CH_2CH_2Br$.

With minor modifications, we may expect the same behavior to occur in most pyrolyses of alkyl chlorides and bromides.

5.12 SELECTIVITY OF RADICALS

Because exothermic radical metathesis reactions have very small activation energies, we do not expect them to be very selective in the exothermic direction. For example, the radical chlorination of a hydrocarbon will proceed via

$$Cl + RH \rightarrow HCl + \dot{R}.$$

This is a reasonably exothermic reaction, having $\Delta H_r^\circ = -5$ kcal for primary C—H bonds; $\Delta H_r^\circ = -9$ kcal for secondary, -12 kcal for tertiary C—H, and $\Delta H_r^\circ \sim -22$ kcal for allylic or benzylic C—H bonds. However, the activation energy is already so small (~ 1 kcal) for the primary C—H-bond metatheses that there can be no appreciable variation in activation energy with bond strength. We thus expect Cl atoms to abstract H atoms, almost indiscriminately from organic compounds with little concern for their bond strengths. An example of this is provided by the gas phase, free-radical chlorination of butane that gives (sec-butyl chloride)/(primary butyl chloride) ratios of 2.4, almost independent of temperature. Correcting by the statistical number of abstractable atoms, the per-atom, metathesis rate ratio becomes 3.6.

A similar example to that given above is provided by the liquid phase chlorination of n-butyl benzene in dilute CCl_4 solution.[28] The statistically corrected, relative rate ratios[29] are indicated in the following:

$$C_6H_5 \underset{5.9}{-\overset{\alpha}{CH_2}} \underset{:\quad 2.7}{-\overset{\beta}{CH_2}} \underset{:\quad 4.0}{-\overset{\gamma}{CH_2}} \underset{:\quad 1.0}{-\overset{\delta}{CH_3}}.$$

[28] G. A. Russell, A. Ito, and D. G. Hendry, *J. Am. Chem. Soc.*, **85**, 2976 (1963).

[29] To make explicit the counting, the actual ratios of $\alpha:\beta:\gamma:\delta$ chlorides, which are observed, are 5.9:2.7:4.0:1.5.

By way of contrast, we should then expect to see a very strong effect of bond strength on selectivity in atom abstraction reactions which occur in the endothermic direction. Abstraction of H atoms by Br atoms will be generally endothermic with endothermicities of 17 kcal for CH_4, 11 kcal for primary C—H, 7.5 kcal for secondary C—H, and 4.5 kcal for tertiary C—H. The exothermic reactions of all of these radicals with HBr have similar activation energies of about 1.5 ± 0.5 kcal.

We expect to see the full differences in bond strengths reflected in the activation energies of Br atom attack, and this is indeed the case. As an example, the ratio of secondary/primary bromides produced by vapor-phase bromination of n-butane at 160°C is about 55, to be compared to the quoted 2.4 for Cl atom attack.

One has, however, to exercise some caution in interpreting such results as those gained above because the final products may be affected by unimolecular abstraction reactions. Thus, in the case of n-pentane, a primary radical may isomerize as follows:

$$\dot{C}H_2CH_2CH_2CH_2CH_3 \xrightarrow{1} CH_3CH_2CH_2\dot{C}HCH_3$$

$$+ Cl_2 \xrightarrow{2} ClCH_2CH_2CH_2CH_2CH_3 + Cl.$$

The ratio of these two paths is given by

$$\frac{R_1}{R_2} = \frac{k_1}{k_2(Cl_2)} \approx \frac{10^{11-15/\theta}}{10^{9.5-1/\theta}(10^{-2.5})}$$

$$= 10^{4-14/\theta}$$

The ratio is very small below 180°C (10^{-3} at 450°K) but becomes significant above 700°K, or at lower Cl_2 concentrations.

The A factor for the internal abstraction is estimated from our A factors for five-membered ring reactions, and the activation energy is obtained by adding to the expected 8 kcal for H-atom abstraction, 7 kcal of ring strain in the five-membered ring.[30]

5.13 OXIDATION

One of the most intriguing of the complex chain reactions is the oxidation of organic molecules. At low temperatures (<190°C) the main products are hydroperoxides and oxygenated species obtained from their secondary reactions. These include alcohols, ketones, aldehydes, and acids, as well as CO and CO_2.

---

[30] A more extensive discussion is given by G. A. Russell, *Free Radicals*, Vol. I, Ch. 7, J. K. Kochi (ed.), Wiley, New York, 1973.

A simple chain mechanism can account for the species listed above.

Chain: $\begin{cases} R\dot{O}_2 + RH \xrightarrow{1} RO_2H + \dot{R}; \\ \dot{R} + O_2 \xrightarrow{2} R\dot{O}_2 \end{cases}$

Transfer: $\begin{cases} 2R\dot{O}_2 \xrightarrow{3} 2R\dot{O} + O_2 \\ R\dot{O} + RH \xrightarrow{4} ROH + \dot{R} \\ R\dot{O} \xrightarrow{5} R''R'C{=}O + \dot{R}'''; \\ \dot{R}''' + O_2 \xrightarrow{6} R'''O_2 \end{cases}$

and so on.

The complexity in these oxidations is introduced by the unusual Step 3, which is only slightly exothermic ($\Delta H_3 \sim -6$ kcal) and which proceeds through the formation of a weakly bonded tetraoxide.[31] It is estimated that the concerted split of the tetraoxide to give $2R\dot{O} + O_2$ is about thermoneutral.

The $RO\cdot$ radicals can undergo fission to produce stable ketones or aldehydes and a smaller radical. Thus with $t\text{-BuO}\cdot$, t-amyl $O\cdot$, or cumyl $O\overset{\cdot}{}$:

$$(CH_3)_3C{-}\dot{O} \to (CH_3)_2CO + \dot{C}H_3,$$

$$(CH_3)_2(C_2H_5)C{-}\dot{O} \to (CH_3)_2CO + \dot{C}_2H_5,$$

$$\phi(CH_3)_2C{-}\dot{O} \to \phi COCH_3 + \dot{C}H_3.$$

In general the weakest bond, as may be expected, is one that undergoes fission.

Near 100°C, with an excess of O_2, it is possible to get very long chains and almost quantitative production of the RO_2H from RH. However for saturated secondary or primary hydroperoxides, the yield is limited by the very slow abstraction step

$$R\dot{O}_2 + RH \underset{-1}{\overset{1}{\rightleftharpoons}} RO_2H + \dot{R},$$

and the competing side paths via Step 3 can become dominant.[32] The RO_2H is, itself, protected from attack on its weak O—H bond by the excess O_2 that converts all $R\cdot$ radicals to $RO_2^{\cdot}$ with extremely high rate constants ($10^{9.5}$ liter/mole-s) with no activation energy. $RO\cdot$ radicals are generally scavenged by the large excess of RH or by metathesis with O_2 to form HO_2.

---

[31] K. Adamic, J. A. Howard, and K. U. Ingold, *Can. J. Chem.*, **47**, 3803 (1969).

[32] In addition there seems to be a fairly rapid termination process, at least in solution [$2RCH_2O_2^{\cdot} \to RCHO + RCH_2OH + O_2$], if primary or secondary $RO_2^{\cdot}$ radicals are involved.

The initiation in the systems discussed above is generally provided by sensitizers. Spontaneous thermal initiation of saturated alkanes with O_2 is too slow to be of importance below 150°C. The same is true of olefins $+O_2$. For the latter, both addition or abstraction reactions will have activation energies in excess of 35 kcal.

As examples, consider $C_2H_4 + O_2$ addition reactions:

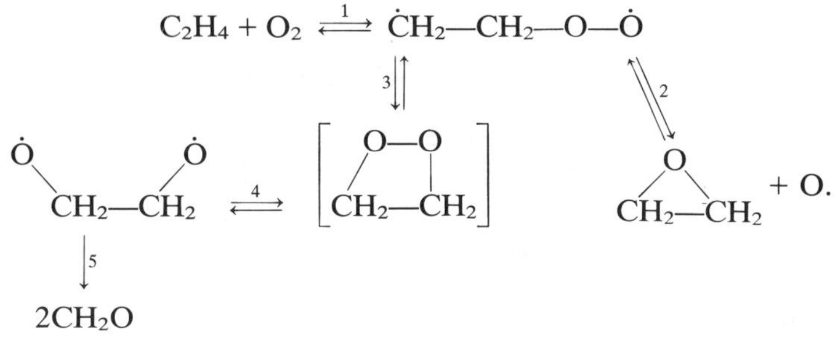

$$C_2H_4 + O_2 \overset{1}{\rightleftharpoons} \dot{C}H_2\text{---}CH_2\text{---}O\text{---}\dot{O}$$

$$\overset{\dot{O}}{\underset{CH_2\text{---}CH_2}{\diagdown}}\overset{\dot{O}}{\diagup} \overset{4}{\rightleftharpoons} \left[\underset{CH_2\text{---}CH_2}{\overset{O\text{---}O}{\diagup \diagdown}} \right] \qquad \underset{CH_2\text{---}CH_2}{\overset{O}{\diagup \diagdown}} + O.$$

$$\downarrow 5$$

$$2CH_2O$$

Step 1 is endothermic by about 32 kcal and reverses quickly unless followed by the exothermic Step 3, which is expected to have an activation energy of another 5 kcal and then lead rapidly to formaldehyde. Step 2 is estimated to have an activation energy of about 16 kcal. It is the only one which leads to a propagating radical.

Abstraction of allylic, or benzylic, H is more favorable. With 3-methyl butene-1, we would expect

$$(CH_3)_2CH\text{---}CH\text{=}CH_2 + O_2 \rightarrow (CH_3)_2\dot{C}\text{---}CH\text{=}CH_2 + H\dot{O}_2$$

with an activation energy just equal to the endothermicity of 31 kcal. Above 100°K in neat liquid phase, this would be a quite significant radical source, even at low O_2 concentrations (10^{-4} M).

The overall net reaction, which is the production of RO_2H from $RH + O_2$, is exothermic by about 20 kcal and can become a source of self ignition only at very high O_2 concentrations, or at temperatures near 300°C where gas mixtures exhibit the well-studied phenomenon of "cool" flames.

Above 160°C the hydroperoxides themselves begin to undergo homogeneous decompositions into $RO + OH$ radicals and so can act as secondary initiation centers for further oxidation. Above 300°C with estimated rate constants for this fission of about $10^{15-43/\theta}$ s^{-1}, the lifetimes of RO_2H are of the order of 10 s and their slow formation and rapid decomposition may be shown to account for the long induction periods and periodic flames characterizing the cool-flame region. The overall phenomenon is described as "degenerate" chain branching because a product in the reaction is giving rise to chain center multiplication.

"Spontaneous" oxidation of hydrocarbons at $T < 300°C$ in either the gas or liquid phases appears to be companied by induction periods of unknown origin, and then a steady-state oxidation. It is presumed and appears likely that this induction period gives rise to a steady-state concentration of hydroperoxides RCH_2OOH or R_2CHOOH, which can continue the oxidation by a rapid, bimolecular isomerization reaction:

$$RCH_2\dot{O_2} + RCH_2OOH \xrightarrow{\ 4\ } RCH_2OOH + R\dot{C}HOOH$$

$$R\dot{C}HOOH \xrightarrow{\ 5\ } RCHO + OH.$$

Reaction 5 is expected to be extremely rapid and perhaps almost coincident with Step 4. It gives rise to the very active OH radical which is a nonselective, very rapid chain carrier. Since Steps 4 and 5 lead to no net destruction of RO_2H, we see that instead they constitute a very fast path for the net reaction:

$$RCH_2\dot{O_2} \rightarrow RCHO + OH$$

The aldehyde produced in this process is itself very easily attacked by selective radicals, such as $R\dot{O_2}$, since it has much weaker bond, $DH°(RCO—H) = 87$ kcal, than is found in aliphatic hydrocarbons.

Because of the RO_2H pyrolysis or catalytic isomerization, the rate of reaction of hydrocarbons with O_2 shows an auto-acceleration from about 200 to 350°C. At the latter temperature the homogeneous gas-phase rate reaches a maximum, and then declines to the point where it is almost immeasurably slow at 400°C in a flow system.[33] In a static system this region of negative temperature coefficient is observed to be accompanied by a change over in products from oxygen-containing compounds, such as RO_2H, and its secondary products to olefins and $H_2O + H_2O_2$.

All of the results described above are accounted for by a very simple scheme:

$$\dot{R} + O_2 \underset{-1}{\overset{1}{\rightleftharpoons}} R\dot{O_2}$$

$$\underset{-1'}{\overset{1'}{\rightleftharpoons}} H\dot{O_2} + \text{olefin};$$

$$R\dot{O_2} + RH \xrightarrow{\ 2\ } RO_2H + \dot{R};$$

$$H\dot{O_2} + RH \xrightarrow{\ 2'\ } HO_2H + \dot{R}.$$

The abstraction of H from R· to give $H\dot{O_2} + $ olefin (Step 1') is at least 200 times slower than the addition of O_2 to R· to give $R\dot{O_2}$. The latter is

[33] K. C. Salooja, *Combustion and Flame*, **6**, 275 (1962); **8**, 311 (1964); **9**, 219 (1965). Y. H. Chung and S. Sandler, *Combustion and Flame*, **6**, 295 (1962).

a radical recombination with a rate constant measured at $10^{9.5}$ liter/mole-s in its high-pressure region. The competing abstraction step, though exothermic by about 7 kcal/mole or more, has a rate constant estimated at $10^{9.2-6/\theta}$ liter/mole-s. It can never be significant compared to addition.

However, Step 1, which below 200°C is the rate determining step in oxidation, becomes significantly reversible above 250°C, so that Step 2 becomes the slow, rate-determining oxidation step. This is indicated in the overall oxidation rate

$$\frac{d(RO_2H)}{dt} = \frac{k_1 k_2 (RH)(\dot{R})(O_2)}{k_{-1} + k_2(RH)}$$ (5.38)

$$\xrightarrow{T<200°C} k_1(\dot{R})(O_2)$$

$$\xrightarrow[T>250°C]{} K_1 k_2(RH)(O_2)(\dot{R}).$$ (5.39)

The turnover point in the rate law will occur when $k_{-1} = k_2(RH)$. With $k_{-1} \sim 10^{14.5-28/\theta}$, $k_2 \sim 10^{8.5-14/\theta}$, and E_2 depending somewhat on the nature of the hydrocarbon, turnover will occur at about 400–500°K for $(RH) = 1$ atm. Let us compare this with the rate law for olefin production via exothermic Step 1′, which can be taken as effectively irreversible in the system,

$$\frac{d(\text{olefin})}{dt} = k_1'(\dot{R})(O_2).$$

The ratio of rates is given by

$$\frac{R(\text{olefin})}{R(RO_2H)} = \frac{k_1'[k_{-1} + k_2(RH)]}{k_1 k_2(RH)} \underset{\substack{\nearrow \\ T<200°C}}{\overset{\dfrac{k_1'}{k_1} \ll 1}{}} $$

$$\underset{\substack{\searrow \\ T>250°C}}{} \frac{k_1'}{K_1 k_2(RH)}$$

Using the values of the assigned rate constants, we find that above 250°C, the ratio becomes

$$\frac{R(\text{olefin})}{R(RO_2H)} \approx \frac{10^{5.7-20/\theta}}{(RH)}.$$

This relation fits the observed data for a number of hydrocarbon oxidations under very varied conditions.[34] In particular, the inverse dependence on (RH) has been observed as well.

[34] See S. W. Benson, *J. A. Chem. Soc.*, **87**, 972 (1965) for a summary; also see S. W. Benson, *Adv. Chem.*, **75, 76**, (1968).

Above 400°C the rates of oxidation in flow systems begin to rise again as the product H_2O_2 begins to decompose homogeneously to provide a new source of radicals $(M + H_2O_2 \rightarrow 2OH + M)$ and play the role of the missing RO_2H. Above 500°C this is sufficiently rapid that secondary oxidation of the product olefin provides sufficient heat to cause self-ignition and explosion. Note that the overall reaction to produce H_2O_2 is nearly thermoneutral and involves a very small entropy change:

$$O_2 + RH \rightarrow \text{olefin} + H_2O_2 + 2 \text{ kcal.}$$

Olefins turn out to be more easily oxidized than alkanes. The mechanism involves addition to the double bond and/or abstraction of allylic H atoms:

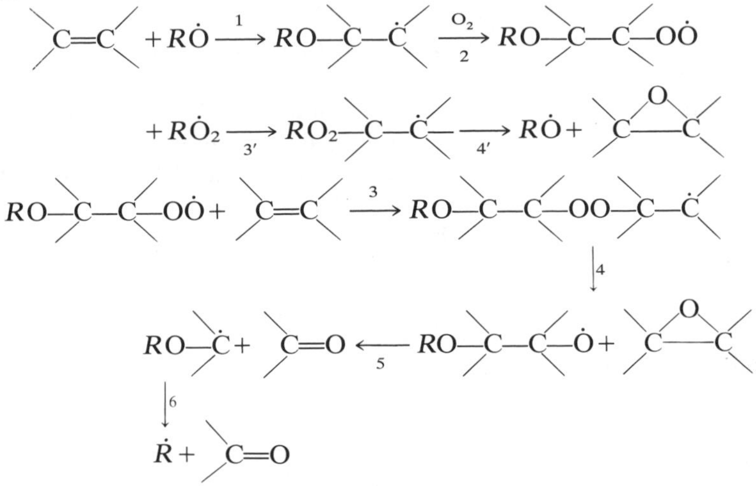

The sequence above provides a chain mechanism for conversion of original olefins into epoxides and aldehydes, via $R\cdot$, $RO\cdot$, and RO_2.

Allylic H atom abstraction leads to

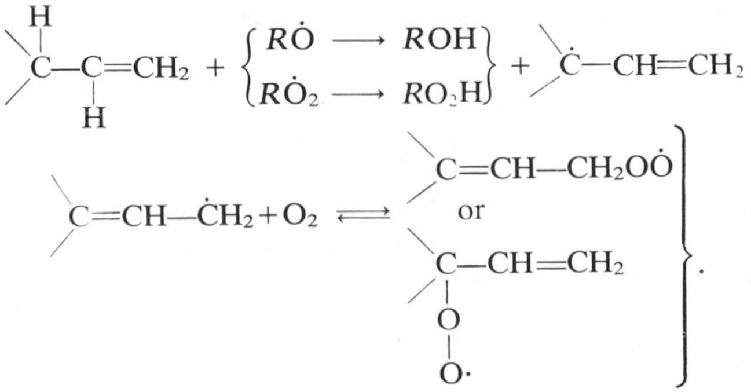

The allyl peroxy radicals are even more weakly bound than the alkyl peroxy, and at $O_2 = 0.1$ atm will be in labile equilibrium with an equal concentration of allyl radicals above 500°K. This tends to make diene formation more likely by the HO_2' mechanism at the higher temperatures for C_4 or higher olefins. It also lowers the turnover temperature at which RO_2H production starts to decrease.

5.14   INERT INITIATION RADICALS

We have seen how in complex compounds, abstraction from a wrong position can lead to an inert radical that cannot react rapidly to propagate the chain. These inert radicals can in many instances arise from initiation reactions. An example arises in the pyrolysis of n-propyl benzene, where the fastest initiation is fission of the weak C—C bond to produce benzyl + ethyl radicals:

$$\phi CH_2 - CH_2 - CH_3 \rightarrow \phi C\dot{H}_2 + \dot{C}_2H_5.$$

Whereas $\cdot C_2H_5$ can abstract with relatively low activation energy from all 3C—H bonds, the corresponding reaction for benzyl radical is expected to have about a 4–6 kcal higher activation energy for the αC—H, and 19 to 22 kcal total activation energy for β and γ—C—H, and hence be very slow. In this system, competing chains exist by virtue of the different C—H bond abstractions:

$$\dot{R} + \phi CH_2 - CH_2 - CH_3 \xrightarrow{\ 1\ } RH + \phi \dot{C}H - CH_2 - CH_3,$$
$$\xrightarrow{\ 1'\ } RH + \phi CH_2 - \dot{C}H - CH_3,$$
$$\xrightarrow{\ 1''\ } RH + \phi CH_2 - CH_2 - \dot{C}H_2,$$
$$\phi \dot{C}H - CH_2 - CH_3 \xrightarrow{\ 2\ } \phi CH{=}CH_2 + \dot{C}H_3,$$
$$\phi CH_2 - \dot{C}H - CH_3 \xrightarrow{\ 2'\ } \phi CH{=}CH - CH_3 + \cdot H,$$
$$\phi CH_2 - CH_2 - \dot{C}H_2 \xrightarrow{\ 2''\ } \phi C\dot{H}_2 + C_2H_4.$$

Of the three possible fission steps, Step 2″ is the fastest, with an expected activation of about 20 kcal, but a low A factor of about $10^{12.8}\ s^{-1}$. However, it produces an inert $\phi C\dot{H}_2$ radical, which can propagate slowly and only at the α—C—H bond. The next fastest fission step is Step 2′ with an expected activation energy of 38 kcal and a low A factor of about $10^{12.8}\ s^{-1}$ also. It gives rise, however, to a very nonselective H atom, which can attack rapidly at any of the C—H positions, and hence will not produce long chains.

The slowest fission step is Step 2 with an expected activation energy of about 50 kcal and an A factor of about $10^{14} \, s^{-1}$. It produces a CH_3, which again is not very selective. It can be seen that there are no long chains expected for n-propyl benzene because only inert, or wrong, radicals are produced in one, or both, of the propagation steps.

Another example comes from the pyrolysis of neopentane (NpH). The initiation produces $\cdot CH_3 +$ the inert t-butyl radical:

$$(CH_3)_4C \xrightarrow{\ i\ } \dot{C}H_3 + \dot{C}(CH_3)_3.$$

The activation energy for t-butyl attack on the primary C—H bond, which is endothermic by 6 kcal, is expected to be about 16 kcal, making this a very slow step. The alternative route is fission of t-Butyl to give $H + $isobutene with an expected activation energy of about 42 kcal ($\Delta H° = 40$ kcal/mole). This will be far more rapid than the H atom abstraction, so that the sequence will be

$$(CH_3)_3\dot{C} \xrightarrow{\ 1\ } H + CH_2 = C(CH_3)_2.$$

The stationary concentration of t-Bu$\cdot$ will be

$$(t\text{-Bu}) = \frac{k_i(\text{NpH})}{k_1} \approx \frac{10^{16.5-80/\theta}(10^{-2.5})}{10^{14.0-42/\theta}}$$

$$= 10^{-38/\theta}.$$

At 750°K ($\theta = 3.5$ kcal/mole) this is $10^{-10.6} \, M$, and relatively minor compared to the concentration of propagating radical CH_3, which is about 100-fold higher ($\sim 10^{-8} \, M$ at 750°K) at all temperatures. This generally turns out to be the case for inert radicals formed in initiation, but one must analyze the system carefully to be certain.

5.15 SECONDARY REACTIONS

We have already seen that the initial stages of the pyrolysis of molecules with more than three polyvalent atoms is complicated by competing chain reactions. Such complexity is further compounded in the later stages of the decomposition when secondary reactions with product molecules become appreciable. These secondary reactions can also produce inhibition, or acceleration of the initial chain and in extreme cases, completely obscure the nature of the initial products.

We can expect secondary reactions to become important in depleting products when the rates of radical reactions with these products are comparable in magnitude to the parallel rates in the initial chain. Consider, for example, the pyrolysis of dimethyl ether. The initial chain is the very simple pyrolytic, two-center scheme:

Chain:
$$\begin{cases} \dot{C}H_3 + CH_3OCH_3 \xrightarrow{1} CH_4 + \dot{C}H_2OCH_3 \text{ (slow)} \\ \dot{C}H_2OCH_3 + M \xrightarrow{2} CH_2O + \dot{C}H_3 + M \text{ (fast)}; \end{cases}$$

Net reaction: $\qquad CH_3OCH_3 \longrightarrow CH_2O + CH_4.$

CH_2O is an excellent source of abstractable H atoms, so that it will compete along with the original ether for the CH_3 radicals. The activation energy appears to be about 2 kcal lower for $CH_2O + CH_3$ than that for CH_3 abstraction of H from ether, so that if the A factors can be taken as equal, we expect a $10^{2/\theta}$ faster attack on CH_2O by $\cdot CH_3$ than on Me_2O. At 800°K ($\theta = 3.6$ kcal) in the middle of the range where the reaction has been studied, this amounts to a factor of about 4, in excellent agreement with the observation that, during the pyrolysis, the CH_2O concentration builds up to a steady-state ratio, relative to Me_2O of about $\frac{1}{4}$.

More quantitatively, if CH_2O is destroyed by $\cdot CH_3$ attack,

Secondary chain:
$$\begin{cases} \dot{C}H_3 + CH_2O \xrightarrow{1'} CH_4 + \dot{C}HO \\ \dot{C}HO + M \xrightarrow{2'} CO + \dot{H} + M, \end{cases}$$

it should approach a stationary state given by

$$\frac{d(CH_2O)}{dt} \simeq k_1(CH_3)(Me_2O) - k_1'(CH_3)(CH_2O).$$

When the steady state is reached and $d(CH_2O)/dt \approx 0$, then

$$\frac{(CH_2O)}{(Me_2O)} \approx \frac{k_1}{k_1'} \approx \frac{1}{4}.$$

The agreement needs to be modified to allow for H atoms, which are also active carriers in the system, produced by the secondary chain. They can attack both Me_2O and CH_2O. The data suggest that the same ratio of about 4 applies to the relative rates of H atom attack on CH_2O and Me_2O:

$$H + (CH_3)_2O \rightarrow H_2 + \dot{C}H_2OCH_3,$$
$$H + CH_2O \rightarrow H_2 + \dot{C}HO.$$

In the ether–CH_2O system just discussed the secondary reactions would not be important at 1% reaction, but they would become appreciable at 20% reaction. The reason for this is that the two chain carriers are the not very selective radicals H and CH_3. We would expect much more significant secondary reactions in chains carried by less active, hence more

selective, radicals. An example would be provided by the pyrolysis of
t-butyl bromide, where the carriers would be the fairly inert Br atoms.
The chain steps are:

Chain:
$$\begin{cases} Br + (CH_3)_3C\!-\!Br \overset{1}{\rightleftharpoons} HBr + \dot{C}H_2(CH_3)_2C\!-\!Br, \\ \dot{C}H_2(CH_3)_2\!-\!Br \overset{2}{\rightleftharpoons} CH_2\!=\!C(CH_3)_2 + Br \end{cases}$$

Secondary
reaction: $\quad Br + CH_2\!=\!C(CH_3)_2 \overset{3}{\rightleftharpoons} HBr + CH_2\!=\!C(CH_3)\!-\!\dot{C}H_2$

The slow chain carrying step is the endothermic ($\Delta H° = 11$ kcal)
abstraction of H by Br atoms (Step 1). The secondary reaction competing
with this is the thermoneutral abstraction of H from the isobutylene
product to give the stable methyl allyl radical (Step 3). Even with A
factors favoring Step 1 by about a factor of 10, the expected difference of
10 kcal in the activation energies of Steps 1 and 3 will make the rate
constant for the latter 10^3-fold faster at 300°C. This means that even at
0.1% decomposition, Br attack on the olefin is as fast as on the parent
molecule. As a consequence, Reaction 3 very quickly reaches a state of
equilibrium with the (Br)/(methyl allyl) ratio given by

$$\frac{(Br)}{(\text{methyl allyl})} = K_{-3}\frac{(HBr)}{(\text{olefin})} \approx \frac{1}{40}$$

The estimate is made by noting that $\Delta H_3 = 0$ and $\Delta S_3 \sim 7.4$ eu and that
$(HBr) = (\text{olefin})$ by stoichiometry. Because the other radical is negligibly
small at 580°K, compared to Br, this immediately establishes Br and
methyl allyl as the terminating species. Further analysis will also show
that the chain is extremely slow with an overall rate given by

$$\frac{-d(RBr)}{dt} = k_1\left(\frac{k_i K_{-3}}{k_t}\right)^{1/2}\left(\frac{HBr}{\text{olefin}}\right)^{1/2}(RBr)^{3/2}$$

$$\approx k_1\left(\frac{k_i K_{-3}}{k_t}\right)^{1/2}(RBr)^{3/2}.$$

The initiation and termination steps are

$$t\text{-BuBr} \overset{i}{\longrightarrow} t\text{-B\dot{u}} + Br,$$

$$\text{methyl allyl} + Br \overset{t}{\longrightarrow} CH_2\!=\!C(CH_3)CH_2Br.$$

We have neglected the termination of two methylallyl radicals, which is
expected to have a rate constant about 10-fold smaller than k_t. This
introduces only a small error. Estimating $k_i = 10^{14.5 - 68/\theta}$, $k_t = 10^{10}$, and
$k_1 = 10^{10.5 - 13.5/\theta}$, we find for the $\frac{3}{2}$ order rate constant

$$k_{3/2} = k_1(k_i K_{-3}/k_t)^{1/2} \simeq 10^{12 - 48/\theta} \text{ (liter/mole)}^{1/2}\text{ s}^{-1}.$$

At 550°K and 100 torr of t-BuBr, this predicts a half-life of 10 years, which is negligible compared to the observed rate and which can thus be properly ascribed to the unimolecular, four-center elimination of HBr (Table 3.10). Similar considerations will show that none of the secondary or tertiary bromides can decompose by chain reactions at the temperatures where they have been studied. This is in agreement with the observations, which show no acceleration of the rate by Br_2 and no effect of radical inhibitors, such as NO or propylene on these reactions.[35]

As already noted, such conclusions do not hold for the chlorides, where the active radical, Cl, is nonselective. They do, however, predict substantial chain inhibition in these systems. We would similarly expect that the pyrolysis of amines and alcohols, which give rise to the nonselective $\dot{N}H_2$ and $\dot{O}H$ chain carriers, can also show chain character with inhibition by products and also by "wrong" radicals.

A last, and somewhat unusual example of important secondary reactions, is provided by the chain chlorination of benzene. Because of the high C—H-bond energy in benzene (~110 kcal/mole), very few radicals or atoms can abstract H atoms from benzene. For relatively inactive radicals, such as Br and CH_3, the preferred mode of reaction is addition. With more active radicals like Cl and OH, both modes can occur. We illustrate the addition reaction below for the case of Cl_2. Note that it proceeds through an intermediate cyclohexadiene compound which is unstable to free radical attack at higher temperatures.

The scheme is

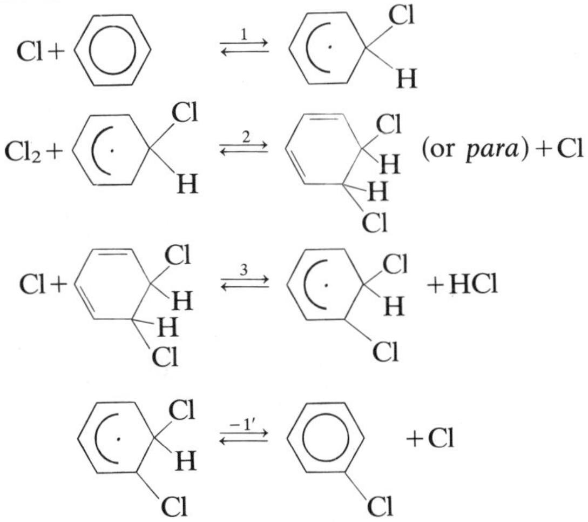

---

[35] Note that as termination products biallyl and allyl bromide build up, they would become competitive initiation sources.

The competing reaction to Step 3 above is Cl atom addition to form trichloro-allyl radical. Because of the unfavorable, though exothermic preequilibrium (Step 1), the overall rate of attack of Cl on adduct (Step 3) is about 500 times faster than the net rate of formation of dichloride, and hence this latter never appears as a significant product of the reaction, despite the fact that all of the reaction proceeds through it. Its net rate of production is given by

$$\frac{d(RCl_2)}{dt} \simeq k_2 K_1 K_{Cl_2}^{1/2}(Cl_2)^{3/2}(\phi H) - k_3 K_{Cl_2}^{1/2}(Cl_2)^{1/2}(RCl_2)$$

where we have neglected back reactions with ϕCl and have assumed that Cl atoms are the only appreciable chain centers in the system.[36] The steady-state concentration of RCl_2 is

$$\frac{(RCl_2)_{ss}}{(\phi H)} = \frac{k_2 K_1(Cl_2)}{k_3}.$$

It is likely that $k_2/k_3 \sim 10^{-1-3/\theta}$, so that with $(Cl_2) = 100$ torr and the assigned value of K_1, $(RCl_2)/(\phi H) = 10^{-8.2+7/\theta} = 10^{-5.7}$ at 550°K.

The present mechanism yields for the rate of chlorination

$$\frac{d(\phi Cl)}{dt} \approx \frac{-d(\phi H)}{dt} = k_2 K_1 K_{Cl_2}^{1/2}(Cl_2)^{3/2}(\phi H)$$

$$\approx 10^{7.0-25.5/\theta}(Cl_2)^{3/2}(\phi H). \tag{5.38}$$

The apparent 5/2 order rate constant is notable for having an activation energy less than half the dissociation energy of Cl_2 (28.5 kcal). This arises from the exothermic preequilibrium Step 1.

If we compare this addition mechanism with the expected rate of the more typical abstraction reaction, we find that they are about equal at 450°K when $(Cl_2) = 1$ atm. At higher temperatures we tend towards the abstraction mechanism while low temperatures favor the addition. For bromination only the addition is important.[37]

5.16   GAS-PHASE POLYMERIZATION

Addition reactions are always subject to a very important species of secondary reactions, polymerization. This is usually not observed in the

---

[36] The ratio $(\cdot RCl)/(Cl) \simeq K_1(\phi H) \approx 10^{-5.8+6/\theta}(\phi H)$ in units of atmospheres. At $(\phi H) = 100$ torr and 550°K, $(\cdot RCl)/(Cl) = 10^{-4.4}$.

[37] For a general discussion of vapor-phase halogenation of aromatics, see E. C. Kooyman, *Advances in Free Radical Chemistry*, Vol. 1, Ch. 4, Academic, New York, 1965. The addition mechanism is discussed in A. S. Rodgers, D. M. Golden, and S. W. Benson, *J. Amer. Chem. Soc.*, **89**, 4578 (1967).

radical catalyzed additions of small molecules, such as the hydrogen halides, or halogens to unsaturates, but is found in the additions (telomerizations) of more complex species, such as, CCl_3Br, CCl_4, CCl_2O, CF_3CN, or as a special case, H_2 to olefins. Addition to acetylenes or conjugated olefins are almost never without side reactions of polymerization, even for the simple molecules.

The reason for the result described above is that the simple chain telomerization is always in competition with the addition of a radical to another molecule of olefin. Let us consider, as an example, the addition of H_2 to C_2H_4 to produce C_2H_6. Above 400°C, where the rate is appreciable, C_4 and higher hydrocarbons are produced, unless $(H_2) \gg (C_2H_4)$. The relevant steps in the chain are

$$\text{Chain:} \quad \begin{cases} H + C_2H_4 \xrightarrow{\;1\;} \cdot C_2H_5 \\ \cdot C_2H_5 + H_2 \xrightarrow{\;2\;} C_2H_6 + H \end{cases}$$

$$\text{Polymer formation:} \quad \cdot C_2H_5 + C_2H_4 \xrightarrow{\;3\;} n\text{-}\dot{C}_4H_9,$$

and so on.

The ratio of rates of production of polymer formation $(\geq C_4)$ to C_2H_6 is given by

$$\frac{R\,(\geq C_4)}{R\,(C_2H_6)} = \frac{k_3}{k_2}\frac{(C_2H_4)}{(H_2)}$$

$$\approx 10^{-1.3+8/\theta}\frac{(C_2H_4)}{(H_2)}$$

where we have used average assignments for A_3 $(10^{8.3})$ and A_2 $(10^{9.6})$. At $(C_2H_4)/(H_2) = 0.10$ and $\theta = 3.5$ $(\sim 750°K)$, this would give $R(\geq C_4)/R(C_2H_6) \sim 1$, in reasonable agreement with observations.[38] We see from this result that it would require $(H_2)/(C_2H_4)$ ratios in excess of 800 to reduce polymerization to 1% of hydrogenation.

Such behavior as that described above is not observed with HCl, HBr, Cl_2, Br_2, and so on, because these species have much lower activation energies than H_2 for radical attack. In contrast, acetylenes and conjugated dienes are much more susceptible to polymerization because the activation energy for radical addition to these species is down to about 4 kcal, rather than 8, as it is with the olefins. This is somewhat compensated for in the case of acetylenes by the fact that the product of the addition, vinyl

[38] H. A. Taylor and A. Van Hook, *J. Phys. Chem.*, **39**, 811 (1935); R. N. Pease, *J. Am. Chem. Soc.*, **54**, 1877 (1932).

radicals are extremely reactive, having appreciably lower activation energies for atom abstraction than alkyl radicals.

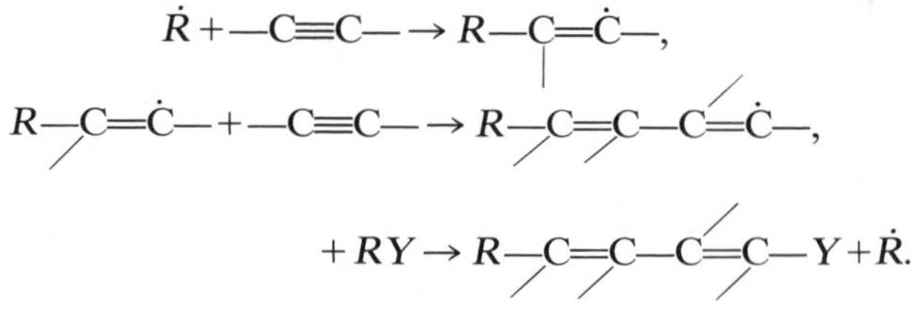

$$\dot{R} + -C \equiv C- \rightarrow R-C \overset{|}{=} \dot{C}-,$$

$$R-\overset{\diagup}{C} = \dot{C}- + -C \equiv C- \rightarrow R-\overset{\diagup}{C} = C-C \overset{\diagup}{=} \dot{C}-,$$

$$+ RY \rightarrow R-\overset{\diagup}{C} = C-C \overset{\diagup}{=} C-Y + \dot{R}.$$

By way of contrast, 1,3-dienes show enhanced polymerization tendencies because of the slowness of the chain-carrier allyl radical in abstraction reactions.

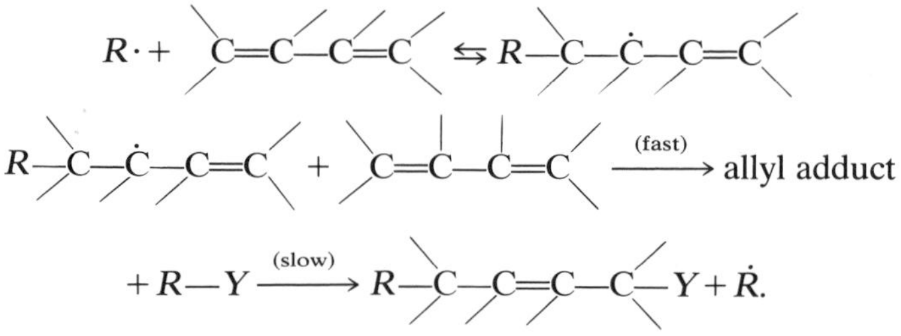

$$R \cdot + \quad C=C-C=C \quad \rightleftharpoons \quad R-C-\dot{C}-C=C$$

$$R-C-\dot{C}-C=C \quad + \quad C=C-C=C \xrightarrow{\text{(fast)}} \text{allyl adduct}$$

$$+ R-Y \xrightarrow{\text{(slow)}} R-C-C=C-C-Y + \dot{R}.$$

It is difficult to make quantitative studies of gas-phase polymerizations or telomerizations because the higher molecular weight products may easily exceed their vapor pressures at the reaction temperatures and condense on the walls. This tends to make for a second phase at the walls and a heterogeneous system. Even without product condensation, the higher molecular weight radicals tend to be adsorbed at the walls and continue the reaction at fixed surface sites. In commercial polymer plants, such a disastrous episode is called "popcorn" polymerization.

5.17 HOMOGENEOUS CATALYSIS

It is in principle possible to catalyze a chain reaction in two ways. One is to increase the stationary supply of radicals by increasing the rate of initiation. This in essence is the method of chain sensitization by species, such as ethylene oxide or peroxides. It does not, however, represent true catalysis because the sensitizers are used up as the reaction proceeds.

The second method is to use a catalyst to speed up the chain itself by providing an alternative, faster path. An excellent example of this is to be found in the catalytic effect produced by H_2 gas on the pyrolyses of dimethyl ether on C_2H_6 and on almost all hydrocarbon pyrolyses. (We should use some caution in interpreting the results because in many cases, the products are affected and H_2 is consumed.) In the case of $(CH_3)_2O$ the effect is truly catalytic, whereas in the case of C_3H_8, large concentrations of H_2 inhibit C_2H_4 and C_3H_6 production, as well as other minor species, and lead to $C_2H_6 + CH_4$ as almost quantitative products.[39]

Because H_2 is a product of many of these pyrolyses, this also leads to some autocatalytic effect on the rate. In the case of $(CH_3)_2O$ the chain mechanism in the presence of large amounts of H_2 is

$$\text{Catalysis:} \left\{ \begin{array}{l} \cdot CH_3 + H_2 \overset{1}{\rightleftharpoons} CH_4 + H \\ H + CH_3OCH_3 \overset{2}{\rightleftharpoons} H_2 + \cdot CH_2OCH_3 \end{array} \right.$$

$$\text{Chain:} \qquad \cdot CH_2OCH_3 \overset{3}{\rightleftharpoons} CH_2O + \cdot CH_3.$$

The effect of added H_2 is to replace the normal chain step, $\cdot CH_3 + CH_3OCH_3 \overset{1'}{\longrightarrow} CH_4 + \cdot CH_2OCH_3$, by two faster propagation steps, Steps 1 and 2. Steps 1 and 1′ have about the same activation energy, but Step 1 has an A factor higher by about a factor of 2. The rate constant for Step 2 is extremely fast compared to that for Step 1, owing to a still higher A factor by about a factor of 10 and also a lower activation energy by about 4 kcal. This results in a very low stationary state concentration of H atoms and makes Step 1 essentially irreversible.

The combined pair, Steps 1 and 2, regenerate the H_2 catalyst, and can be looked upon as catalysis by chain transfer. The ratio of rates of $(CH_3)_2O$ pyrolysis with and without H_2 are

$$\frac{R(H_2)}{R \text{ (normal)}} \approx \frac{k_1(CH_3)(H_2)}{k_1'(CH_3)(Me_2O)} = \frac{k_1(H_2)}{k_1'(Me_2O)} \tag{5.39}$$

This is calculated for the initial stages of the reaction before H_2 (from the secondary chain) and CH_2O have accumulated in the system. The maximum catalysis is then a factor of 2 in the initial rate when H_2 and Me_2O are equal—not a very large effect.

A much more pronounced catalysis can be obtained via the use of HCl. Reactions of radicals with HCl tend to have activation energies just slightly (~1 kcal) in excess of the endothermicity of the reaction. The second catalytic chain step, involving Cl atom attack on RH to regenerate HCl, is extremely rapid, so that small amounts of HCl can produce very pronounced catalysis. In the $(CH_3)_2O$ pyrolysis 20 mole % of HCl is observed to give an eightfold increase in rate.[40] The catalytic transfer

[39] The mechanism in such cases is known as hydrogenolysis.

[40] S. W. Benson and K. H. Anderson, *J. Chem. Phys.*, **39**, 1677 (1963).

steps are

$$
\text{Catalysis:} \quad \left\{ \begin{array}{l} \dot{C}H_3 + HCl \overset{1}{\rightleftharpoons} CH_4 + Cl \\ Cl + (CH_3)_2O \overset{2}{\rightleftharpoons} HCl + \dot{C}H_2OCH_3; \end{array} \right.
$$

$$
\text{Net reaction:} \quad \dot{C}H_3 + (CH_3)_2O \overset{1'}{\rightleftharpoons} CH_4 + \dot{C}H_2OCH_3.
$$

Here the rate ratios are initially

$$
\frac{R(HCl)}{R \text{ (normal)}} \simeq \frac{k_1(\dot{C}H_3)(HCl)}{k_1'(\dot{C}H_3)(Me_2O)} = \frac{k_1(HCl)}{k_1'(Me_2O)}. \tag{5.40}
$$

A_1 is about threefold larger than A_1', but E_1 is only 2.5 kcal, as against 9.5 kcal for E_1', so that at 750°K the net acceleration at (HCl) = (Me$_2$O) is about 300-fold. This effect is rapidly diminished as CH$_4$ accumulates, for it tends to compete with ether for Cl via Step -1. Including this in the rate expression yields

$$
\frac{R(HCl)}{R \text{ (normal)}} = \frac{k_1(HCl)}{k_1'(Me_2O)[1 + k_{-1}(CH_4)/k_2(Me_2O)]}. \tag{5.41}
$$

Because k_{-1} and k_2 do not differ very much at 750°K, the initial acceleration by HCl is steadily diminished as CH$_4$ accumulates.

Such catalytic effects as those described above are very useful in providing a very important diagnostic tool for examining chains. In a number of instances a priori estimates are not sufficiently reliable to establish which of two chain steps is the faster in a simple scheme. Let us consider the Me$_2$O pyrolysis from this point of view.

The normal chain is

$$
\dot{C}H_3 + CH_3OCH_3 \overset{1'}{\longrightarrow} CH_4 + \dot{C}H_2OCH_3;
$$

$$
\dot{C}H_2OCH_3 \overset{2}{\longrightarrow} CH_2O + \dot{C}H_3.
$$

If no other reactions are important, the steady-state ratio of $\cdot CH_2OCH_3$ to $\cdot CH_3$ is given by

$$
\frac{(\dot{C}H_2OCH_3)}{(\dot{C}H_3)} = \frac{k_1'(Me_2O)}{k_2}. \tag{5.42}
$$

The addition of HCl cannot affect the rate of the fission Step 2, but it can speed up the metathesis Step 1'. If it does, it effectively shifts the steady ratio in favor of higher $(\dot{C}H_2OCH)/(\dot{C}H_3)$ ratios by replacing $k_1'(Me_2O)$ in (5.40) by the much larger $k_1(HCl)$ (5.42). In the limit when this ratio is very large, these are relatively few $\dot{C}H_3$ radicals in the system, and termination is now predominantly by $\dot{C}H_2OCH_3$. Step 2 in the chain has now become the slow rate-determining step. The time of the chain cycle is now effectively τ_2, whereas it was $\tau_1' + \tau_2$. The maximum acceleration in rate is

$$\frac{(\tau_1' + \tau_2)}{\tau_2} = 1 + \frac{\tau_1'}{\tau_2}$$

if we assume that termination rate constants are about the same for the two different radicals. Let us suppose that initially $\tau_1' \sim \tau_2$, implying that $(\dot{C}H_2OCH_3)/(\dot{C}H_3) \sim 1$. In such a case, the maximum acceleration we could observe would be a factor of 2.[41]

In the case of Me_2O small amounts of HCl produced 10-fold accelerations in the rate. This clearly signified that initially $\tau_1' \gg \tau_2$, so that $(\dot{C}H_2OCH_3) \ll (\dot{C}H_3)$, and consequently, $\dot{C}H_2OCH_3$ could be ignored in termination reactions. It also provided lower limits on a direct measure of the rate constant k_2, which was not otherwise measurable.

A similar study of the neopentane pyrolysis[42] showed that large accelerations in the initial rate could be produced by small amounts of HCl, again indicating that the pyrolytic step rate constant was much faster than the specific rate of the metathesis step. The mechanism is

$$\begin{array}{l} \text{Catalytic} \\ \text{Transfer:} \\ \text{Chain:} \end{array} \left\{ \begin{array}{l} \cdot CH_3 + HCl \overset{1}{\rightleftharpoons} CH_4 + Cl \\[2mm] Cl + NpH \overset{2}{\longrightarrow} HCl + N\dot{p} \\[2mm] N\dot{p} \overset{3}{\longrightarrow} (CH_3)_2C{=}CH_2 + \dot{C}H_3 \end{array} \right.$$

and the conclusion was that $(N\dot{p}) \ll (\dot{C}H_3)$ and that $k_3 > k_1'(NpH)$, where k_1' is for the step

$$CH_3 + NpH \overset{1'}{\longrightarrow} CH_4 + N\dot{p}.$$

A very substantial catalysis of many orders of magnitude is observed in the effect of HCl on the pyrolysis of peroxides. In the case of $t\text{-}Bu_2O_2$, very small amounts of HCl produce a catalytic effect independent of HCl and rapidly quenched by products.[43] The products are, in addition, altered:

$$t\text{-}Bu_2O_2 \overset{i}{\longrightarrow} 2t\text{-}Bu\dot{O},$$

$$t\text{-}Bu\dot{O} + HCl \overset{1}{\longrightarrow} t\text{-}BuOH + Cl,$$

$$t\text{-}Bu\dot{O} \overset{2}{\longrightarrow} CH_3COCH_3 + \dot{C}H_3,$$

---

[41] This ratio depends on CH_3OCH_3 concentration, (5.42).

[42] S. W. Benson and K. C. Anderson, *J. Chem. Phys.*, **39**, 1677 (1963); **40**, 3747 (1964).

[43] M. Flowers, L. Batt, and S. W. Benson, *J. Chem. Phys.*, **37**, 2662 (1962).

$$Cl + t\text{-}Bu_2O_2 \xrightarrow{3} HCl + \dot{C}H_2C(CH_3)_2OO\text{-}t\text{-}Bu$$

$$\downarrow 4$$

$$t\text{-}BuO + H_2C\overset{O}{\underset{\diagdown}{\overset{\diagup}{}}}C(Me)_2$$

$$\dot{C}H_3 + HCl \underset{5}{\rightleftharpoons} CH_4 + Cl,$$

$$Cl + CH_3COCH_3 \xrightarrow{t} HCl + \dot{C}H_2COCH_3,$$

$$2\dot{C}H_2COCH_3 \xrightarrow{t'} (CH_2COCH_3)_2.$$

The $\dot{C}H_2COCH_3$ is essentially an inert radical in this system, incapable of reacting rapidly with HCl ($\Delta H_t \simeq -6$ kcal), and so can only terminate. In consequence Step t can be considered the virtual termination. The overall rate of pyrolysis becomes

$$\frac{-d(t\text{-}Bu_2O_2)}{dt} = k_i(t\text{-}Bu_2O_2) + \frac{2k_3k_i(t\text{-}Bu_2O_2)^2}{k_t(CH_3COCH_3)}. \tag{5.43}$$

Below 100°C only the second term in (5.43), is important. It is an approximation that applies only after a few tenths of a percent of decomposition and above some small concentration of HCl, which can be $<10^{-3}$ mole fraction. Because k_3 and k_t are comparable, the catalysis dies out very rapidly as CH_3COCH_3 approaches peroxide in concentration.

Equation (5.43) is also imprecise to the extent that it ignores contributions to "dead" radical production from the other products, isobutylene oxide and t-BuOH.

We can expect similar effects to those described above from HBr and H_2S for reactants that have weak C—H bonds. Thus HBr is a true catalyst for the gas-phase oxidation of i-C_4H_{10}. The mechanism is[44]

$$t\text{-}Bu\cdot + O_2 \xrightarrow{1} t\text{-}BuO_2\cdot,$$

$$t\text{-}BuO_2\cdot + HBr \underset{2}{\rightleftharpoons} t\text{-}BuO_2H + Br,$$

$$Br + i\text{-}C_4H_{10} \underset{3}{\rightleftharpoons} t\text{-}Bu + HBr.$$

The rapid Reaction 1 keeps the t-Bu· concentration low and thereby inhibits the rapid back reaction, Step -3. Step 2 is exothermic by about 3 kcal, so that the equilibrium is well over toward t-BuO$_2$H at 150°C.

A similar behavior of H_2S has been explored in the case of Me_2O[45] and

[44] F. F. Rust and W. E. Vaughn, *Ind. Eng. Chem.*, **41**, 2595 (1949).
[45] N. Imai and O. Toyama, *Bull. Chem. Soc. Japan*, **34**, 328 (1961).

$CH_3CHO.^{46}$ The energetics of $\cdot SH$ radical and $Br\cdot$ atom are very similar.

Species such as I_2 and HI, which were among the first catalysts to be investigated, operate essentially by altering the initiation mechanism and increasing the radical supply, as well as changing the chain. In the case of the I_2-catalyzed decomposition of H_2CO, for example,[47] the I_2 is almost instantaneously consumed to form an I_2–HI mixture, and the remaining small concentration of I_2 reacts as follows:

$$I_2 \overset{1}{\rightleftharpoons} 2I \text{ (rapid, equilibrium)},$$

$$I + H_2CO \overset{2}{\rightleftharpoons} HI + H\dot{C}O,$$

$$H\dot{C}O + I_2 \overset{3}{\longrightarrow} HICO + I,$$

$$HICO \underset{\text{(wall?)}}{\longrightarrow} HI + CO,$$

$$2HI \longrightarrow H_2 + I_2.$$

In this system catalysis can only occur if the temperature is sufficiently high ($>350°C$), so that the HI reaction can regenerate I_2. A similarly complex mechanism is found to be operative in the I_2-catalyzed pyrolysis of CH_3CHO.

5.18 HOMOGENEOUS CATALYTIC RECOMBINATION

An unusual instance of catalysis occurs in the recombination reactions involving atoms and small radicals. Reactions such as $H + H \rightarrow H_2$ or $O + NO \rightarrow NO_2$ all require third bodies to remove the energy, and hence are termolecular. This makes them fairly slow reactions compared to recombinations of larger radicals, such as $2Me \rightarrow C_2H_6$. At 1 atm pressure the termolecular recombination rates are about 1/25 that of the bimolecular recombinations, and this factor increases proportionately with the reciprocal pressure. At 1 torr, the ratio is about $10^{-4.3}$.

Such termolecular recombination can be speeded up by suitable catalysts C, which can engage in the following cycle:

$$\text{Chain recombination:} \quad \left\{ \begin{array}{l} X + \dot{C} \overset{1}{\rightleftharpoons} CX^* \\ CX^* + M \overset{2}{\longrightarrow} CX + M; \\ X + CX \overset{3}{\longrightarrow} X_2 + C \end{array} \right.$$

$$\text{Net reaction:} \quad 2X \longrightarrow X_2.$$

[46] N. Imai and O. Toyama, *Bull. Chem. Soc. Japan*, **33**, 1120, 1408 (1960).

[47] R. F. Faull and G. K. Rollefson, *J. Am. Chem. Soc.*, **58**, 1755 (1936); **59**, 625 (1937); R. Walsh and S. W. Benson, *J. Am. Chem. Soc.*, **88**, 4570 (1966).

The requirements that such a catalyst be effective are that k_3, the metathesis step, be fast (that is, have a low activation energy) and that the lifetime of the hot species, CX^*, be sufficiently long under the ambient conditions that a reasonable fraction of them be stabilized by collisional deactivation (Step 2).

Analysis[48] has shown that above 1000°K, $\dot{C}H_3$ can act as an extremely effective catalyst for H atom recombination. At room temperature almost any hydrocarbon, ketone, or aldehyde, can perform a similar role. Thus 0.1 mole percent of acetone in H atoms can increase the effective recombination rate at 1 torr total pressure by a factor of about 100-fold. Such effects suggest that impurities can be very important in systems containing H atoms.

NO_2 plays a similar, but less spectacular, role in catalyzing the recombination of O atoms via

$$O + NO_2 \rightarrow O_2 + NO,$$

$$O + NO + M \rightarrow NO_2 + M.$$

O_2 appears to do something comparable in the recombination of Cl atoms, the mechanism being

$$Cl + O_2 + M \rightleftarrows Cl\dot{O}\dot{O} + M,$$

$$Cl + Cl\dot{O}\dot{O} \rightarrow Cl_2 + O_2.$$

An interesting consequence of such catalysis as those described above is observed in the pyrolysis of N_2O. The initial step is the production of O atoms by direct split:

$$N_2O \xrightarrow{i} N_2 + O.$$

In the initial stages of the reaction, this is followed by secondary attack of the O on N_2O, leading to

$$N_2O + O \begin{array}{c} \nearrow^{1} 2NO \\ \\ \searrow_{1'} N_2 + O_2 \end{array}$$

Both of these paths have about the same rate constants, and the initial stoichiometry is 2 moles of N_2O decomposed for each O atom produced.[49] However, both these secondary reactions are quickly arrested by the catalytic effect of NO in decreasing the O atom concentration, and the

[48] S. W. Benson, *J. Chem. Phys.*, **38**, 2285 (1963).
[49] F. Kaufman, *Progress in Reaction Kinetics*, **1**, 1 (1961).

apparent rate of decomposition of N_2O falls by a factor of 2 as the stoichiometry changes to 1 mole N_2O per O atom produced. If NO is added initially, no further NO is produced, and the stoichiometry remains constant at unity, all the O atoms going to O_2 via $NO-NO_2$ catalysis.

5.19 ISOMERIZATION STEPS IN CHAINS

Radicals containing more than two polyvalent atoms may in principle isomerize by a unimolecular process.[50] This can take the form of atom abstraction, cyclization, displacement, or our already discussed fission reaction. We may expect that our usual rules apply to these steps. For example, we do not expect to find displacements to occur at saturated C, N, or O atoms, except with very large activation energies, and we may expect to find the activation energies for atom abstraction to parallel those for the more usual bimolecular metatheses.

A new feature of these unimolecular processes is that they must go through a cyclic transition state and hence, should have low A factors, depending on the number of hindered internal rotations, which become inhibited by the ring formation. In addition, ring formation involves strain energy, and we should expect to see this reflected in correspondingly higher activation energies. As an example, we should expect those processes to be the fastest that involve the formation of strain-free, six-membered rings. Five- and seven-membered rings have about 7 kcal of strain, and these should be less favorable by this amount. Four- and three-membered rings have strain energies of about 27 kcal for atoms from the second row of the periodic table, and this should make such transition states extremely rare.

In line with such expectations as those presented above are the observations that internal H-atom abstraction occurs with facility between carbon atoms n and $n+5$ in alkyl radicals, and more slowly between $n \leftrightarrow n+4$ and $n \leftrightarrow n+6$. n-butyl radicals can exchange D atoms via such a path:

$$CD_3CH_2CH_2\dot{C}H_2 \qquad\qquad \dot{C}D_2CH_2CH_2CH_2D$$

$$CD_3\dot{C}H_2 + C_2H_4 \quad \begin{bmatrix} \overset{\textstyle .}{C}D_2 \overset{\textstyle D}{} \overset{\textstyle .}{C}H_2 \\ CH_2{-}CH_2 \end{bmatrix}^{\ddagger} \quad CD_2{=}CH_2 + \dot{C}H_2CH_2D.$$

---

[50] A recent review is provided by J. W. Wilt in *Free Radicals*, Vol. 1, Ed. by J. K. Kochi, John Wiley and Sons, 1973.

Such a 1–4 shift has been observed in n-pentyl radicals:[51]

$$\dot{C}H_3 + C_2H_4 \rightarrow n\dot{C}_3H_7$$
$$\dot{C}_3H_7 + C_2H_4 \rightarrow n\dot{C}_5H_{11} \;\rightleftharpoons\; \left[\begin{array}{c} \overset{\displaystyle \cdot\; H\; \cdot}{} \\ \dot{C}H_2 \qquad CH{-}CH_2 \\ CH{-}CH_3 \end{array} \right]^{\ddagger}$$

$$\| \|$$

$$\dot{C}_2H_5 + C_3H_6 \leftarrow CH_3{-}CH_2{-}CH_2{-}\dot{C}H{-}CH_3$$

No evidence exists for 1,2 or 1,3 H atom shifts in alkyl radicals.[52]

The isomerization of n-propyl radical to i-propyl radical would have to take place via a three-membered ring:

$$\dot{C}H_2{-}CH_2{-}CH_3 \rightleftharpoons \left[\overset{\displaystyle \cdot\; H\; \cdot}{CH_2{-}CH{-}CH_3} \right]^{\ddagger} \rightleftharpoons CH_3{-}\dot{C}H{-}CH_3.$$

If this ring has the same strain energy as cyclopropane or ethylene oxide, namely 27 kcal, and the normal activation energy for such abstraction is about 8 kcal (e.g., $\cdot C_2H_5$ abstracting secondary H atoms), we might expect a net activation energy of about 35 kcal. The A factor should be about $10^{13.3} \, s^{-1}$ because we are losing a low barrier, $\cdot CH_2$ hindered rotation. The competing steps could be external H-atom abstraction, recombination, or elimination of CH_3. This latter has an A factor of about $10^{14.0} \, s^{-1}$ and an activation energy of 33 kcal; thus it would be about 30 to 100 times faster at most temperatures, and we would not expect to see such isomerizations.

The same analysis and conclusions made above should also extend to the n-butyl radical, which could only isomerize by a 1,2 or 1,3 H-atom shift. The four-membered ring has only 1 kcal less strain than the three-membered ring, and the A factor should be lower by another factor of 5 because it involves loss of an additional hindered rotor.

It has been shown by isotope-labeling experiments that the exothermic ($\Delta H^\circ = -3.5$ kcal) isomerization of n-butyl radical to s-butyl radical must have an activation energy in excess of 36 kcal to account for the absence of s-butyl products,[53] or intermediates which would form $C_3H_6 + CH_3$.

[51] A. S. Gordon and J. R. McNesby, *J. Chem. Phys.*, **31**, 853 (1959); **33**, 1882 (1960); L. Endrenyi and D. L. LeRoy, *J. Phys. Chem.*, **70**, 4081 (1966).

[52] From data on competing reactions, it is estimated that for 1,2-H atom shifts $E_{act} \geq 38$ kcal and $E_{act} \geq 36$ kcal for 1,3-H atom shifts.

[53] A. S. Gordon and J. R. McNesby, *J. Chem. Phys.*, **33**, 1882 (1960).

In surprising contrast to the result above, a considerable body of evidence has accumulated to indicate that Cl atoms can undergo 1,2 transfer in radicals. Thus the radical addition of Br_2 to trichloropropene leads to rearranged product[54]

$$Br + CH_2\!\!=\!\!CH\!-\!CCl_3 \rightleftharpoons BrCH_2\!-\!\dot{C}H\!-\!CCl_3$$

$$\updownarrow$$

$$CH_2Br\!-\!CHCl\!-\!CCl_2Br \xleftarrow{\ Br_2\ } BrCH_2\!-\!CHCl\!-\!\dot{C}Cl_2$$

The C—Cl bond in these radicals is only about 16 kcal, so that it is not impossible that the abstraction takes place via dissociation in a solvent cage, followed by readdition of the Cl atom to the short-lived olefin.[55] It would require an activation energy at least this small to account for the results. By way of partial explanation, however, it is observed that third-row atoms, like S, have strain energies in three- and four-membered rings about 8 kcal lower than that for second row atoms.

Another piece of evidence on H atom internal abstraction comes from data on the biradicals produced in cyclopropane pyrolysis. Analysis of these data has shown that the 1,2 H-atom shift in the trimethylene radical, which produces C_3H_6 with $\Delta H° = -62$ kcal, has an activation energy of 11 kcal and the expected A factor of about $10^{13}\,s^{-1}$. The driving force for the reaction is the double-bond formation, and it appears from the low activation energy as though it is about 40% developed in the transition state:

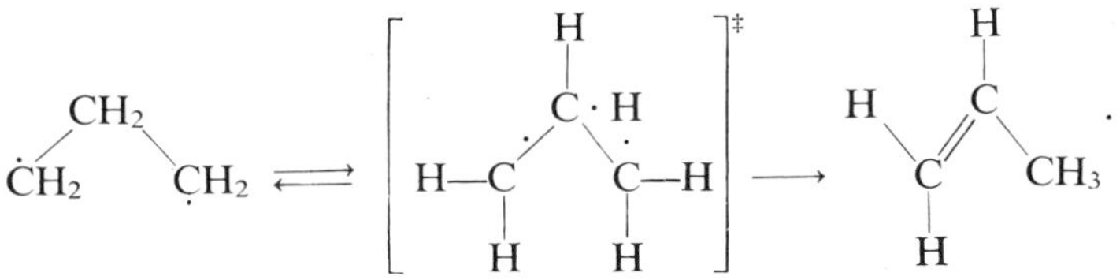

One consequence of this facile 1,2-H atom shift for olefins is that it presents a significant competing pathway for isomerization of olefins. Thus cis-butene-2, which can isomerize by simple double bond rotation, as already noted, can also isomerize at a competing rate by biradical

[54] A. N. Nesmeyanov, R. K. Freidlina, and V. N. Kost, *Tetra. Lett.*, **1**, 241 (1957).
[55] Note also that in polar solvents, heterolytic fission of the radical into the ion-pair; olefin$^+$ Cl$^-$ can be favored.

formation:[56]

$$\underset{CH_3}{\overset{CH=CH}{\diagdown}}\underset{CH_3}{\diagup} \quad \overset{1}{\rightleftharpoons} \quad \underset{\cdot CH_2}{\overset{CH_2-\dot{C}H}{\diagdown}}\underset{CH_3}{\diagup} \quad \overset{2}{\rightleftharpoons} \quad \underset{CH_2}{\overset{CH-CH_2}{\diagup\diagdown}}\underset{CH_3}{}$$

(cis)   3$\updownarrow$

$$\underset{CH_3}{\overset{CH=CH}{\diagdown}}\underset{}{\overset{CH_3}{\diagup}} \quad \overset{1'}{\rightleftharpoons} \quad \underset{\cdot CH_2}{\overset{CH_3}{}} CH_2-\dot{C}H \qquad \overset{2'}{\leftarrow}$$

(trans)   4$\updownarrow$

$$\underset{CH_2}{\overset{CH_2}{\diagdown}}\underset{CH}{\diagup}\overset{CH_3}{}$$

The rate determining step is expected to be k_1 with estimated parameters, $A_{700} = 10^{14.5}$ s^{-1} and $E = 71$ kcal. At 700°K, $k_1 = 10^{-7.7}$ s^{-1}, which is negligible compared to the observed rate, but at 1100°K in shock tubes the two rates are comparable. At both temperatures the secondary products, butene-1 and methyl cyclopropane would be unstable.

Oxidation of hydrocarbons in the liquid state provides a very interesting example of internal H-atom abstraction. In the sensitized oxidation of 2,4-dimethyl pentane,[57] it is observed that very high yields of a di-OOH are formed. The mechanism is

$$R\dot{O}_2 + Me_2\underset{}{\overset{H}{C}}\underset{CH_2}{\diagdown}\underset{}{\overset{H}{C}Me_2} \quad \overset{1}{\longrightarrow} \quad Me_2\dot{C}-CH_2-\overset{H}{C}Me_2 + RO_2H,$$

$$Me_2\dot{C}-CH_2-\overset{H}{C}Me_2 + O_2 \quad \overset{2}{\longrightarrow} \quad Me_2C\underset{OOH}{\overset{O-\dot{O}\ H}{\diagup}}CH_2-CMe_2,$$

$$Me_2C\underset{CH_2-CMe_2}{\overset{O-\dot{O}\quad H}{\diagdown}} \quad \overset{3}{\longrightarrow} \quad Me_2C\underset{CH_2-\dot{C}Me_2}{\overset{OOH}{\diagup}}$$

$$\downarrow +O_2$$

$$Me_2C\underset{+\dot{R}}{\overset{O_2H\quad O_2H}{}}CH_2-CMe_2 \quad \overset{+RH}{\longleftarrow} \quad Me_2C\underset{CH_2-CMe_2}{\overset{OOH}{\diagup}}\dot{O}_2$$

---

[56] The detailed mechanism, whether biradical or concerted, is irrelevant since the isomerization proceeds via the known, reverse reaction to methyl cyclopropane which can then go back to butene-1 or butene-2 (cis and trans).

[57] F. F. Rust, J. Am. Chem. Soc., **79**, 4000 (1957).

Here a six-membered ring is involved, so that the activation energies for internal and external abstraction should be equal. The A factor for internal abstraction involves tightening of three internal rotors and therefore should be about 10^{11} s^{-1}, which is still larger than the expected $10^{8.5}$ (RH) for external abstraction even in pure liquid where $(RH) = 10$ M. If the isomer 2,3-dimethyl pentane is used, then only monohydroperoxide is produced in agreement with our simple expectation of an additional 7 kcal activation energy for H atom abstraction in the five-membered ring. A faster process would involve H-atom abstraction via a six-membered ring, but this is about 3 kcal endothermic, compared to tertiary H-atom abstraction, and might have an activation energy 4 kcal higher. This would make it about fivefold slower than external abstraction at 150°C, and there is some evidence for comparable amounts of such a product with the isomeric hydrocarbon.

It is interesting to consider, in the light of this analysis, a commonly assumed reaction of alkyl radicals with O_2:

$$R\dot{C}H_2 + O_2 \rightarrow RCHO + \dot{O}H.$$

Such a reaction is about 57 kcal exothermic. However it must proceed via the peroxy intermediate with liberation of about 28 kcal:

$$R\dot{C}H_2 + O_2 \rightleftarrows (RCH_2O\dot{O})^* + 28 \text{ kcal}.$$

The peroxy radical would then rearrange to a hydroperoxyl radical via a four-membered ring:

$$RCH_2O\dot{O} \rightleftarrows \left[\begin{matrix} RCH\!\!-\!\!O \\ \cdotO \\ H\cdot \end{matrix} \right]^{\ddagger} \rightleftarrows RCHO_2H.$$

This reaction is about 3.5 kcal endothermic and would have an estimated activation energy of about 26 (strain energy) + 7 (normal abstractions) + 3.5 (endothermicity) = 36.5 kcal. The initial excitation is insufficient to provide this, so the reaction rate should not be faster for the initially hot radical, which will instead be rapidly thermalized.

However, if the hydroperoxy radical is formed, we may expect a rapid, exothermic fission into $\dot{O}H + RCHO$ ($\Delta H° \sim -26$ kcal). This step should still have some small activation energy (~ 5 kcal). We thus see that despite the large exothermicity of the radical oxidation, the overall rate is expected to be negligibly slow, for it has to proceed as a bimolecular reaction with about 40 kcal of activation energy.[58] There is, in fact, very little experimental evidence for this step.

[58] Note also that the competing step is redissociation back into $O_2 + R\dot{C}H_2$ with a lower activation energy and a higher A factor.

About the first investigators to suggest the importance of internal abstraction in hydrocarbon pyrolyses were Benson and Kistiakowsky.[59] This was immediately applied by Rice and Kossiakoff,[60] who showed that in the pyrolysis of long-chain hydrocarbons, $C_{16}H_{34}$, and so on, the product distribution was in reasonable accord with a statistical abstraction of secondary H atoms, followed by very rapid isomerization equilibrium via H-atom abstraction across mainly six-, but also five- and seven-membered rings. At pyrolysis temperatures of 700–1100°K, these isomerizations are indeed rapid, compared to competing bimolecular and fission steps.

There is one point worth emphasizing in regard to these internal rearrangements described above. Generally, only exothermic processes ($\Delta H_r < 0$) are important. The reason for this is that any endothermic process always has a reverse rate, which is exothermic and hence faster by the factor $10^{-\Delta H_r/\theta}$. Thus, if we consider the endothermic reaction

$$CH_3CH_2CH_2O\dot{O} \rightleftarrows \dot{C}H_2CH_2CH_2O_2H,$$

the reverse reaction is exothermic by about 8 kcal and hence faster by about 10^3 at 600°K. The result is a negligible steady-state concentration of the endothermic species. We may make a general rule that we expect internal rearrangements to be significant only if the reaction is exothermic or nearly thermoneutral. However we must always consider competitive reactions before summarily neglecting such steps.[61]

5.20   HOT-MOLECULE REACTION—CHEMICAL ACTIVATION

Many chain reactions are exothermic. This means that one or more of the chain-propagation steps must reflect this exothermicity and hence produce radical or molecule species that jointly contain this heat of reaction. If the exothermic step had an activation energy in addition, this too must appear in the product species. The following are some representative, fairly exothermic steps:

$$H + Cl_2 \xrightarrow{1} HCl + Cl + 44 \text{ kcal},$$

$$H + F_2 \xrightarrow{2} HF + F + 96 \text{ kcal},$$

$$O + O_3 \xrightarrow{3} 2O_2 + 93 \text{ kcal},$$

[59] S. W. Benson and G. B. Kistiakowsky, *J. Amer. Chem. Soc.*, **64**, 80 (1942).

[60] A. Kossiakoff and F. O. Rice, *J. Am. Chem. Soc.*, **65**, 590 (1943).

[61] S. W. Benson, in *The Mechanisms of Pyrolysis, Oxidation, and Burning Materials*, NBS Special Publication 357, L. Wall (ed.), 1972.

$$H + C_2\dot{H}_4 \xrightarrow{4} C_2H_5 + 39 \text{ kcal},$$

$$\dot{C}H_2CH_2C\dot{H}_2 \xrightarrow{5} CH_3CH{=}CH_2 + 62 \text{ kcal},$$

$$2CH_3\dot{O} \xrightarrow{6} CH_2O + CH_3OH + 80 \text{ kcal},$$

$$O + NO_2 \xrightarrow{7} O_2 + NO + 46 \text{ kcal},$$

$$H + NO_2 \xrightarrow{8} OH + NO + 30 \text{ kcal},$$

$$H + O_3 \xrightarrow{9} OH + O_2 + 77 \text{ kcal}.$$

The energy of the reaction can be distributed among all of the degrees of freedom of the products, including translation, vibration, rotation, and electronic. Recent studies of excited HCl produced in Step 1, show that up to 90% of the heat of the reaction can appear in vibrational excitation.[62] On the average, about 60% of the heat of reaction appears in vibration.

The translationally excited species produced in exothermic reactions can have abnormally high probabilities for bimolecular reaction and thus give rise to what can appear to be an abnormal rate constant. Such effects arise, typically, in photolysis reactions which fragment bonds. Thus, the H—I photolysis at 2537 Å gives rise to an H atom with 99.2% of the excess energy of the reaction, namely 41 kcal. It is abnormally reactive.[63]

When the reaction is an association or is a unimolecular isomerization, all of the excess energy is localized in the vibrational modes of the product. The unimolecular isomerization of cyclopropane to propylene is a good example. The reaction is 7.5 kcal exothermic and has an activation energy of 65 kcal. Hence the product C_3H_6 molecules are formed with an excess internal energy of 72.5 kcal.[64] This qualifies them as hot molecules. This energy will be dissipated in consecutive collisions, but it can also participate in other competitive internal reactions. As an example, the isomerization of methyl cyclopropane produces butene-2 molecules with virtually the same 72 kcal excess energy. Because the cis–trans isomerization of butene-2 requires only 63 kcal (Table 3.8) of activation energy, we might expect to obtain cis–trans equilibrium in this pyrolysis. This is not the case (Table 3.15). The cis is produced about

---

[62] J. C. Polanyi, *J. Quant. Spec. Radiat. Trans.*, **3**, 471 (1963). The HBr* case is discussed in J. R. Airey, P. D. Pacey, and J. C. Polanyi, *Eleventh International Symposium on Combustion*, **85**, Combustion Institute, Pittsburgh, 1967.

[63] Also, because of its light mass, it requires many collisions before it is "thermalized."

[64] This is above the normal thermal energy that they already have at the reaction temperature, $E_v = \int_0^{T_C} C_{\text{vib.}} \, dT$.

twice as fast as the thermodynamically more stable *trans*. This implies that collisional deactivation is faster than reaction of the hot species.

We can use the RRK theory to make a rough estimate of the result described above. The rate of reaction of the hot cis species will be given by

$$k_{c \to t} = A \left(\frac{E - E^*}{E} \right)^{s-1} \tag{5.44}$$

$$= 10^{13.8} \left(\frac{E - 63}{E} \right)^{s-1} s^{-1}. \tag{5.45}$$

when the A factor and activation energy are for the high-pressure reaction parameters (Table 3.8). The thermal energy is estimated from

$$\int_0^{700} (C_p - 8) \, dT = E_{th.} = 11.0 \, \text{kcal}$$

with C_p taken from our tables in the appendix. The molecule C_4H_8 has 30 internal modes, of which we can take $\frac{2}{3} = 20$, or more accurately use $S_{eff.} = (C_p - 8)/R = 15$ at 700°K. Then, with $E = \Delta H^\circ + E_{th.}$, we have

$$k_{c \to t} = 10^{13.8} \left(\tfrac{20}{83} \right)^{14} \approx 10^{5.4} \, s^{-1}.$$

This is so slow compared to collision frequencies, even at 1 torr $(10^{6.5} \, s^{-1})$, that we can conclude there is no significant effect of the excess energy.

An entirely different case is presented by the pyrolysis of ethylene oxide, isoelectronic with cyclopropane. We can expect a parallel mechanism of decomposition to acetaldehyde, which is in fact, observed:

$$\overline{CH_2 - CH_2 - O} \rightleftarrows \cdot CH_2 - CH_2 - O \cdot \to CH_3 - CHO^*.$$

The overall reaction has $\Delta H^\circ = -28$ kcal/mole and an observed activation energy of 57 kcal. The product CH_3CHO is thus excited by 85 kcal of excess energy, and because ΔH° for decomposition to $\cdot CH_3 + \dot{C}HO$ radicals is 82 kcal[65] at 700°K, we can write

$$k(CH_3CHO^*) \approx 10^{16} \left(\frac{10.5}{90} \right)^9 \sim 10^{7.5} \, s^{-1}$$

where $E_{th} \sim 5$ kcal and $S_{eff} \sim 10$, $S_{max} = 3n - 6 = 15$. Assuming a deactivation efficiency of $\frac{1}{3}$, we can estimate that half of the CH_3CHO^* will decompose at 10 torr and only 5% at 1 atm, in good agreement with observation (Reference 9, Table 3.15). At 1 atm, about 10% of the product species are dissociated into radicals.

[65] We actually use $\Delta E^\circ - RT$ instead of ΔH° on the assumption that the mean rotational energy of RT is effectively utilized in bond breaking.

Comparable types of reactions are seen in photochemical systems where recombinations can produce excited species with sufficient energy to undergo further reaction. The recombination of two $\cdot CH_2I$ radicals will be about 88 kcal exothermic and produce hot $C_2H_4I_2$, which requires only 36 kcal to split into I_2 and C_2H_4.

$$k(C_2H_4I_2^*) = 10^{13.5}(\tfrac{52}{88})^{13} \approx 10^{10.9} \text{ s}^{-1}$$

where we have neglected the very small amount of thermal energy. Decomposition is about 20 times faster than collision at 1 atm, so we can anticipate almost nothing but $I_2 + C_2H_4$ as products of the recombination. Similar effects have been observed with the recombination of $\cdot CH_3 + \cdot CH_2Cl$, $2 \cdot CH_2Cl$, and $\cdot CHCl_2 + \cdot CH_3$, and the corresponding fluoro radicals, all of which have enough excess energy to split out $HF + $ olefin in the range 300°K to 500°K.[66]

Even more energetic species are obtained from the carbenes (CH_2, etc.) and isoelectronic O-atom reactions. A typical CH_2 reaction is addition to a double bond. The reaction of CH_2 with C_2H_4 leads to hot trimethylene, or hot cyclopropane. The latter has 86 kcal of excess energy calculated for ground state (triplet) CH_2. Because only 65 kcal are required for chemical isomerization to propylene, we can write (at 300°K)

$$k(c\text{-}C_3H_6^*) \simeq 10^{15.2}(\tfrac{21}{86})^{s-1} \text{ s}^{-1}.$$

To estimate S_{eff}, we must calculate it for a temperature at which the internal energy would be 24 kcal. This is given approximately by the temperature T where

$$\int_0^T (C_p - 8)\, dT \sim 21 \text{ kcal.}$$

At about 1250°K

$$\int_0^T (C_p - 8)\, dT \simeq 21 \text{ kcal, \quad and \quad } (C_p - 8)/R \approx 15(\sim 2/3 S_{max}).$$

Thus it seems reasonable to use S_{eff} ($\leq 3n - 6$) = 15.

$$k \sim 10^{15.2}\left(\frac{1}{4.2}\right)^{14} \approx 10^{6.4} \text{ s}^{-1}.$$

It is interesting to make a calculation similar to that made above for the cis–trans isomerization of dimethyl cyclopropane, produced by the addition of CH_2 to cis-butene-2. The reaction is about 89 kcal exothermic

---

[66] S. W. Benson and G. R. Haugen, *J. Phys. Chem.*, **69**, 3898 (1965). For later work see C. F. Larson and B. S. Rabinovitch, *J. Chem. Phys.*, **52**, 5181 (1970) and references therein.

and the Arrhenius parameters from Table 3.15 give (using $S_{eff} = \frac{2}{3} S_{max} = 26$)

$$k^* = 10^{15.25} (\tfrac{30}{89})^{25}$$
$$\approx 10^{3.2} \text{ s}^{-1},$$

so that isomerization should be almost negligibly small compared to quenching in either gas or liquid states.[67] The implication of this calculation is that any observed isomerization in the system cannot come after the ring has formed. It has to come, instead, from an initially open biradical with 38 kcal of excess energy and with three competing paths open to it[68]: rotation to produce geometric isomer, ring closing to form cyclopropane, and H migration to yield 2-methyl butenes-1 (or 2). The relative rates for rotation R_{rot}, for example, relative to ring closure $R_{(c)}$ is

$$\frac{R_{(c)}}{R_{(rot)}} = \frac{10^{13.5}}{10^{11.5}} \left(\frac{38-10}{38-5} \right)^{25} \approx 10^{2.0-1.8} \approx 1.5$$

so that, depending on the rate of quenching and the precise energy of the system, we get varying amounts of isomerization. These calculations militate very strongly against the initial formation of a ring compound. They do not rule out triplet state intervention, which would also lead only to biradical, however.

5.21   REACTIONS OF OZONE WITH OLEFINS

The amazing and not too stable compound O_3 undergoes very unusual reactions of addition to olefins, both at low temperatures ($-70°C$) in nonpolar solvents and also at higher temperatures in the gas phase. Many of the adducts of such reactions are quite stable species, but show the unexpected structure

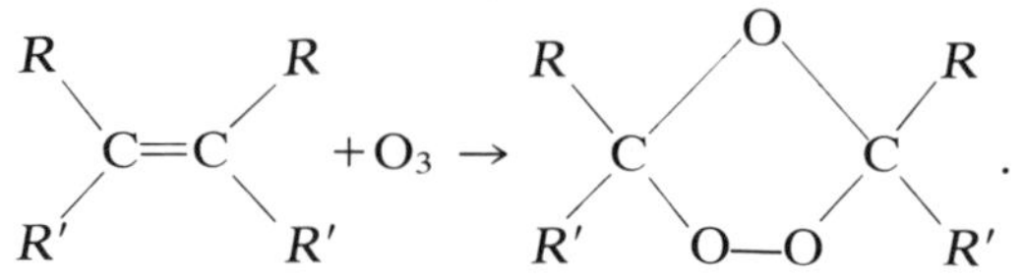

The unusual finding is that there has been a scission of the $C{=}C$ double bond. Let us consider these reactions from an energetic point of view. For

---

[67] For detailed discussion, see J. W. Simons and B. S. Rabinovitch, *J. Phys. Chem.*, **68**, 1322 (1964).

[68] This could be the triplet biradical which doesn't react until it converts to singlet.

the simplest olefin, $C_2H_4 + O_3$, the production of the expected product would be 45 kcal exothermic:

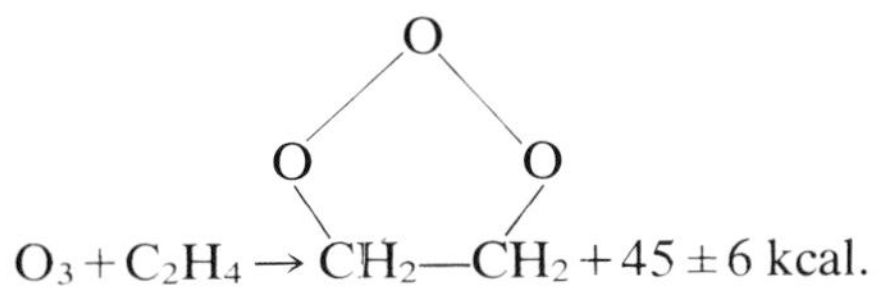

$$O_3 + C_2H_4 \rightarrow \underset{\substack{| \\ CH_2—CH_2}}{\overset{\substack{O \\ \diagup \diagdown \\ O \qquad O}}{}} + 45 \pm 6 \text{ kcal.}$$

This is a sufficiently energetic reaction to produce a consequently "hot" species. However even if this initial hot adduct is stabilized, the energy to open the ring and form the peroxy biradical is only 13 ± 2 kcal. Even at $-70°$ with $\theta = 0.92$ kcal, the half-life for ring opening with an estimated A factor of $10^{14.5} \text{ s}^{-1}$ is ~0.1 s:

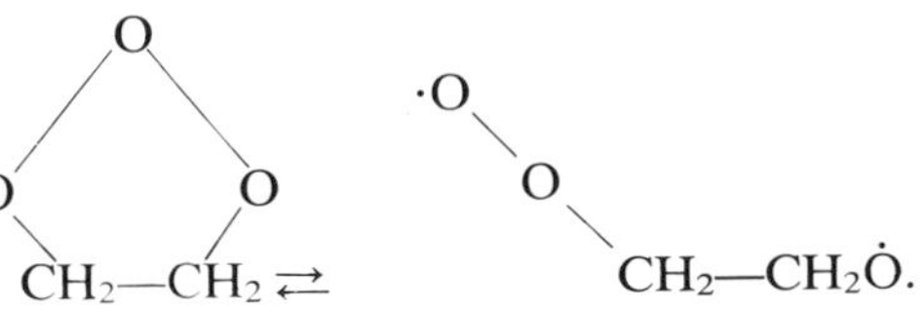

$$\underset{\substack{| \\ CH_2—CH_2}}{\overset{\substack{O \\ \diagup \diagdown \\ O \qquad O}}{}} \rightleftarrows \underset{\substack{| \\ CH_2—CH_2\dot{O}}}{\overset{\substack{\cdot O \\ \diagdown \\ O}}{}}.$$

A secondary split into two fragments has an enthalpy change ≤ 23 kcal, the upper limit being set by the assumption of no interaction energy aside from σ bonding in the Criegee biradical $\dot{C}H_2O\dot{O}$:

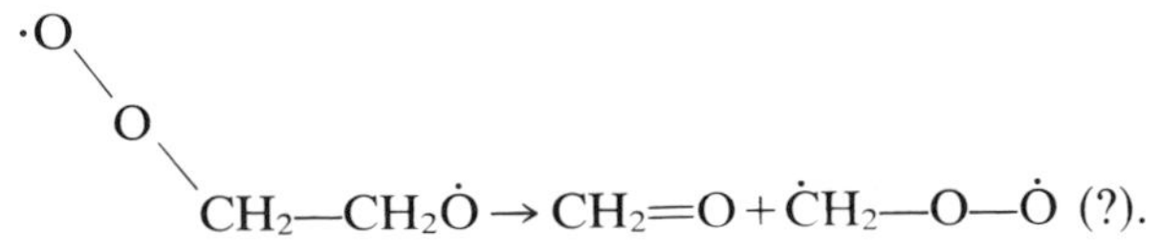

$$\underset{\substack{| \\ CH_2—CH_2\dot{O}}}{\overset{\substack{\cdot O \\ \diagdown \\ O}}{}} \rightarrow CH_2{=}O + \dot{C}H_2—O—\dot{O} \ (?).$$

If the biradical has the structure of an aldehyde-oxide, the reaction above is actually more than 9 kcal exothermic:

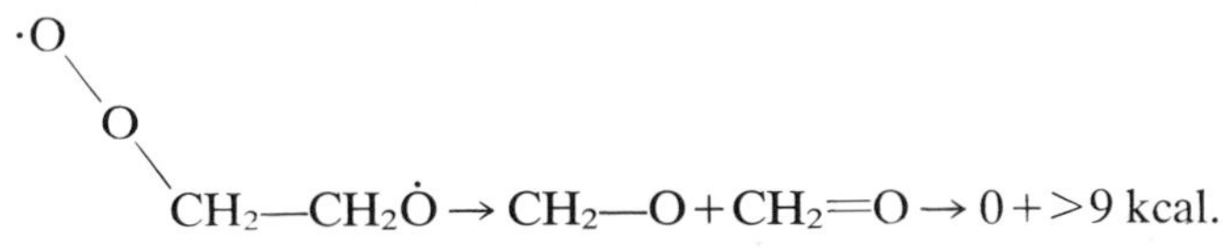

$$\underset{\substack{| \\ CH_2—CH_2\dot{O}}}{\overset{\substack{\cdot O \\ \diagdown \\ O}}{}} \rightarrow CH_2—O + CH_2{=}O \rightarrow 0 + {>}9 \text{ kcal.}$$

The formation of such a pair of species in a cage would now begin to make sense of the subsequent reactions of true ozonide formation and exchange of species with dissolved aldehydes or ketones.[69]

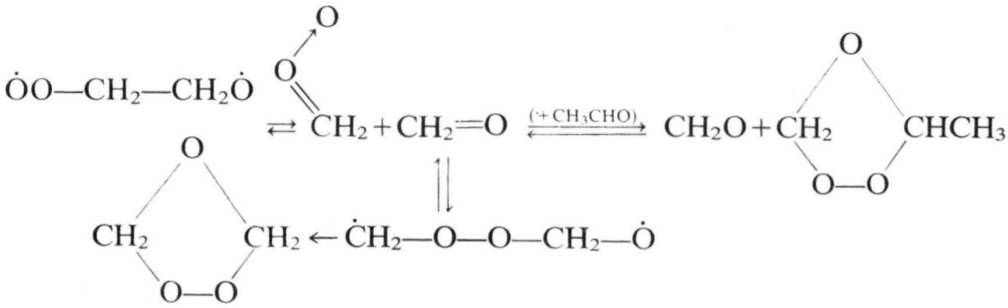

$$\dot{O}O-CH_2-CH_2\dot{O}$$

$$\rightleftarrows \dot{C}H_2 + CH_2 {=} O \xrightarrow{(+CH_3CHO)} CH_2O + \dot{C}H_2 \quad CHCH_3$$

$$CH_2 \quad \dot{C}H_2 \leftarrow \dot{C}H_2-O-O-CH_2-\dot{O}$$

The overall rearrangement of initial trioxacyclopentane to true ozonide is exothermic by an estimated 61 kcal, so that the entire reaction starting with $C_2H_4 + O_3$ is exothermic by 106 kcal/mole! The oxonide is now capable of a further ring opening to form biradicals with an estimated activation energy of ~30 kcal. Such a reaction, however, leads to no net products without a further split into the dioxo biradicals:

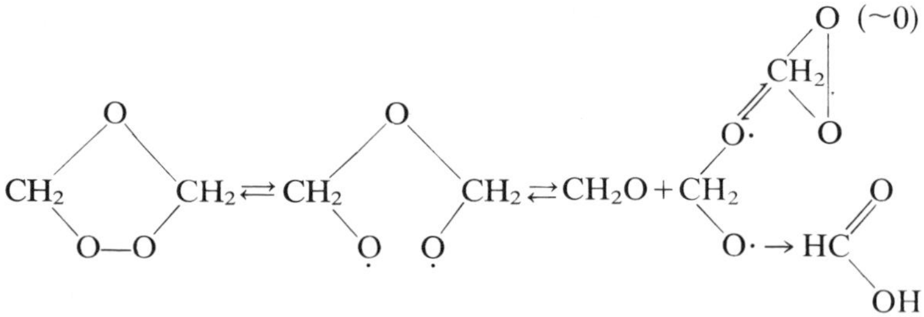

$$CH_2 \quad CH_2 \rightleftarrows \dot{C}H_2 \quad \dot{C}H_2 \rightleftarrows CH_2O + \dot{C}H_2$$

$$O \cdot \rightarrow HC\begin{smallmatrix}O\\\\OH\end{smallmatrix}$$

At the moment we can only speculate about the existence of the aldehyde oxide species, but there is evidence from the chain oxidation of C_2F_4 to indicate that the active chain carrier is $CF_2{=}O \rightarrow O$; furthermore, a very stable diperoxide has been obtained from the photolysis of diphenyl ketene in O_2[70]:

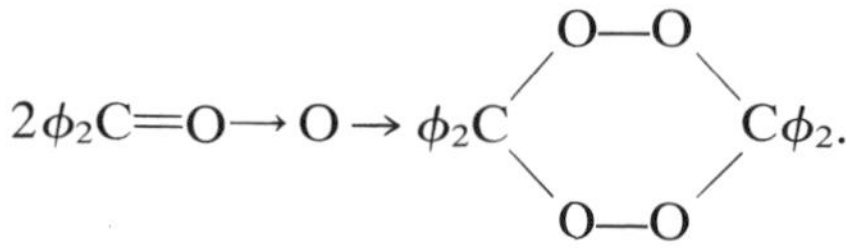

$$2\phi_2C{=}O \rightarrow O \rightarrow \phi_2C\begin{smallmatrix}O-O\\\\O-O\end{smallmatrix}C\phi_2.$$

Despite the weak peroxide linkage, this turns out to be a very stable

[69] See detailed discussion in *Oxidation of Organic Compounds*, Vol. III, *Advances in Chem. Series 77*, American Chemical Society Publication, 1968.

[70] P. D. Bartlett and T. G. Traylor, *J. Am. Chem. Soc.*, **84**, 3408 (1962). See also H. E. O'Neal and C. Blumstein, *Int. J. Chem. Kinet.*, **5**, 397 (1973).

peroxide, for a second, fairly energetic bond has to be ruptured before a net reaction can occur:

The overall activation energy can be in excess of 45 kcal/mole.

Niki et al.[71] have determined a number of the ozone-olefin products in gas phase studies.

[71] H. Niki et al., *J. Am. Chem. Soc.*, in press (1976).

Appendix: Tables

The following data all (unless otherwise noted) refer to standard states of ideal gas at 1 atm and 25°C. ΔH_f° are in kcal/mole; S° and C_{pT}° are in cal/mole-°K.

Much of the data is discussed in an article by S. W. Benson, F. R. Cruickshank, D. M. Golden, G. R. Haugen, H. E. O'Neal, A. S. Rodgers, R. Shaw, R. Waish, *Chem. Rev.*, **69,** 279 (1969), and a subsequent article by H. K. Eigenmann, S. W. Benson, and D. M. Golden, *J. Phys. Chem.*, **77,** 1687 (1973) on the ΔH_f° of oxygen-containing compounds. One of the most important sources have been the *JANAF* Thermochemical tables (2nd ed.) edited by D. R. Stull and H. Prophet et al., NSDRS, NBS-37, U.S. Government Printing Office, Washington, D.C. (1971). Two other valuable sources have been, J. D. Cox and G. Pilcher, *Thermochemistry of Organic and Organometallic Compounds*, Academic Press, London (1970), and D. R. Stull, E. F. Westrum, Jr., and G. C. Sinke, *The Chemical Thermodynamics of Organic Compounds*, Wiley, New York, 1969.

Table A.1. Group Values for ΔH_f°, S_{int}°, and C_{pT}°, Hydrocarbons

Group	ΔH_f° 298	S_{int}° 298	C_p° 300	400	500	600	800	1000	1500
C—(H)₃(C)	−10.20	30.41	6.19	7.84	9.40	10.79	13.02	14.77	17.58
C—(H)₂(C)₂	−4.93	9.42	5.50	6.95	8.25	9.35	11.07	12.34	14.25
C—(H)(C)₃	−1.90	−12.07	4.54	6.00	7.17	8.05	9.31	10.05	11.17
C—(C)₄	0.50	−35.10	4.37	6.13	7.36	8.12	8.77	8.76	8.12
Cd—(H)₂	6.26	27.61	5.10	6.36	7.51	8.50	10.07	11.27	13.19
Cd—(H)(C)	8.59	7.97	4.16	5.03	5.81	6.50	7.65	8.45	9.62
Cd—(C)₂	10.34	−12.70	4.10	4.61	4.99	5.26	5.80	6.08	6.36
Cd—(Cd)(H)	6.78	6.38	4.46	5.79	6.75	7.42	8.35	8.99	9.98
Cd—(Cd)(C)	8.88	−14.6	(4.40)	(5.37)	(5.93)	(6.18)	(6.50)	(6.62)	(6.72
[Cd—(CB)(H)]	6.78	6.38	4.46	5.79	6.75	7.42	8.35	8.99	9.98
Cd—(CB)(C)	8.64	(−14.6)	(4.40)	(5.37)	(5.93)	(6.18)	(6.50)	(6.62)	(6.72
[Cd—(Ct)(H)]	6.78	6.38	4.46	5.79	6.75	7.42	8.35	8.99	9.98
Cd—(CB)₂	8.0								
Cd—(Cd)₂	4.6								
C—(Cd)(C)(H)₂	−4.76	9.80	5.12	6.86	8.32	9.49	11.22	12.48	14.36
C—(Cd)₂(H)₂	−4.29	(10.2)	(4.7)	(6.8)	(8.4)	(9.6)	(11.3)	(12.6)	(14.4
C—(Cd)(CB)(H)₂	−4.29	(10.2)	(4.7)	(6.8)	(8.4)	(9.6)	(11.3)	(12.6)	(14.4
C—(Ct)(C)(H)₂	−4.73	10.30	4.95	6.56	7.93	9.08	10.86	12.19	14.26
C—(CB)(C)(H)₂	−4.86	9.34	5.84	7.61	8.98	10.01	11.49	12.54	13.76
C—(Cd)(C)₂(H)	−1.48	(−11.69)	(4.16)	(5.91)	(7.34)	(8.19)	(9.46)	(10.19)	(11.28
C—(Ct)(C)₂(H)	−1.72	(−11.19)	(3.99)	(5.61)	(6.85)	(7.78)	(9.10)	(9.90)	(11.1
C—(CB)(C)₂(H)	−0.98	(−12.15)	(4.88)	(6.66)	(7.90)	(8.75)	(9.73)	(10.25)	(10.68
C—(Cd)(C)₃	1.68	(−34.72)	(3.99)	(6.04)	(7.43)	(8.26)	(8.92)	(8.96)	(8.2
C—(CB)(C)₃	2.81	(−35.18)	(4.37)	(6.79)	(8.09)	(8.78)	(9.19)	(8.96)	(7.6
Ct—(H)	26.93	24.7	5.27	5.99	6.49	6.87	7.47	7.96	8.8
Ct—(C)	27.55	6.35	3.13	3.48	3.81	4.09	4.60	4.92	6.35
Ct—(Cd)	29.20	(6.43)	(2.57)	(3.54)	(3.50)	(4.92)	(5.34)	(5.50)	(5.86
Ct—(CB)	(29.20)	6.43	2.57	3.54	3.50	4.92	5.34	5.50	5.80
CB—(H)	3.30	11.53	3.24	4.44	5.46	6.30	7.54	8.41	9.73
CB—(C)	5.51	−7.69	2.67	3.14	3.68	4.15	4.96	5.44	5.98
CB—(Cd)	5.68	−7.80	3.59	3.97	4.38	4.72	5.28	5.61	5.72
[CB—(Ct)]	5.68	−7.80	3.59	3.97	4.38	4.72	5.28	5.61	5.75
CB—(CB)	4.96	−8.64	3.33	4.22	4.89	5.27	5.76	5.95	(6.05
Ca	34.20	6.0	3.9	4.4	4.7	5.0	5.3	5.5	5.7
CBF—(CB)₂(CBF)	4.8	−5.0	3.0	3.7	4.2	4.6	5.2	5.5	—
CBF—(CB)(CBF)₂	3.7	−5.0	3.0	3.7	4.2	4.6	5.2	5.5	—
CBF—(CBF)₃	1.5	1.4	2.0	2.9	3.5	4.0	4.7	5.1	—

Cd represents double-bonded C atom, Ct the triple bonded C-atom, CB the C atom in a benzene ring and Ca allenic C atom. By convention group values for C—(X)(H)₃ will always be taken as those for C—(C)(H)₃ when is any other polyvalent atom such as Cd, Ct, CB, O, and S. CBF represents a carbon atom in a fused ring syst such as naphthalene, anthracene, etc. CBF—(CBF)₃ represents the group in graphite.

Table A.1. (*Contd.*)
Non-Next-nearest Neighbor Corrections

Group	ΔH_f° 298	S_{int}° 298	C_p° 300	400	500	600	800	1000	1500
Alkane *gauche* correction	0.80								
Alkene *gauche* correction	0.50								
cis-Correction	1.00[a]	[b]	−1.34	−1.09	−0.81	−0.61	−0.39	−0.26	0
ortho Correction	0.57	−1.61	1.12	1.35	1.30	1.17	0.88	0.66	−0.05
1,5 H repulsion[c]	1.5								

[a] When one of the groups is *t*-butyl the *cis* correction = 4.00, when both are *t*-butyl, *cis* correction = ~10.00, and when there are two *cis* corrections around one double bond, the total correction is 3.00.

[b] +1.2 for but-2-ene, 0 for all other 2-enes, and −0.6 for 3-enes.

[c] These refer to repulsions between the H atoms attached to the 1,5 C atoms in such compounds as 2,2,4,4-tetramethyl pentane, and then only to the methyls close to each other (see text, p. 31).

Corrections to be Applied to Ring-compound Estimates[a]

Ring (σ)	ΔH_f° 298	S_{int}° 298	C_p° 300	400	500	600	800	1000	1500
Cyclopropane (6)	27.6	32.1	−3.05	−2.53	−2.10	−1.90	−1.77	−1.62	(−1.52)
Cyclopropene (2)	53.7	33.6							
Cyclobutane (8)	26.2	29.8	−4.61	−3.89	−3.14	−2.64	−1.88	−1.38	−0.67
Cyclobutene (2)	29.8	29.0	−2.53	−2.19	−1.89	−1.68	−1.48	−1.33	−1.22
Cyclopentane (10)	6.3	27.3	−6.50	−5.5	−4.5	−3.8	−2.8	−1.93	−0.37
Cyclopentene (2)	5.9	25.8	−5.98	−5.35	−4.89	−4.14	−2.93	−2.26	−1.08
Cyclopentadiene (2)	6.0	28.0	−4.3						
Cyclohexane (6)	0	18.8	−5.8	−4.1	−2.9	−1.3	1.1	2.2	3.3
Cyclohexene (2)	1.4	21.5	−4.28	−3.04	−1.98	−1.43	−0.29	0.08	0.81
Cyclohexadiene 1,3	4.8								
Cyclohexadiene 1,4	0.5								
Cycloheptane (1)	6.4	15.9							
Cycloheptene	5.4								
Cycloheptadiene, 1,3	6.6								
Cycloheptatriene 1,3,5 (1)	4.7	23.7							
Cycloöctane (8)	9.9	16.5							
cis-Cycloöctene	6.0								
trans-Cycloöctene	15.3								
Cycloöctatriene 1,3,5	8.9								
Cycloöctatetraene	17.1								
Cyclononane	12.8								
cis-Cyclononene	9.9								
trans-Cyclononene	12.8								
Cyclodecane	12.6								
Cyclododecane	4.4								
Spiropentane (4)	63.5	67.6							
Bicycloheptadiene	31.6								
Biphenylene	58.8								

[a] Note that in most cases the ΔH_f° correction equals ring-strain energy.

Corrections to be Applied to Ring-compound Estimates[a] (Contd.)

Ring (σ)	ΔH_f° 298	S_{int}° 298	C_p° 300	400	500	600	800	1000	150(0)
Bicycloheptane (2,2,1)	16.2								
Bicyclo-(1,1,0)-butane (2)	67.0	69.2							
Bicyclo-(2,1,0)-pentane	55.3								
Bicyclo-(3,1,0)-hexane	32.7								
Bicyclo-(4,1,0)-heptane	28.9								
Bicyclo-(5,1,0)-octane	29.6								
Bicyclo-(6,1,0)-nonane	31.1								
Methylene cyclopropane	41								

Table A.2. Oxygen-containing Compounds

Group	$\Delta H_{f\ 298}^\circ$	$S_{int\ 298}^\circ$	C_p° 300	400	500	600	800	1000	1500
O(H$_2$)	−57.8	45.1	8.0	8.4	9.2	9.9	11.2		
O(H)(C)	−37.9	29.07	4.3	4.4	4.8	5.2	6.0	6.6	
O(H)(C$_B$)	−37.9	29.1	4.3	4.5	4.8	5.2	6.0	6.6	
O(H)(O)	−16.3	27.85	5.2	5.8	6.3	6.7	7.2	7.5	8.2
O(H)(CO)	−58.1	24.5	3.8	5.0	5.8	6.3	7.2	7.8	
O(C)$_2$	−23.2	8.68	3.4	3.7	3.7	3.8	4.4	4.6	
O(C)(C$_d$)	−30.5	9.7							
O(C)(C$_B$)^a	−23.0								
O(C)(O)	−4.5	[9.4]	3.7	3.7	3.7	3.7	4.2	4.2	4.8
O(C)(CO)	−43.1	8.4							
O(C$_d$)$_2$	−33.0	10.1							
O(C$_B$)$_2$	−21.1								
O(C$_d$)(CO)	−45.2								
O(C$_B$)(CO)	−36.7								
O(O)(CO)	−19.0								
O(CO)$_2$	−46.5								
O(O)$_2$	[19.0]	[9.4]	[3.7]	[3.7]	[3.7]	[3.7]	[4.2]	[4.2]	[4.8]
CO(H)$_2$	−26.0	52.3	8.5	10.5	13.4	14.8	17.0		
CO(H)(C)	−29.1	34.9	7.0	7.8	8.8	9.7	11.2	12.2	
CO(H)(C$_B$)^b	−29.1								
CO(H)(C$_d$)	−29.1								
CO(H)(C$_t$)	−29.1								
CO(H)(CO)	−25.3								
CO(H)(O)	−32.1	34.9	7.0	7.9	8.8	9.7	11.2	12.2	

^{a} O(H)(C$_d$) ≡ O(H)(C$_B$) ≡ O(H)(C$_t$) ≡ O(H)(C), assigned.
^{b} CO(H)(O) ≡ CO(H)(C), assigned.

274

Table A.2. (Contd.)

Group	$\Delta H^{\circ}_{f\ 298}$	$S^{\circ}_{int\ 298}$	C°_p 300	400	500	600	800	1000	1500
CO(C)$_2$	−31.4	15.0	5.6	6.3	7.1	7.8	8.9	9.6	
CO(C)(C$_B$)	−30.9								
CO(C$_B$)$_2$	−25.8								
CO(C)(O)	−35.1	14.8	6.0	6.7	7.3	8.0	8.9	9.4	
CO(C)(CO)	−29.2								
CO(C$_d$)(O)^c	−32.0								
CO(C$_B$)(O)	−36.6								
CO(C$_B$)(CO)	−26.8								
CO(O$_2$)	−29.9								
CO(O)(CO)	−29.3								
C(H)$_3$(O)^d	−10.08	30.41	6.19	7.84	9.40	10.79	13.03	14.77	17.58
C(H)$_2$(O)(C)	−8.1	9.8	4.99	6.85	8.30	9.43	11.11	12.33	
C(H)$_2$(O)(C$_d$)	−6.5								
C(H)$_2$(O)(C$_B$)	−8.1	9.7							
C(H)$_2$(O)(C$_t$)	−6.5								
C(H)$_2$(O)(CO)									
C(H)$_2$(O)$_2$	−16.1								
C(H)(O)(C)$_2$	−7.2	−11.0	4.80	6.64	8.10	8.73	9.81	10.40	
C(H)(O)$_2$(C)	−16.3								
C(O)(C)$_3$	−6.6	−33.56	4.33	6.19	7.25	7.70	8.20	8.24	
C(O)$_2$(C)$_2$	−18.6								
C(H)$_3$(CO)^e	−10.08	30.41	6.19	7.84	9.40	10.79	13.02	14.77	17.58
C(H)$_2$(CO)(C)	−5.2	9.6	6.2	7.7	8.7	9.5	11.1	12.2	
C(H)$_2$(CO)(C$_d$)	−3.8								
C(H)$_2$(CO)(C$_B$)	−5.4								
C(H)$_2$(CO)(C$_t$)	−5.4								
C(H)$_2$(CO)$_2$	−7.6								
C(H)(CO)(C)$_2$^{f}	−1.7	−12.0							
C(CO)(C)$_3$	1.4								
C$_d$(O)(H)^g	8.6	8.0	4.2	5.0	5.8	6.5	7.6	8.4	9.6
C$_d$(O)(C)^h	10.3								
C$_d$(O)(C$_d$)^i	8.9								
C$_d$(O)(CO)	11.6								
C$_d$(H)(CO)	5.0								
C$_d$(CO)(C)	7.5								
C$_B$(O)	−0.9	−10.2	3.9	5.3	6.2	6.6	6.9	6.9	
C$_B$(CO)	3.7								

^{c} CO(C)(C$_d$)≡CO(C$_B$)(O), assigned.
^{d} C(H)$_3$(O)≡C(H)$_3$(C), assigned.
^{e} C(H)$_3$(CO)≡C(H)$_3$(C), assigned.
^{f} C(H)(CO)(C$_2$), estimated.
^{g} C$_d$(H)(O)≡C$_d$(H)(C), assigned.
^{h} C$_d$(O)(C)≡C$_d$(C)(C$_d$).
^{i} C$_d$(O)(C$_d$)≡C$_d$(C)(C$_d$).

Table A.2. (*Contd.*)
Non-Nearest Neighbor and Ring Corrections

Strain	$\Delta H^\circ_{f\,298}$	$S^\circ_{int\,298}$	C°_p						
			300	400	500	600	800	1000	1500
Ether–oxygen *gauche*	0.5								
Di-tertiary ethers	7.8								
Oxygen *gauche*	0								
Oxygen *ortho*	0								
	26.9	30.5	−2.0	−2.8	−3.0	−2.6	−2.3	−2.3	
	25.7	27.7	−4.6	−5.0	−4.2	−3.5	−2.6	+0.2	
	5.9								
	0.5								
	0.2								
	3.3								
	6.6								
	4.7								
	6.0								
	−5.8								
	1.2								
	4.5								

Table A.2. (*Contd.*)
Non-Nearest Neighbor and Ring Corrections

Strain	$\Delta H^\circ_{f\,298}$	$S^\circ_{int\,298}$	C°_p						
			300	400	500	600	800	1000	1500
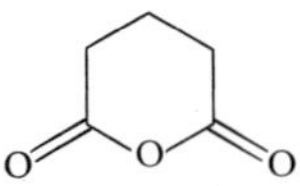	0.8								
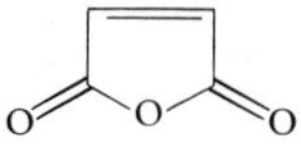	3.6								

Ring Correction	$\Delta H^\circ_{f\,298}$	Ring Correction	$\Delta H^\circ_{f\,298}$
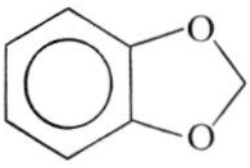	16.6	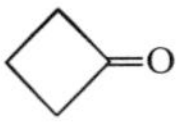	22.6
		Cyclopentanone	5.2
		Cyclohexanone	2.2
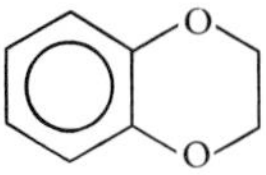		Cycloheptanone	2.3
	2.0	Cyclooctanone	1.5
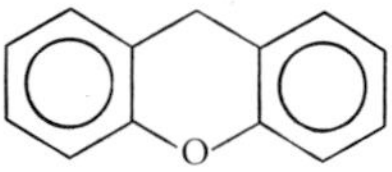	2.3	Cyclononanone	4.7
		Cyclodecanone	3.6
		Cycloundecanone	4.4
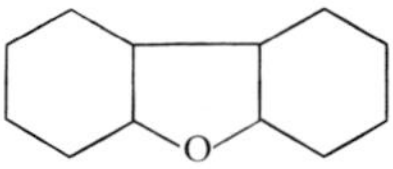	11.4	Cyclododecanone	3.0
	cis 15.3 / *trans* 20.9	Cyclo(C_{15})anone / Cyclo(C_{17})anone	2.1 / 1.1
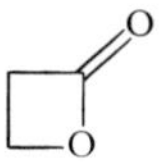	23.9		
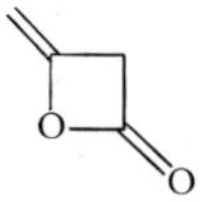	22.0		

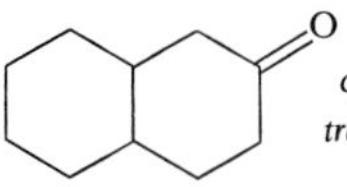

Table A.3. Group Contributions to C_{pT}°, S°, and ΔH_f° at 25°C and 1 atm for Nitrogen-containing Compounds

Group	$\Delta H_{f\,298}^\circ$	$S_{int\,298}^\circ$	C_p° 300	400	500	600	800	1000	1500
C—(N)(H)$_3$	−10.08	30.41	6.19	7.84	9.40	10.79	13.02	14.77	17.58
C—(N)(C)(H)$_2$	−6.6	9.8^a	5.25^a	6.90^a	8.28^a	9.39^a	11.09^a	12.34^a	
C—(N)(C)$_2$(H)	−5.2	−11.7^a	4.67^a	6.32^a	7.64^a	8.39^a	9.56^a	10.23^a	
C—(N)(C)$_3$	−3.2	−34.1^a	4.35^a	6.16^a	7.31^a	7.91^a	8.49^a	8.50^a	
N—(C)(H)$_2$	4.8	29.71	5.72	6.51	7.32	8.07	9.41	10.47	12.28
N—(C)$_2$(H)	15.4	8.94	4.20	5.21	6.13	6.83	7.90	8.65	9.55
N—(C)$_3$	24.4	−13.46	3.48	4.56	5.43	5.97	6.56	6.67	6.50
N—(N)(H)$_2$	11.4	29.13	6.10	7.38	8.43	9.27	10.54	11.52	13.19
N—(N)(C)(H)	20.9	9.61	4.82	5.8	6.5	7.0	7.8	8.3	9.0
N—(N)(C)$_2$	29.2	−13.80							
N—(N)(C$_B$)(H)	22.1								
N$_I$—(H)	16.3								
N$_I$—(C)	21.3								
N$_I$—(C$_B$)^b	16.7								
N$_A$—(H)	25.1	26.8	4.38	4.89	5.44	5.94	6.77	7.42	8.44
N$_A$—(C)	27								
N—(C$_d$)(C)(H)	15.4								
N—(C$_d$)(C)(N)	30								
N—(N$_I$)(C)(H)	21								
N—(C$_d$)(H)$_2$	4.8								
N—(C$_d$)(C)$_2$	24.4								
N—(C$_d$)(H)(N)	21.5								
N—(C$_B$)(H)$_2$	4.8	29.71	5.72	6.51	7.32	8.07	9.41	10.47	12.28
N—(C$_B$)(C)(H)	14.9								
N—(C$_B$)(C)$_2$	26.2								
N—(C$_B$)$_2$(H)	16.3								
N$_A$—(N)	23.0								
C$_B$—(N)	−0.5	−9.69	3.95	5.21	5.94	6.32	6.53	6.56	
CO—(N)(H)	−29.6	34.93	7.03	7.87	8.82	9.68	11.16	12.20	
CO—(N)(C)	−32.8	16.2	5.37	6.17	7.07	7.66	9.62	11.19	
N—(CO)(H)$_2$	−14.9	24.69	4.07	5.74	7.13	8.29	9.96	11.22	
N—(CO)(C)(H)	−4.4	3.9^a							
N—(CO)(C)$_2$									
N—(CO)(C$_B$)(H)	+0.4								
N—(CO)$_2$(H)	−18.5								
N—(CO)$_2$(C)	−5.9								
N—(CO)$_2$(C$_B$)	−0.5								
C—(N$_A$)(C)(H)$_2$	−6.0								
C—(N$_A$)(C)$_2$(H)	−3.4								
C—(N$_A$)(C)$_3$	−3.0								
C—(CN)(C)(H)$_2$	22.5	40.20	11.10	13.40	15.50	17.20	19.7	21.30	
C—(CN)(C)$_2$(H)	25.8	19.80	11.00	12.70	14.10	15.40	17.30	18.60	
C—(CN)(C)$_3$	29.0	−2.80							
C—(CN)$_2$(C)$_2$		28.40							
C$_d$—(CN)(H)	37.4	36.58	9.80	11.70	13.30	14.50	16.30	17.30	
C$_d$—(CN)$_2$	34.1								

Table A.3. (Contd.)

Group	$\Delta H^\circ_{f\ 298}$	$S^\circ_{int\ 298}$	C°_P 300	400	500	600	800	1000	1500
C_d—(NO₂)(H)		44.4	12.3	15.1	17.4	19.2	21.6	23.2	25.3
C_B—(CN)	35.8	20.50	9.8	11.2	12.3	13.1	14.2	14.9	
C_t—(CN)	63.8	35.40	10.30	11.30	12.10	12.70	13.60	14.30	15.30
C—(NCO)	−10.2	48.9	15.4						
C—(NO₂)(C)(H)₂	−15.1	48.4[a]							
C—(NO₂)(C)₂(H)	−15.8	26.9[a]							
C—(NO₂)(C)₃		3.9[a]							
C—(NO₂)₂(C)(H)	−14.9								
O—(NO)(C)	−5.9	41.9	9.10	10.30	11.2	12.0	13.3	13.9	14.5
O—(NO₂)(C)	−19.4	48.50							
O—(C)(CN)	2.0	39.5	10.0						
O—(C_d)(CN)	7.5	43.1	13.0						
O—(C_B)(CN)	7.0	29.2	8.3						

Corrections to be Applied to Ring-compound Estimates

Ethyleneimine	27.7	31.6[a]							

| Azetidine | 26.2[a] | 29.3[a] | | | | | | | |

| Pyrrolidine | 6.8 | 26.7 | −6.17 | −5.58 | −4.80 | −4.00 | −2.87 | −2.17 | |

| Piperidine | 1.0 | | | | | | | | |

| | 3.4 | | | | | | | | |

| | 8.5 | | | | | | | | |

[a] Estimates by authors.
[b] For *ortho* or *para* substitution in pyridine add −1.5 kcal/mole per group.
N_I stands for imino N atom; N_A represents azo N atom.

C—(N_I)(C)(H)₂ ≡ C—(N)(C)(H)₂; assigned
C—(N_I)(C)₂(H) = C—(N)(C)₂(H); C_d—(N_I)(H) ≡ C_d—(C_d)(H)

Table A.4. Halogen-containing Compounds. Group Contribution to ΔH_f° 298, S° 298, and C_{pT}°, Ideal Gas at 1 atm

Group	ΔH_f° 298	S_{int}° 298	C_p° 300	400	500	600	800	1000
C—(F)₃(C)	−158.4	42.5	12.7	15.0	16.4	17.9	19.3	20.0
C—(F)₂(H)(C)	(−102.3)	39.1	9.9	12.0		15.1		
C—(F)(H)₂(C)	−51.5	35.4	8.1	10.0	12.0	13.0	15.2	16.6
C—(F)₂(C)₂	−97.0	17.8	9.9	11.8	13.5			
C—(F)(H)(C)₂	−49.0	(14.0)						
C—(F)(C)₃	−48.5							
C—(F)₂(Cl)(C)	−106.3	40.5	13.7	16.1	17.5			
C—(Cl)₃(C)	−20.7	50.4	16.3	18.0	19.1	19.8	20.6	21.0
C—(Cl)₂(H)(C)	(−18.9)	43.7	12.1	14.0	15.4	16.5	17.9	18.7
C—(Cl)(H)₂(C)	−16.5	37.8	8.9	10.7	12.3	13.4	15.3	16.7
C—(Cl)₂(C)₂	−22.0	22.4	12.2					
C—(Cl)(H)(C)₂	−14.8	17.6	9.0	9.9	10.5	11.2		
C—(Cl)(C)₃	−12.8	5.4	9.3	10.5	11.0	11.3		
C—(Br)₃(C)		55.7	16.7	18.0	18.8	19.4	19.9	20.3
C—(Br)(H)₂(C)	−5.4	40.8	9.1	11.0	12.6	13.7	15.5	16.8
C—(Br)(H)(C)₂	−3.4							
C—(Br)(C)₃	−0.4	−2.0	9.3	11.0				
C—(I)(H)₂(C)	8.0	43.0	9.2	11.0	12.9	13.9	15.8	17.2
C—(I)(H)(C)₂	10.5	21.3	9.2	10.9	12.2	13.0	14.2	14.8
C—(I)(C)₃	13.0	0.0	9.7					
C—(I)₂(C)(H)	(26.0)	(54.6)	(12.2)	—	(16.4)	(17.0)		
C—(Cl)(Br)(H)(C)		45.7	12.4	14.0	15.6	16.3	17.9	19.0
N—(F)₂(C)	−7.8							
C—(Cl)(C)(O)(H)	−21.6	15.9	(9.0)	(9.9)	(10.5)	(11.2)		
C—(I)(O)(H)₂	3.8	40.7						
Cd—(C)(Cl)	−2.1	15.0						
Cd—(F)₂	−77.5	37.3	9.7	11.0	12.0	12.7	13.8	14.5
Cd—(Cl)₂	−1.8	42.1	11.4	12.5	13.3	13.9	14.6	15.0
Cd—(Br)₂		47.6	12.3	13.2	13.9	14.3	14.9	15.2
Cd—(F)(Cl)		39.8	10.3	11.7	12.6	13.3	14.2	14.7
Cd—(F)(Br)		42.5	10.8	12.0	12.8	13.5	14.3	14.7
Cd—(Cl)(Br)		45.1	12.1	12.7	13.5	14.1	14.7	14.7
Cd—(F)(H)	−37.6	32.8	6.8	8.4	9.5	10.5	11.8	12.7
Cd—(Cl)(H)	−1.2	35.4	7.9	9.2	10.3	11.2	12.3	13.1
Cd—(Br)(H)	11.0	38.3	8.1	9.5	10.6	11.4	12.4	13.2
Cd—(I)(H)	24.5	40.5	8.8	10.0	10.9	11.6	12.6	13.3
Ct—(Cl)		33.4	7.9	8.4	8.7	9.0	9.4	9.6
Ct—(Br)		36.1	8.3	8.7	9.0	9.2	9.5	9.7
Ct—(I)		37.9	8.4	8.8	9.1	9.3	9.6	9.8

Table A.4. (*Contd.*)

Group	ΔH_f° 298	S_{int}° 298	C_p° 300	400	500	600	800	1000
Arenes								
C_B—(F)	−42.8	16.1	6.3	7.6	8.5	9.1	9.8	10.2
C_B—(Cl)	−3.8	18.9	7.4	8.4	9.2	9.7	10.2	10.4
C_B—(Br)	10.7	21.6	7.8	8.7	9.4	9.9	10.3	10.5
C_B—(I)	24.0	23.7	8.0	8.9	9.6	9.9	10.3	10.5
C—$(C_B)(F)_3$	−162.7	42.8	12.5	15.3	17.2	18.5	20.1	21.0
C—$(C_B)(Br)(H)_2$	−5.1							
C—$(C_B)(I)(H)_2$	8.4							
Corrections for Non-Next-nearest Neighbors								
ortho (F)(F)	5.0	0	0	0	0	0	0	0
ortho (Cl)(Cl)	2.2							
ortho (alk)(halogen)[a]	0.6							
cis (halogen)(halogen)	−0.3							
cis (halogen)(alk)	−0.8							

[a] Halogen = Cl, Br, I only.
The *gauche* correction = 1.0 kcal for Cl, Br, I; none for X—Me and none for F—halogen.

Table A.5. Sulfur-containing Compounds. Group Contributions to ΔH_f° 298, S° 298, and C_{pT}°.

Group	ΔH_f° 298	S_{int}° 298	C_p° 300	400	500	600	800	1000
C—$(H)_3(S)$[a]	−10.08	30.41	6.19	7.84	9.40	10.79	13.02	14.77
C—$(C)(H)_2(S)$	−5.65	9.88	5.38	7.08	8.60	9.97	12.26	14.15
C—$(C)_2(H)(S)$	−2.64	−11.32	4.85	6.51	7.78	8.69	9.90	10.57
C—$(C)_3(S)$	−0.55	−34.41	4.57	6.27	7.45	8.15	8.72	8.10
C—$(C_B)(H)_2(S)$	−4.73							
C—$(C_d)(H)_2(S)$	−6.45							
C—$(H)_2(S)_2$	−6.0±3							
C_B—(S)[b]	−1.8	10.20	3.90	5.30	6.20	6.60	6.90	6.90
C_d—(H)(S)[c]	8.56	8.0	4.16	5.03	5.81	6.50	7.65	8.45
C_d—(C)(S)	10.93	−12.41	3.50	3.57	3.83	4.09	4.41	5.00
S—(C)(H)	4.62	32.73	5.86	6.20	6.51	6.78	7.30	7.71
S—$(C_B)(H)$	11.96	12.66	5.12	5.26	5.57	6.03	6.99	7.84
S—$(C)_2$	11.51	13.15	4.99	4.96	5.02	5.07	5.41	5.73
S—$(C)(C_d)$	9.97							

Table A.5. (*Contd.*)

Group	ΔH_f° 298	S_{int}° 298	C_p°					
			300	400	500	600	800	1000
S—$(C_d)_2$	−4.54	16.48	4.79	5.58	5.53	6.29	7.94	9.73
S—$(C_B)(C)$	19.16							
S—$(C_B)_2$	25.90							
S—(S)(C)	7.05	12.37	5.23	5.42	5.51	5.51	5.38	5.12
S—$(S)(C_B)$	14.5							
S—$(S)_2$	3.01	13.4	4.7	5.0	5.1	5.2	5.3	5.4
C—$(SO)(H)_3{}^d$	−10.08	30.41	6.19	7.84	9.40	10.79	13.02	14.77
C—$(C)(SO)(H)_2$	−7.72							
C—$(C)_3(SO)$	−3.05							
C—$(C_d)(SO)(H)_2$	−7.35							
C_B—$(SO)^e$	2.3							
SO—$(C)_2$	−14.41	18.10	8.88	10.03	10.50	10.79	10.98	11.17
SO—$(C_B)_2$	−12.0							
C—$(SO_2)(H)_3{}^f$	−10.08	30.41	6.19	7.84	9.40	10.79	13.02	14.77
C—$(C)(SO_2)(H)_2$	−7.68							
C—$(C)_2(SO_2)(H)$	−2.62							
C—$(C)_3(SO_2)$	−0.61							
C—$(C_d)(SO_2)(H)_2$	−7.14							
C—$(C_B)(SO_2)(H)_2$	−5.54							
C_B—$(SO_2)^g$	2.3							
C_d—$(H)(SO_2)$	12.5							
C_d—$(C)(SO_2)$	14.5							
SO_2—$(C_d)(C_B)$	−68.6							
SO_2—$(C_d)_2$	−73.6							
SO_2—$(C)_2$	−69.74	20.90	11.52					
SO_2—$(C)(C_B)$	−72.29							
SO_2—$(C_B)_2$	−68.58							
SO_2—$(SO_2)(C_B)$	−76.25							
CO—$(S)(C)^h$	−31.56	15.43	5.59	6.32	7.09	7.76	8.89	9.61
S—(H)(CO)	−1.41	31.20	7.63	8.09	8.12	8.17	8.50	8.24
C—$(S)(F)_3$		38.9						
CS—$(N)_2{}^i$	−31.56	15.43	5.59	6.32	7.09	7.76	8.89	9.61
N—$(CS)(H)_2$	12.78	29.19	6.07	7.28	8.18	8.91	10.09	10.98

[a] C—$(S)(H)_3 \equiv$ C−$(C)(H)_3$, assigned.
[b] C_B—$(S) \equiv C_B$—(O), assigned.
[c] C_d—$(S)(H) \equiv C_d$—(O)(H), assigned.
[d] C—$(SO)(H)_3 \equiv$ C—$(CO)(H)_3$, assigned.
[e] C_B—$(SO) \equiv C_B$—(CO), assigned.
[f] C—$(SO_2)(H)_3 \equiv$ C—$(SO)(H)_3$.
[g] C_B—$(SO_2) \equiv C_B$—(CO), assigned.
[h] CO—$(S)(C) \equiv$ CO—$(C)_2$, assigned.
[i] CS—$(N)_2 \equiv$ CO—$(C)_2$, assigned.
[j] S—$(S)(N) \equiv$ O—(O)(C), assigned.
[k] SO—$(N)_2 \equiv$ CO—$(C)_2$, assigned.
[l] SO_2—$(N)_2 \equiv$ SO—$(N)_2$, assigned.

Table A.5. *(Contd.)*

Group	ΔH_f° 298	S_{int}° 298	C_p° 300	400	500	600	800	1000
S—(S)(N)[j]	−4.90							
N—(S)(C)$_2$	29.9							
SO—(N)$_2$[k]	−31.56							
N—(SO)(C)$_2$	16.0							
SO$_2$—(N)$_2$[l]	−31.56							
N—(SO$_2$)(C)$_2$	−20.4							

Corrections to be Applied to Organosulfur Ring Compounds

Ring (σ)	ΔH_f° 298	S_{int}° 298	C_p° 300	400	500	600	800	1000
 (3-membered S ring) (2)	17.7	29.5	−2.9	−2.6	−2.7	−3.0	−4.3	−5.8
 (4-membered S ring) (2)	19.4	27.2	−4.6	−4.2	−3.9	−3.9	−4.6	−5.7
 (5-membered S ring) (2)	1.7	23.6	−4.9	−4.7	−3.7	−3.7	−4.4	−5.6
 (6-membered S ring) (1)	0	16.1	−6.2	−4.3	−2.8	−0.7	0.9	−1.3
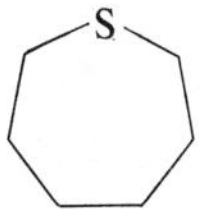 (7-membered S ring) (1)	3.9							
 (5-membered S ring, unsaturated) (2)	5.0							

[m] Assume ring corrections for (5-membered S ring with one double bond) and (5-membered S ring, saturated) are the same.

[n] Assume ring corrections for (4-membered S ring with double bond) and (5-membered S ring) are the same.

283

Table A.5. (*Contd.*)
Corrections to be Applied to Organosulfur Ring Compounds

Ring (σ)		ΔH_f° 298	S_{int}° 298	C_p° 300	400	500	600	800	1000
	(1)	5.0							
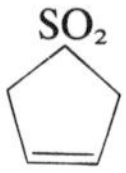	(2)	5.7							
	(2)	1.7	23.6	−4.9	−4.7	−3.7	−3.7	−4.4	−5.6

Table A.6. Organometallic Compounds[a]

Metal	Groups	ΔH_f° 298	Remarks
Tin	C—(Sn)(H)$_3$	−10.08	C—(Sn)(H)$_3$≡C—(C)(H)$_3$, assigned
	C—(Sn)(C)(H)$_2$	−2.18	
	C—(Sn)(C)$_2$(H)	3.38	
	C—(Sn)(C)$_3$	8.16	
	C—(Sn)(C$_B$)(H)$_2$	−7.77	
	C$_B$—(Sn)	5.51	C$_B$—(Sn)≡C$_B$—(C), assigned
	C$_d$—(Sn)(H)	8.77	C$_d$—(Sn)(H)≡C$_d$—(C)(H), assigned
	Sn—(C)$_4$	36.2	
	Sn—(C)$_3$(Cl)	−9.8	
	Sn—(C)$_2$(Cl)$_2$	−49.2	
	Sn—(C)(Cl)$_3$	−89.5	
	Sn—(C)$_3$(Br)	−1.8	
	Sn—(C)$_3$(I)	9.9	
	Sn—(C)$_3$(H)	34.8	
	Sn—(C$_d$)$_4$	36.2	Sn—(C$_d$)$_4$≡Sn—(C)$_4$, assigned
	Sn—(C$_d$)$_3$(Cl)	−8.2	
	Sn—(C$_d$)$_2$(Cl)$_2$	−50.7	
	Sn—(C$_d$)(Cl)$_3$	−82.2	
	Sn—(C)$_3$(C$_d$)	37.6	
	Sn—(C$_B$)$_4$	26.2	
	Sn—(C)$_3$(C$_B$)	34.9	
	Sn—(C)$_3$(Sn)	26.4	
Lead	C—(Pb)(H)$_3$	−10.08	C—(Pb)(H)$_3$≡C—(C)(H)$_3$, assigned
	C—(Pb)(C)(H$_2$)	−1.7	
	Pb—(C)$_4$	72.9	

Table A.6. (*Contd.*)

Metal	Groups	$\Delta H_f^\circ 298$	Remarks
Chromium	O—(Cr)(C)	−23.5	O—(Cr)(C)≡O—(Ti)(C), assigned
	Cr—(O)$_4$	−64.0	
Zinc	C—(Zn)(H)$_3$	−10.08	C—(Zn)(H)$_3$≡C—(C)(H)$_3$, assigned
	C—(Zn)(C)(H)$_2$	−1.8	
	Zn—(C)$_2$	33.3	
Titanium	O—(Ti)(C)	−23.5	O—(Ti)(C)≡O—(P)(C), assigned
	Ti—(O)$_4$	−157	
	N—(Ti)(C)$_2$	39.1	N—(Ti)(C)$_2$≡N—(P)(C)$_2$, assigned
	Ti—(N)$_4$	−123	
Vanadium	O—(V)(C)	−23.5	O—(V)(C)≡O—(Ti)(C), assigned
	V—(O)$_4$	−87.0	
Cadmium	C—(Cd)(H)$_3$	−10.08	C—(Cd)(H)$_3$≡C—(C)(H)$_3$, assigned
	C—(Cd)(C)(H)$_2$	−0.3	
	Cd—(C)$_2$	46.4	
Aluminum	C—(Al)(H)$_3$	−10.08	C—(Al)(H)$_3$≡C—(C)(H)$_3$, assigned
	C—(Al)(C)(H)$_2$	0.7	
	Al—(C)$_3$	9.2	
Germanium	C—(Ge)(C)(H)$_2$	−7.7	
	Ge—(C)$_4$	36.2	Ge—(C)$_4$≡Sn—(C)$_4$, assigned
	Ge—(Ge)(C)$_3$	15.6	
Mercury	C—(Hg)(H)$_3$	−10.08	C—(Hg)(H)$_3$≡C—(C)(H)$_3$, assigned
	C—(Hg)(C)(H)$_2$	−2.7	
	C—(Hg)(C)$_2$(H)	3.6	
	C$_B$—(Hg)	−1.8	C$_B$—(Hg)≡C$_B$—(O), assigned
	Hg—(C)$_2$	42.5	
	Hg—(C)(Cl)	−2.8	
	Hg—(C)(Br)	4.9	
	Hg—(C)(I)	15.8	
	Hg—(C$_B$)$_2$	64.4	
	Hg—(C$_B$)(Cl)	9.9	
	Hg—(C$_B$)(Br)	18.1	
	Hg—(C$_B$)(I)	27.9	

[a] No *gauche* corrections across C—*M* bond.

Table A.6. (*Contd.*)

Organophosphorus Groups[a]

Group	$\Delta H_f^\circ\,298$	$S_{int}^\circ\,298$	Remarks
C—(P)(H)$_3$	−10.08	30.4	C—(P)(H)$_3$≡C—(C)(H)$_3$, assigned
C—(P)(C)(H)$_2$	−2.47		
C—(PO)(H)$_3$	−10.08	30.4	C—(PO)(H)$_3$≡C—(C)(H)$_3$, assigned
C—(PO)(C)(H)$_2$	−3.4		
C—(P:N)(H)$_3$	−10.08	30.4	C—(P:N)(H)$_3$≡C—(C)(H)$_3$, assigned
C—(N:P)(C)(H)$_2$	19.4		
C$_B$—(P)	−1.8		C$_B$—(P)≡C$_B$—(O), assigned
C$_B$—(PO)	2.3		C$_B$—(PO)≡C$_B$—(CO), assigned
C$_B$—(P:N)	2.3		C$_B$—(P:N)≡C$_B$—(CO), assigned
P—(C)$_3$	7.0		
P—(C)(Cl)$_2$	−50.1		
P—(C$_B$)$_3$	28.3		
P—(O)$_3$	−66.8		
P—(N)$_3$	−66.8		P—(N)$_3$≡P—(O)$_3$, assigned
PO—(C)$_3$	−72.8		
PO—(C)(F)$_2$		46.7	
PO—(C)(Cl)(F)		50.8	
PO—(C)(Cl)$_2$	−123.0	53.0	
PO—(C)(O)(Cl)	−112.6		
PO—(C)(O)$_2$	−99.5		
PO—(O)$_3$	−104.6		
PO—(O)$_2$(F)	−167.7		
PO—(C$_B$)$_3$	−52.9		

[a] No *gauche* corrections across the X—P, X—PO, and X—P:N bonds (X represents C, O, N).

Group	$\Delta H_f^\circ\,298$	Remarks
PO—(N)$_3$	−104.6	PO—(N)$_3$≡PO—(O)$_3$, assigned
O—(C)(P)	−23.5	O—(C)(P)≡O—(C)$_2$, assigned
O—(H)(P)	−58.7	
O—(C)(PO)	−40.7	O—(C)(PO)≡O—(C)(CO), assigned
O—(H)(PO)	−65.0	
O—(PO)$_2$	−54.5	
O—(P:N)(C)	−40.7	O—(P:N)(C)≡O—(C)(CO), assigned
N—(P)(C)$_2$	32.2	
N—(PO)(C)$_2$	17.8	
P:N—(C)$_3$(C)	0.50	P:N—(C)$_3$(C)≡C—(C)$_4$, assigned
P:N—(C$_B$)$_3$(C)	−25.7	
P:N—(N:P)(C)$_2$(P:N)	−15.5	
P:N—(N:P)(C$_B$)$_2$(P:N)	−22.9	
P:N—(N:P)(Cl)$_2$(P:N)	−58.2	
P:N—(N:P)(O)$_2$(P:N)	−43.4	

Table A.6. *(Contd.)*

Organoboron Groups

Group	$\Delta H_f^\circ 298$	Remarks
C—(B)(H)$_3$	-10.1	C—(B)(H)$_3\equiv$C—(C)(H)$_3$, assigned
C—(B)(C)(H)$_2$	-2.22	
C—(B)(C)$_2$(H)	1.1	
C—(BO$_3$)(H)$_3$	-10.1	C—(BO$_3$)(H)$_3\equiv$C—(C)(H)$_3$, assigned
C—(BO$_3$)(C)(H)$_2$	-2.2	
C$_d$—(B)(H)	15.6	
B—(C)$_3$	0.9	
B—(C)(F)$_2$	-187.9	
B—(C)$_2$(Cl)	-42.7	
B—(C)$_2$(Br)	-26.9	
B—(C)$_2$(I)	-8.9	
B—(C)$_2$(O)	29.3	B—(C)$_2$(O)$\equiv$N—(C)$_2$(N), assigned
B—(C$_d$)(F)$_2$	-192.9	B—(C$_d$)(F)$_2\equiv$B—(C)(F)$_2$, assigned
B—(O)$_3$	24.4	B—(O)$_3\equiv$B—(N)$_3$, assigned
B—(O)$_2$(Cl)	-19.7	
B—(O)(Cl)$_2$	-61.2	
B—(O)$_2$(H)	19.9	
B—(N)$_3$	24.4	B—(N)$_3\equiv$N—(C)$_3$, assigned
B—(N)$_2$(Cl)	-23.8	
B—(N)(Cl)$_2$	-67.9	
BO$_3$—(C)$_3$	-208.7	
O—(B)(H)	-115.5	
O—(B)(C)	-69.4	
N—(B)(C)$_2$	-9.9	
B—(S)$_3$	24.4	
S—(B)(C)	-14.5	
S—(B)(C$_B$)	-7.8	

The *gauche* corrections across the C—B bond are $+0.8$ kcal/mole.

Table A.7. Values of C_p° for the Common Elements in Their Standard States at Different Temperatures

	Temperature (°K)							
Element	298	400	500	600	800	1000	1200	1500
$H_2(g)$	6.9	7.0	7.0	7.0	7.1	7.2	7.4	7.7
$O_2(g)$	7.0	7.2	7.4	7.7	8.1	8.3	8.5	8.7
B(cr)	2.7	3.7	4.5	5.0	5.6	6.0	6.3	6.7
C(cr)graphite	2.0	2.9	3.5	4.0	4.7	5.1	5.4	5.7
$N_2(g)$	7.0	7.0	7.1	7.2	7.5	7.8	8.1	8.3
$F_2(g)$	7.5	7.9	8.2	8.4	8.7	8.9	9.0	9.1
S(cr)(l)/(g)(S_2)	5.4(cr)	7.7(l)	9.1(l)	8.21(l)	8.8(g)	8.8(g)	9.0(g)	9.0(g)
Na(cr)/(l)/(g)	6.7(cr)	7.5(l)	7.3(l)	7.1(l)	6.9(l)	6.9(l)	5.0	5.0
Mg(cr)/(l)/(g)	6.0(cr)	6.3(cr)	6.6(cr)	6.8(cr)	7.4(cr)	7.9(l)	8.4(l)	5.0
Al(cr)/(l)	5.8(cr)	6.1(cr)	6.4(cr)	6.7(cr)	7.3(cr)	7.0(l)	7.0(l)	7.0(l)
Hg(l)/(g)	6.7(l)	6.6(l)	6.5(l)	6.5(l)	5.0	5.0	5.0	5.0
$Cl_2(g)$	8.1	8.4	8.6	8.7	8.9	9.0	9.0	9.1
$Br_2(l)(g)$	18.1(l)	8.8	8.9	8.9	9.0	9.0	9.0	9.1
$I_2(cr)/(l)$	13.0(cr)	19.3(l)	9.0	9.0	9.0	9.1	9.1	9.1
K(cr)/(l)(g)	7.1(cr)	7.5(l)	7.3(l)	7.2(l)	7.1(l)	7.3(l)	5.0	5.0
Pb(cr)/(l)	6.4	6.6	6.8	7.0	7.2(l)	7.0(l)	6.9(l)	6.9(l)
Si(cr)	4.8	5.3	5.6	5.8	6.1	6.3	6.4	6.5
Fe(cr)	6.0	6.5	7.0	7.6	9.2	13.6	8.2^a	8.6^a
W(cr)	5.8	6.0	6.1	6.2	6.4	6.5	6.7	6.9
Ti(cr)	6.0	6.4	6.6	6.8	7.2	7.5	7.7	8.1
Be(cr)	3.9	4.8	5.3	5.6	6.1	6.5	7.0	7.7
P(cr)/$P_2(g)$	5.1	5.5	5.9	6.2	8.6(g)	8.8(g)	8.8(g)	9.0(g)

^{a} Alpha phase to 1184°K; gamma phase to 1665°K; (cr) = crystal; (l) = liquid; (g) = gas.

Table A.8. Thermochemical Data for Some Gas-phase Atomic Species

				C_p°				
Element	(degeneracy)	$\Delta H^\circ_{f\,300}$	S°_{300}	300°K	500°K	800°K	1000°K	1500°K
Ag	(2)	68.4	41.3	5.0				5.0
Al	(6)	78.0	39.3	5.1				5.0
Ar	(1)	0	37.0	5.0				5.0
Au	(2)	87.3	43.1	5.0				5.0
B	(6)	132.8	36.6	5.0				5.0
Be	(1)	78.3	32.5	5.0				5.0
Br	(4)	26.7	41.8	5.0			5.1	5.3
C	(9)	170.9	37.8	5.0				5.0
Ca	(1)	42.2	37.0	5.0				5.0
Cd	(1)	26.8	40.1					
Cl	(4)	28.9	39.5	5.2	5.4		5.3	5.2
Co	(~8)	101.6	42.9					

288

Table A.8. (*Contd.*)
Thermochemical Data for Some Gas-phase Atomic Species

Element	(degeneracy)	$\Delta H^\circ_{f\,300}$	S°_{300}	C°_p 300°K	500°K	800°K	1000°K	1500°K
Cr	(~9)	95	41.6					
Cs	(2)	18.3	41.9					
Cu	(2)	81.0	39.7	5.0				5.0
e^-	(2)	0	5.0	5.0				5.0
F	(6)	18.9	37.9	5.4	5.3	5.1	5.0	5.0
Fe	(9–25)	99.5	43.1	6.1	5.9	5.5	5.4	5.3
H^1	(2)	52.1	27.4	5.0				5.0
H^2(D)	(2)	53.0	29.5	5.0				5.0
He	(1)	0	30.1	5.0				5.0
Hg	(1)	14.7	41.8	5.0				5.0
I	(4)	25.5	43.2	5.0				5.0
K	(2)	21.3	38.3	5.0				5.0
Kr	(1)	0	39.2	5.0				5.0
Li	(2)	38.4	33.1	5.0				5.0
Mg	(1)	35.3	35.5	5.0				5.0
Mn	(5)	67.2	41.5					
N	(4)	113.0	36.6	5.0				5.0
Na	(2)	25.9	36.7	5.0				5.0
Ne	(1)	0	35.0	5.0				5.0
O	(~7)	59.6	38.5	5.2	5.1	5.0		5.0
P^a	(4)	78.8	39.0	5.0				5.0
Pb	(1)	46.7	41.9	5.0			5.0	5.2
S	(~7)	66.7	40.1	5.7	5.4	5.2	5.1	5.1
Se	(~5)	49.2	42.2	5.0				
Si	(~9)	107.7	40.1	5.3	5.1			5.1
Sn	(3)	72.2	40.2	5.1				
Ti	(~21)	113.0	43.1	5.8	5.3	5.1	5.1	5.3
W	(1)	203.4	41.5	5.1	6.3	9.0	9.9	9.0
Xe	(1)	0	40.5	5.0				5.0
Zn	(1)	31.2	38.5	5.0			5.0	

Thermochemical Data for Some Gas-phase Diatomic Species

Substance	$\Delta H^\circ_{f\,300}$	S°_{300}	C°_p 300	500	800	1000	1500
O_2	0	49.0	7.0	7.4	8.1	8.3	8.7
H_2	0	31.2	6.9	7.0	7.1	7.2	7.7
D_2	0	34.6	7.0	7.1	7.2	7.3	7.8
HD	0.1	34.3	7.0	7.1	7.1	7.2	7.7
HO	9.4	43.9	7.2	7.1	7.2	7.3	7.9
DO	8.7	45.3	7.1				

^{a} Standard State is white phosphorous. JANAF uses red phosphorous as standard.

Table A.8. (*Contd.*)
Thermochemical Data for Some Gas-phase Diatomic Species

Substance	$\Delta H^\circ_{f\,300}$	S°_{300}	C°_p				
			300	500	800	1000	1500
F_2	0	48.4	7.5	8.2	8.7	8.9	9.1
HF	−64.8	41.5	7.0	7.0	7.1	7.2	7.7
FO	26 ± 1	51.8	7.3	8.0	8.6	8.8	9.0
Cl_2	0	53.3	8.1	8.6	8.9	9.0	9.1
HCl	−22.0	44.6	7.0	7.0	7.3	7.6	8.1
ClO	24.2	54.1	7.5	8.2	8.7	8.8	9.0
ClF	−12.1	52.1	7.7	8.3	8.7	8.9	9.0
Br_2	7.4	58.6	8.6	8.9	9.0	9.0	9.1
HBr	−8.7	47.5	7.0	7.0	7.4	7.7	8.3
BrF	−14.0	54.7	7.9	8.5	8.8	8.9	9.1
BrCl	3.5	57.3	8.4	8.7	8.9	9.0	9.1
BrO	30	56.8	7.7				
I_2	14.9	62.3	8.8	8.9	9.0	9.1	9.1
HI	6.3	49.4	7.0	7.1	7.6	7.9	8.5
IF	−22.7	56.4	8.0	8.6	8.9	9.0	9.2
ICl	4.2	59.1	8.5	8.8	9.0	9.0	9.1
IBr	9.8	61.8	8.7	8.9	9.0	9.1	9.2
S_2	30.7	54.5	7.8	8.4	8.7	8.8	9.0
SCl		57.3	8.2	8.7	8.9	9.0	
SF		54.3	7.6	8.2	8.7	8.8	
SH	34	46.7	7.7	7.5	7.6	7.9	8.4
SO	1.5	53.0	7.2	7.8	8.4	8.6	8.8
N_2	0	45.8	7.0	7.0	7.5	7.8	8.3
NH	90 ± 5	43.3	7.0	7.0	7.2	7.5	8.1
NO	21.6	50.3	7.1	7.3	7.8	8.1	8.6
NF	60 ± 8	51.4	7.3	7.9	8.5	8.7	8.9
P_2	40.6	52.1	7.7	8.3	8.7	8.8	8.9
PH	58	46.9	7.0	7.1	7.6	7.9	8.4
PO	−2.5 ± 2	53.2	7.6	7.9	8.4	8.6	8.8
PN	24	50.4	7.1	7.6	8.3	8.5	8.8
C_2	199 ± 2	47.6	10.3	8.9	8.5	8.6	8.9
CH	142	43.7	7.0	7.0	7.4	7.7	8.3
CO	−26.4	47.2	7.0	7.1	7.6	7.9	8.4
CF	61 ± 2	50.9	7.2	7.7	8.3	8.5	8.8
CCl	120 ± 5	53.8	7.7	8.3	8.7	8.8	8.9
CN	101	48.4	7.0	7.2	7.7	8.0	8.5
CS	55 ± 5	50.3	7.1	7.7	8.3	8.5	8.8
Si_2	131 ± 5	54.9	8.2	8.7	8.9	9.0	9.0
SiO	−24.2	50.5	7.1	7.7	8.3	8.5	8.8
SiH		47.4	7.0	7.2	8.0	8.1	8.7

Table A.8. (*Contd.*)
Thermochemical Data for Some Gas-phase Diatomic Species

Substance	$\Delta H^\circ_{f\,300}$	S°_{300}	C°_p				
			300	500	800	1000	1500
Pb_2	80 ± 5	67.2	8.8	9.1	9.2	9.3	9.5
B_2	195 ± 6	48.3	7.3	8.0	8.5	8.7	8.9
BH	106 ± 2	41.0	7.0	7.1	7.6	7.9	8.5
BO	10 ± 10	48.6	7.0	7.2	7.8	8.1	8.5
BF	-28 ± 3	47.9	7.1	7.6	8.2	8.5	8.8
BCl	34 ± 4	50.9	7.6	8.2	8.7	8.8	9.0
BC	198 ± 10	49.8	7.1	7.6	8.3	8.5	8.8
AlH	62 ± 5	44.9	7.0	7.4	8.1	8.4	8.8
AlO	22 ± 5	52.2	7.4	8.1	8.6	8.7	8.9
AlF	-62.5	51.4	7.6	8.3	8.7	8.8	9.0
AlCl	-11.2	54.4	8.3	8.7	8.9	9.0	9.1
AlBr	3.6 ± 5	57.3	8.5	8.8	9.0	9.0	9.1
Ag_2	100 ± 3						
AgCl	23						
HgH	57 ± 4	52.5	7.2	7.9	8.6	8.9	9.4
HgCl	18.8 ± 3	62.1	8.7	8.9	9.1	9.1	9.2
Cu_2	116 ± 3	57.7	8.7	8.9	9.0	9.1	9.2
CuCl	21.8	56.7	8.4	8.8	8.9	9.0	9.1
BeF	-50 ± 2	49.2	7.1	7.7	8.3	8.6	8.8
BeCl	14.5 ± 3	52.0	7.6	8.2	8.7	8.8	9.0
BeO	31 ± 3	47.2	7.0	7.5	8.1	8.4	8.7
MgO		52.9	7.5	8.2	8.7	8.8	9.0
MgF	-53.1	52.8	7.8	8.4	8.8	8.9	9.0
MgCl	-10 ± 10	55.8	8.3	8.7	8.9	9.0	9.1
Li_2	50.4	47.1	8.6	8.9	9.1	9.2	9.4
LiF	-79.5 ± 2	47.8	7.5	8.2	8.7	8.8	9.0
LiCl	-47 ± 3	50.9	7.9	8.5	8.8	8.9	9.1
LiI	-21.8 ± 2	55.5	8.3	8.7	9.0	9.0	9.2
Na_2	32.9	55.0	9.0	9.1	9.2	9.3	9.5
NaH	30 ± 5	45.0	7.2	7.9	8.5	8.8	9.1
NaF	-70.1 ± 2	52.0	8.2	8.7	8.9	9.0	9.1
NaCl	-43.4	54.9	8.6	8.9	9.0	9.1	9.2
NaBr	-34.4	57.6	8.7	8.9	9.0	9.1	9.2
K_2	30.4	59.7	9.1	9.2	9.3	9.4	9.7
KH	29 ± 4	47.3	7.4	8.1	8.7	8.9	9.1
KF	-77.9	54.1	8.4	8.8	9.0	9.0	9.2
KCl	-51.3	57.1	8.7	8.9	9.0	9.1	9.2

Table A.9. Thermochemical Data for Some Gas-phase Triatomic Species

Substance	$\Delta H^\circ_{f\,300}$	S°_{300}	C°_p 300	500	800	1000	1500
O_3	34.1	57.1	9.4	11.3	12.7	13.2	13.7
HDO	−58.6	47.7	8.1	8.6	9.7	10.3	11.7
H_2O	−57.8	45.1	8.0	8.4	9.2	9.9	11.2
D_2O	−59.6	47.4	8.2	8.5	10.1	10.9	12.2
HO_2	5.0	54.4	8.3	9.5	10.8	11.4	12.4
F_2O	5.9	59.1	10.4	12.1	13.1	13.3	13.7
F—O—O:	(12±3)	61.9	10.6	11.8	12.8	13.1	13.5
Cl_2O	21.0	64.0	11.4	12.8	13.4	13.6	13.8
O—Cl—O	25±2	61.5	10.0	11.7	13.0	13.3	13.8
ClOO·	23.0	63	11.6				
HOCl	−22±3	56.5	8.9	10.1	11.1	11.6	12.4
HOBr	−19±2	59.2	9.0				
SO_2	−70.9	59.3	9.5	11.1	12.5	13.0	13.6
SOF		62.9	10.0	11.6	12.8	13.2	
SOCl		66.2	10.9	12.1	13.0	13.3	
SF_2	−51.8	60.9	10.4	12.2	13.1	13.4	
H_2S	−4.8	49.2	8.2	8.9	10.2	10.9	12.3
SCl_2	−5.2	67.2	12.1	13.1	13.6	13.7	
S_2O		63.8	10.5	12.0	13.0	13.3	13.6
NO_2	7.9	57.3	8.8	10.3	11.9	12.5	13.2
N_2O	19.6	52.5	9.2	11.0	12.5	13.1	14.0
·NH_2	46±1	46.5	8.0	8.5	9.5	10.2	11.5
HNO	23.8	52.7	8.3	9.3	10.8	11.5	12.5
FNO	−15.7	59.3	9.9	11.2	12.4	12.8	13.3
ClNO	12.4	62.5	10.7	11.8	12.8	13.3	13.9
BrNO	19.6	65.3	10.9	11.8	12.6	13.0	13.4
INO	29.0	67.6	11.2	11.9	12.7	13.0	13.5
·NF_2	8.5±2	59.7	9.8	11.6	12.8	13.2	13.6
I_2F	1						
HIF	−1						
C_3	190±3	56.8	10.2	11.8	13.3	13.6	14.2
CO_2	−94.05	51.1	8.9	10.7	12.3	13.0	14.0
HCN	32.3	48.2	8.6	10.0	11.4	12.2	13.5
FCN	−3±10	53.9	10.1	11.6	12.8	13.3	14.1
ClCN	33.0	56.5	10.8	12.2	13.2	13.6	14.2
BrCN	43.3	59.3	11.2	12.4	13.2	13.6	14.2
ICN	54.6	61.4	11.5	12.5	13.3	13.7	14.3
:NCN		54.2	10.2	12.3	13.6	14.0	14.5
CS_2	28.0	56.8	10.9	12.5	13.6	14.0	14.5
COS	−33.1	55.3	9.9	11.7	13.0	13.6	14.4
·CNO		54.0	9.2				
SiO_2	−76±2	54.7	10.7	12.4	13.7	14.1	14.5
$PbCl_2$	−40.6	76.6	13.4	13.7	13.8	13.9	13.9
·BF_2	−130±10	59.0	9.6	11.1	12.8	12.9	13.4
BOF	−144±3	53.7	9.8	11.5	13.4	13.3	14.2

Table A.9. (Contd.)

Substance	$\Delta H^\circ_{f\,300}$	S°_{300}	C°_p 300	500	800	1000	1500
BO_2	-75	55.0	10.0	12.0	13.5	13.9	
AlOF	-140 ± 3	56.0	10.7	12.8	14.0	14.3	14.6
Al_2O	-31 ± 7	62.0	10.9	12.3	13.2	13.4	13.7
$HgCl_2$	-35 ± 2	70.4	13.9	14.5	14.8	14.8	14.9
$BeCl_2$	-86.1 ± 3	60.3	12.3	13.5	14.2	14.4	14.7
MgF_2	-173 ± 2	61.7	11.6	13.8	13.4	13.0	13.8
$MgCl_2$	-95.5 ± 2	64.9	12.4	13.3	13.6	13.7	13.8
Li_2O	-40 ± 3	54.7	11.9	13.3	14.2	14.4	14.7
XeF_2	-25.6						

Table A.10. Thermochemical Data for Some Gas-phase Tetratomic Species

Substance	$\Delta H^\circ_{f\,300}$	S°_{300}	C°_p 300	500	800	1000	1500
H_2O_2	-32.6	56.0	10.1	12.0	13.8	14.7	16.2
HOOO·	24 ± 2						
F_2O_2	(15 ± 5)						
ClF_3	-38.0	67.3	15.3	17.8	19.0	19.3	19.6
Cl_2O_2	28 ± 6						
BrF_3	-61.1	69.9	15.9	18.1	19.1	19.4	19.7
$SOCl_2$	-50.8	74.0	15.9	17.8	18.9	19.2	
SO_3	-94.6	61.3	12.1	15.1	17.4	18.2	19.1
H_2S_2	3.9		12.3				
SOF_2	-132 ± 5	66.6	13.6	16.4	18.2	18.7	
S_2F_2	-54.5	69.3	15.3	17.7	18.9	19.3	
S_2Cl_2	-4.7	76.5	17.4	18.9	19.5	19.6	
$CSCl_2$		70.8	15.3	17.5	18.8	19.1	
NO_3 (sym)	17.0	60.4	11.2	15.0	17.5	18.3	19.1
ON—OO·	14 ± 3	70	11				
NH_3	-11.0	46.0	8.5	10.0	12.2	13.5	15.9
N_2H_2 (cis)	50 ± 5	52.2	8.7	10.9	13.5	14.8	16.9
HNO_2 (cis)	-18.3	59.6	10.8	13.4	15.6	16.5	17.9
FNO_2	-26 ± 2	62.2	11.9	14.9	17.2	18.0	18.9
$ClNO_2$	2.9	65.0	12.7	15.4	17.5	18.2	19.1
INO_2	14.4 ± 1	70.3 ± 1.5					
NO_2—OF	2.5	70.0	15.6	19.7	22.7	23.6	24.8
N_2F_2 (cis)	17.9	62.1	11.9	15.3	17.6	18.3	19.1
NF_3	-31.4 ± 1	62.3	12.8	16.1	18.2	18.7	19.3
HN_3	70.3	57.1	10.4				
HNCS	30.0	59.1					

Table A.10. (*Contd.*)

Substance	$\Delta H^\circ_{f\,300}$	S°_{300}	C°_p				
			300	500	800	1000	1500
P_4	14.0	66.9	16.1	18.3	19.2	19.4	19.7
PBr_3	−28.9						
PCl_3	−68.6	74.5	17.2	18.8	19.4	19.6	19.7
PF_3	−229	65.2	14.0	16.9	18.5	19.0	19.5
PH_3	1.3	50.2	8.9	11.1	14.0	15.4	17.4
CH_2O	−26.0	52.3	8.5	10.5	13.4	14.8	17.0
C_2H_2	54.2	48.0	10.5	13.1	15.2	16.3	18.3
HNCO		56.9	10.7	13.1	15.2	16.1	17.6
HOCN		57	10.7				
COF_2	−151.7±2	61.9	11.3	14.5	16.9	17.8	18.8
$COCl_2$	−52.6	67.8	13.8	16.3	17.9	18.5	19.2
C_2N_2	73.9	57.7	13.6	15.6	17.4	18.2	19.4
BH_3	22±2	44.9	8.7	10.0	12.5	14.0	16.4
BF_3	−271.7	60.7	12.1	15.0	17.3	18.1	19.0
BCl_3	−96.3	69.3	14.9	17.3	18.7	19.1	19.5
OBOH	−134.0	57.3	10.1	12.5	14.9	15.9	17.5
AlF_3	−290±2	66.2	15.0	17.4	18.7	19.1	19.5
$AlCl_3$	−139.7	74.6	17.0	18.6	19.4	19.5	19.7
Li_2F_2	−222±4	63.6	16.1	18.3	19.2	19.4	19.7
Na_2F_2	−196±4	72.1	18.3	19.3	19.6	19.7	19.8
Na_2Cl_2	−135±2	77.8	18.8	19.5	19.7	19.8	19.8
Na_2Br_2	−116.2	83.4	19.2	19.6	19.8	19.8	19.8
K_2Cl_2	−147.6	84.3	19.3	19.7	19.8	19.8	19.8

Table A.11. Thermochemical Data for Some Polyatomic Species

Species	$\Delta H^\circ_{f\,300}$	S°_{300}	C°_p				
			300	500	800	1000	1500
H_2O_3	[−13.5]	67					
H_2O_4	[+5.5]	77					
HO_4	[+43]	78					
$HClO_3$	11±4	67					
$HClO_4$	4±4	69					
Cl_2O_7	65.0						
ClO_3F	−5.1	66.7	15.5	20.0	23.0	23.9	24.9
BrF_5	−102.5	77.3	24.2	28.5	30.4	30.9	31.4
IF_5	−196.4	80.1	24.7	28.7	30.5	30.9	31.4
IF_7	−224±2	83.1	32.4	38.8	41.7	42.4	43.1
Cyclo-S_6	24.5						
Cyclo-S_8	24.5	103.0	37.4				

Table A.11. (*Contd.*)

Species	$\Delta H^\circ_{f\,300}$	S°_{300}	C°_p				
			300	500	800	1000	1500
H$_2$S$_3$	3.6						
H$_2$S$_4$	5.7						
H$_2$SO$_4$	-177 ± 2	69.1	19.3	25.5	30.0	31.6	34.1
SF$_4$	-187 ± 6	70.0	17.5	21.8	24.1	24.6	25.3
SF$_6$	-291.8	69.7	23.4	30.7	34.6	35.7	36.8
SO$_2$F$_2$	-205 ± 6	67.7	15.8	20.1	23.0	23.9	24.9
SO$_2$Cl$_2$	-87.0	74.5	18.4				
N$_2$O$_3$	19.8	73.9	15.7	18.7	21.4	22.4	23.6
N$_2$O$_4$	2.2	72.7	18.5	23.2	27.1	28.5	30.2
N$_2$O$_5$	2.7	82.8	23.0	29.0	32.7	33.8	34.9
N$_2$H$_4$	22.8	57.1	12.2	16.9	21.1	23.0	26.4
NH$_2$OH	-9						
HONO$_2$	-32.1	63.7	12.7	16.9	20.3	21.6	23.4
N$_2$F$_4$	$-5(\pm1)$	72.0	18.9	25.0	28.7	29.7	30.8
NSF$_3$		69.4	17.2	21.2	23.6	24.3	
P$_2$H$_4$	5.0						
PBr$_5$	-53.0						
PF$_5$	-377.2	70.7	20.1	25.9	29.1	30.0	31.0
POF$_3$	-290 ± 4	68.2	16.5	20.7	23.3	24.1	25.0
POCl$_3$	-133.5	77.8	20.3	23.0	24.5	24.9	25.4
PCl$_5$	-89.6	87.0	26.7	29.7	30.9	31.2	31.5
CH$_4$	-17.9	44.5	8.5	11.1	15.0	17.2	20.7
C$_2$H$_4$	12.5	52.4	10.3	14.9	20.0	22.4	26.3
C$_2$H$_6$	-20.2	54.9	12.7	18.7	25.8	29.3	34.9
C$_3$H$_4$ (allene)	45.9	58.3	14.2	19.8	25.4	28.0	32.1
C$_3$H$_6$	4.9	63.8	15.3	22.6	30.7	34.5	40.4
C$_3$H$_8$	-24.8	64.5	17.6	27.0	37.0	41.8	
n-C$_4$H$_{10}$	-30.2	74.1	23.4	35.3	48.2	54.2	
i-C$_4$H$_{10}$	-32.2	70.4	23.3	35.6	48.5	54.4	
c-C$_3$H$_4$	66.6	58.4					
c-C$_3$H$_6$	12.7	56.8	13.3				
c-C$_4$H$_8$	6.3	63.4	17.3				
c-C$_5$H$_{10}$	-18.5	70.0	20.0	35.9	52.4	59.8	70.9
Cyclopentene	8.6	69.2	18.0	31.6	45.8	51.9	61.1
Cyclopentadiene	32.4	65.6	17.8	29.5	40.6	45.4	
Cyclohexane	-29.3	71.3	25.6	45.5	66.8	75.8	88.6
Cyclohexene	-0.8	74.3	25.1	42.8	59.5	66.6	77.3
Cyclohexadiene-1,3	25.9						
Cyclohexadiene-1,4	26.3						
C$_6$H$_6$	19.8	64.3	19.7	32.8	45.1	50.2	
Toluene	12.0	76.6	24.9	41.0	56.6	63.3	

Table A.11. (*Contd.*)

Species	$\Delta H^\circ_{f\,300}$	S°_{300}	C_p°				
			300	500	800	1000	1500
CH$_3$OH	−48.0	57.3	10.5	14.2	19.0	21.4	
C$_2$H$_5$OH	−56.2	67.5	15.7	22.8	30.3	33.8	
CH$_3$OCH$_3$	−44.0	63.7	15.8	22.5	30.4	34.1	
Ethylene oxide	−12.6	58.1	11.4	18.0	24.6	27.5	31.8
C$_6$H$_5$OH	−23.1	75.1	24.8	38.7	50.7	55.5	62.7
C$_6$H$_5$OCH$_3$	−18.0	[86.2]					
CH$_3$OOH	[−31.3]	[67.5]					
CH$_3$OOCH$_3$	−30.0	[74.1]					
CH$_2$CO	−11.4	57.8	12.4	15.7	18.8	20.3	22.6
CH$_3$CHO	−39.7	63.2	13.2	18.2	24.2	27.0	
CH$_3$COCH$_3$	−51.7	70.5	18.0	25.9	34.9	39.2	45.7
HCOOH	−90.5	59.4	10.8	14.6	18.3	19.9	
CH$_3$COOH	−103.8	67.5	16.0		29.1	31.9	36.5
(COOH)$_2$	[−175.0]	[82]					
C$_3$O$_2$	−23.4	66.0	16.0	19.3	22.1	23.3	25.0
(CH$_3$CO)$_2$	−78.6	[91.6]					
HCOOCH$_3$	−83.6	67.5	16.0	22.6	29.1	32.0	
CH$_3$COOCH$_3$	[−98.0]	79.8					
H$_2$C(OH)$_2$	−93.5	[70]					
H$_2$CO$_3$	[−148]	[70]					
C$_6$H$_5$CHO	−8.9	[86]					
CH$_3$NH$_2$	−5.5	58.1	11.9	16.7	22.4	25.3	
(CH$_3$)$_2$NH	−4.5	65.4	16.6	24.9	33.9	38.2	
(CH$_3$)$_3$N	−5.9	69.2	22.1	33.6	45.6	51.0	
C$_2$H$_5$NH$_2$	−11.0	68.1	17.4	25.4	33.9	37.9	
CH$_3$NHNH$_2$	22.6	66.6	17.0		31.3		
Ethylene imine	29.5	60.0	12.6	20.5	28.1	31.5	
CH$_3$CN	21	58.7	12.5	16.6	21.3	23.5	
$\overline{\text{CH}_2\text{—N}\text{=}\text{N}}$	79	56.9	10.2	14.3	18.4	20.0	
CH$_2$=N=N	71	58.0	12.6	15.6	18.7	20.2	
CH$_3$NC	35.9	59.1	12.8	16.7	21.3	23.6	
HCONH$_2$	−44.5	59.6	11.1		21.1		
CH$_3$NO	[16]	[64]					
C$_2$H$_5$NO	[9]	[74]					
CH$_3$NO$_2$	−17.9	65.8	13.8	19.6	25.6	28.2	
CH$_3$ONO	−15.6	68.0	15.4	21.0	26.9	29.5	
CH$_3$ONO$_2$	−28.6	72.2	18.3	25.0	31.5	34.2	
CH(NO$_2$)$_3$	−0.2						
CH$_3$SH	−5.5	61.0	12.0		20.3		

Table A.11. (*Contd.*)

Species	$\Delta H^\circ_{f\,300}$	S°_{300}	C°_p				
			300	500	800	1000	1500
$(CH_3)_2S$	−8.9	68.3	17.8	27.0	31.6		
$(CH_3)_2S_2$	−5.7	80.5	22.6		37.7		
$HN{=}CS$	+30.0	59.2					
$CO(NH_2)_2$	(−60±4)						
$C(NO_2)_4$	(19.4)						
CH_3F	−55±2	53.3	9.0	12.3	16.4	18.4	21.6
CH_3Cl	−19.6	56.0	9.7	13.2	17.0	18.9	21.8
CH_3Br	−9.5±1	58.9	10.2	13.6	17.3	19.0	21.8
CH_3I	3.3	60.5	10.6	14.0	17.5	19.2	21.9
CH_2Cl_2	−22.8	64.6	12.2	16.0	19.4	20.8	22.9
CH_2F_2	−107.2	59.0	10.3	14.1	18.2	20.0	22.5
CH_2Br_2	5	70.1	13.1	16.7	19.7	21.1	23.1
CH_2I_2	+29.2±1	74.0	13.9	17.3	20.1	21.3	23.2
CHF_3	−166.8	62.0	12.2	16.6	20.3	21.7	23.6
$CHCl_3$	−24.6	70.7	15.7	19.4	21.9	22.9	24.2
CF_2Cl_2	−115±2	71.9	17.3	21.3	23.7	24.4	25.2
CF_4	−222±2	62.5	14.6	19.3	22.6	23.6	24.8
CCl_4	−22.9	74.2	19.9	23.0	24.6	25.0	25.5
C_2F_4	−155±2	71.7	19.2	24.0	27.6	28.9	30.4
C_2Cl_4	−3.6	81.3	22.7	26.6	29.2	30.0	
C_2F_6	−317	79.4	25.4	33.3	38.3	39.8	41.4
C_2Cl_6	−35.3	94.8	32.7	38.3	41.3	42.1	
C_2H_5F	−61	63.3	14.1	20.5	26.5	30.8	
C_2H_5Cl	−26.7	66.1	15.1	21.7	27.4	31.5	
C_2H_5Br	−14.8	68.6	15.5	22.0	28.6	31.6	
C_2H_5I	−2.1	70.7	15.4	22.3	28.8	32.0	
$CH_2{=}CHF$	(−31.6)	60.4	11.9	17.0	21.9	24.0	
$CH_2{=}CHNO_2$		70.6	17.4	24.9	31.7	34.5	38.5
$CH_2{=}CHCl$	5.0	63.1	12.9	17.8	22.4	24.3	
$CH_2{=}CHI$	30.8	68.1	13.6	18.4	22.7	24.6	
CH_3COF	−104						
CH_3COCl	−58.9	70.5	16.3	21.2	26.3	28.6	
CH_3COI	−30.3	76.0	17.5	20.9			
CH_3COBr	−45.6						
SiH_4	8.2	49.0	10.3	14.1	18.3	20.2	22.8
Si_2H_6	17.0	65.6	19.0	25.5	31.8	34.5	
SiF_4	−386.0	67.5	17.6	21.4	23.8	24.4	25.2
SiH_3CH_3	−4.0	61.3	15.7				

Table A.11. (*Contd.*)

Species	$\Delta H^\circ_{f\,300}$	S°_{300}	C°_p				
			300	500	800	1000	1500
SiCl₄	−157±2	79.2	21.7	24.0	25.1	25.3	25.6
H₂BOH	−70.0						
HB(OH)₂	−153.8						
H₃SiBr		62.8	12.7		20.0		
H₃SiI		64.5	12.9		—	—	
H₂SiBr₂		73.9	15.7		21.7	—	
SiBr₄	−98.0	90.2	23.2		25.4	25.5	
ClSiBr₃		90.1	22.8		25.3	25.5	
Cl₂SiBr₂		89.5	22.4		25.2	25.5	
Cl₃SiBr		83.5	21.7		25.1	25.5	
ISiBr₃		96.6	23.4	24.8	25.4	25.6	
Si(CH₃)₄	(−26)	85.8	33.5	46.5	60.5	67.3	
(CH₃)₃SiCl		85.6	31.4	41.6	52.1	57.1	
(CH₃)₂SiCl₂	−106.0	86.0	26.2	34.6	42.5	46.1	
(CH₃)₃SiF		83.8	29.4	40.1	51.3	56.6	
(CH₃)₂SiF₂		80.2	25.3	34.0	42.2	45.9	
CH₃SiF₃		75.1	22.7	—	—	—	
[(CH₃)₃Si]₂O		127.8	57.2	78.2	100.0	110.3	
H₃SiGeH₃	27.8	—	—	—	—	—	
SiI₄	−28.0	—	—	—	—	—	
(CH₃)₃SiH	(−21)	79.0	28.4	39.2	50.5	55.9	
(CH₃)₂SiH₂	(−17)	71.7	22.0	30.7	39.8	44.0	
B₂H₆	9±4	55.7	13.9	21.2	29.1	32.6	37.7
B₅H₉	17.5±2	65.9	22.6	38.6	54.3	60.8	69.6
B₂Cl₄	−116.9	85.8	22.7	26.6	28.9	29.5	30.2
B₃F₃O₃	−565.3	81.8	27.5	36.2	42.8	44.9	47.4
Borazole (H₆B₃N₃)	−122±3	69.0	23.2	36.0	47.1	51.8	58.5
B(CH₃)₃	−29.3	75.3	21.2				
Al₂Cl₆	−309.2	113.8	37.9	41.3	42.7	43.1	43.4
Al₂(CH₃)₆	−62±4	125.4					
Al(CH₃)₃	−21±2						
Al(BH₄)₃	3	91					
XeF₄	−49.3						
XeF₆	−66.7						

Table A.12. Thermochemical Data for Some Gas-phase Organic Free Radicals

Radical	(σ, s)	$\Delta H^\circ_{f\,300}$	S°_{300}	C°_p				
				300	500	800	1000	1500
:CO$_3$	(6, 1)	-45 ± 4	61.2	11.3	15.0	17.5	18.3	19.1
HĊO	$(1, \frac{1}{2})$	9.0	53.7	8.3	9.2	10.5	11.2	12.3
HCOO·	$(2, \frac{1}{2})$	-36 ± 2	57.3	9.6	12.4	15.0	15.9	16.8
:C=C=O	(1, 1)	90 ± 4	55.7	10.3	11.7	13.1	13.7	14.6
·COOH	$(1, \frac{1}{2})$	-50 ± 1	61.0	10.8	13.4	15.6	16.5	17.9
:CH$_2$	(2, 1)	86 ± 3	46.3	7.5	8.8	10.3	11.1	12.6
:CF$_2$	(2, 0)	-43.5 ± 2	57.5	9.3	11.1	12.5	13.0	13.5
:CCl$_2$	(2, 0)	47 ± 3	63.7	10.8	12.3	13.1	13.4	13.7
·C≡C—H	$(1, \frac{1}{2})$	122 ± 3	49.6	8.9	10.2	11.5	12.2	13.3
ClĊO	$(1, \frac{1}{2})$	-4.0	63.5	10.8	11.7	12.5	12.9	13.4
·CH$_3$	$(6, \frac{1}{2})$	34.3	46.4	8.8	10.6	13.2	14.5	16.8
·C$_2$H$_3$	$(1, \frac{1}{2})$	69 ± 2	56.3	9.7	13.1	16.8	18.7	21.8
·CF$_3$	$(3, \frac{1}{2})$	-112.5	63.8	12.2	15.2	17.5	18.2	19.1
·CCl$_3$	$(3, \frac{1}{2})$	18.5	70.8	14.9	17.3	18.7	19.1	19.5
·C$_2$H$_5$	$(6, \frac{1}{2})$	26.5	58.0	11.1	16.3	22.2	25.7	30.4
n-Ċ$_3$H$_7$	$(6, \frac{1}{2})$	21.0	68.5	17.1	25.2	33.7	38.1	44.7
i-Ċ$_3$H$_7$	$(18, \frac{1}{2})$	17.6	66.7	17.0	24.9	33.2	37.9	44.6
i-Ċ$_4$H$_9$	$(9, \frac{1}{2})$	13.7	75.2	22.6	33.5	46.0	50.6	59.1
t-C$_4$H$_9$	$(162, \frac{1}{2})$	8.4	72.1	22.6	33.5	46.0	50.5	59.3
Allyl	$(2, \frac{1}{2})$	40.6	62.1	14.1				
Cyclopropyl	$(2, \frac{1}{2})$	66 ± 3	60.4	12.9				
Cyclobutyl	$(2, \frac{1}{2})$	51 ± 1	67.3	16.8				
Phenyl	$(2, \frac{1}{2})$	78.5	69.4	18.8				
Benzyl	$(2, \frac{1}{2})$	45	75.3	25.6				
Cyclohexyl	$(1, \frac{1}{2})$	13	[76]					
Cyclohexen-yl-3	$(2, \frac{1}{2})$	[30]	[75]					
Cyclohexadien, 1,3-yl-5	$(2, \frac{1}{2})$	50	[69]					
CH$_3$O·	$(3, \frac{1}{2})$	3.5	55	9.0				
CH$_3$S·	$(3, \frac{1}{2})$	[32.5]	[59]	9.0				
CH$_3$NH	$(3, \frac{1}{2})$	45	[57]	10.5				
ĊH$_2$CN	$(2, \frac{1}{2})$	59.0	[58]	11.4				
ĊH$_2$NH$_2$	$(2, \frac{1}{2})$	33.5	[57]	10.5				
CH$_3$ĊO	$(3, \frac{1}{2})$	-5.4	64.5	11.3	15.7			
ĊH$_2$COCH$_3$	$(3, \frac{1}{2})$	-6.0	[71.5]	17.6				
ĊH$_2$OCH$_3$	$(6, \frac{1}{2})$	-2.0	[64.7]	15.5				
C$_2$H$_5$Ȯ	$(3, \frac{1}{2})$	-4	[65.3]	14.2				
i-C$_3$H$_7$Ȯ	$(9, \frac{1}{2})$	[-12.6]	[71.8]					
t-BuȮ	$(81, \frac{1}{2})$	-21.6	[75.7]					
CH$_3$OȮ	$(3, \frac{1}{2})$	[$+6.7$]	[65.3]					
C$_2$H$_5$OȮ	$(3, \frac{1}{2})$	[-1.8]	[75]					
t-BuȮ$_2$	$(81, \frac{1}{2})$	[-19.2]	[85]					
·SF$_5$	$(?, \frac{1}{2})$	-218						
ΦĊO	$(2, \frac{1}{2})$	26.5	[78]					

Table A.13. Frequencies Assigned to Normal and Partial Bond Bending and Stretching Motions

Bond Stretches	Frequency (ω cm^{-1})[a]	Bond Bends	Frequency (ω cm^{-1})
C=O	1700	H–C–H (C /\ H H)	1450
C∸C	1400	C /· \ H H	1000
C—O ethers	1100	(C /\ H C) t, w^b	1150
C—O acids, esters	1200		
C·O	710	C ·\ H C	800
C=C	1650	O /\ H C	1200
C∸C	1300	O ·\ H C	840
C—C	1000	C /\\ H C	1150
C·C^c	675	C /·\ H C	1150
C—H; O—H; N—H	3100		
S—H	2600	C /\\ C C	420
C·H	2200	C /·\\ C C	290
C—F	1100	C /·\ C C	420
C·F	820	C /\\ O O	420
C—Cl	650	C /··\ O O	420

300

Bond Stretches	Frequency (ω cm^{-1})^a	Bond Bends	Frequency (ω cm^{-1})
C·Cl	490	C=C=C	850
C—I	500	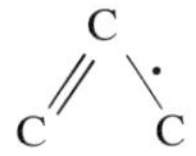 C=C, C···C (C above)	635
C·I	375	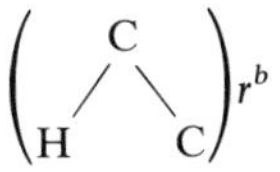 (H—C—C)r^b	700
C—Br	560	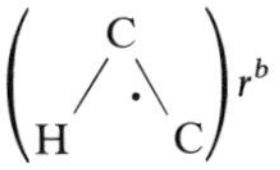 (H—C···C)r^b	700
C·Br	420	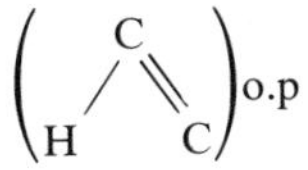 (H—C=C)o.p	700
		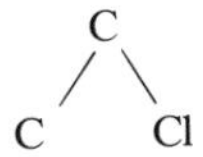 C—C—Cl	400
		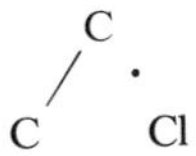 C—C···Cl	280
		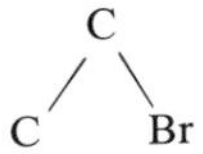 C—C—Br	360
		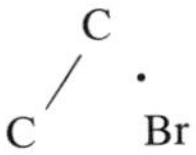 C—C···Br	250
		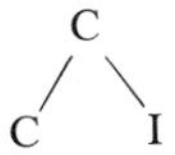 C—C—I	320
		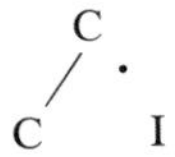 C—C···I	220
		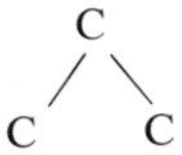 C—C—C	420
		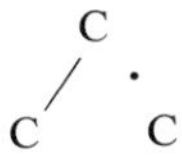 C—C···C	300
		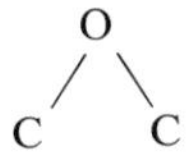 C—O—C (O above)	400

Table A.13. (*Contd.*)

Bond Stretches	Frequency (ω cm^{-1})^a	Bond Bends	Frequency (ω cm^{-1})
		C–O···C bend	280
		C–C–O bend	400
		C–C···O bend	280
		O–C–O bend	400

^{a} Note that bending frequencies are surprisingly consistent with the relation, $\omega_1/\omega_2 = (\mu_2/\mu_1)^{1/2}$. Deviations from this relation seldom exceed 50 cm^{-1}. Here the reduced mass $\mu = [M_A M_B/(M_A + M_B)]$ for the bend (A-R-B).

^{b} Methyl and methylene wags and twists, whose frequencies range within 1000–1300 cm^{-1}, have been equated with $\left(\begin{smallmatrix} & C & \\ / & & \backslash \\ H & & C \end{smallmatrix}\right)$ t, w bends and assigned a mean value of 1150 cm^{-1}.

Methylene rocks have lower frequencies (i.e., ~700 cm^{-1}), which correspond closely to the out of plane $\left(\begin{smallmatrix} & C & \\ / & & \backslash\backslash \\ H & & C \end{smallmatrix}\right)$ bends in olefins.

^{c} Single dots, as in C·C, are meant to represent one-electron bonds.

Table A.14. Some Average Covalent Radii, Bond Lengths^a

Atom	Radius (Å)	Atom	Radius (Å)	Bond	Bond Lengths	Bond	Bond Lengths
H	0.32	P	1.10	C–H	1.10±0.02 (range)	C–I	2.13
C	0.77	P$_d$	1.00	O–H	0.96	C–N	1.47
C$_d$	0.67	B	0.81	N–H	1.01	C≡N	1.16
C$_t$	0.60	Hg	1.48	S–H	1.34	C–O	1.43
N	0.74	F	0.72	B–H	1.18	C=O	1.23
N$_d$	0.62	Cl	0.99	C–C	1.54	C–S	1.81
N$_t$	0.55	Br	1.11	C=C	1.34	C=S	1.17
O	0.74	I	1.28	C≡C	1.20	N=O	1.25
O$_d$	0.62	Si	1.17	C–F	1.33	C≡N	1.28
S	1.04	Si$_d$	1.07	C–Cl	1.77		
S$_d$	0.94	Sn	1.40	C–Br	1.94		

^{a} Subscripts refer to (d) double and (t) triple bonds. Van der Waals' Radii are obtained from above by adding 0.95Å.

302

Table A.15. Contributions of Harmonic Oscillator to C_p as a Function of Frequency and Temperature

Frequency (cm^{-1})	Temperature (°K)						
	300	400	500	600	800	1000	1500
100	1.95	1.97	1.97	1.98	1.98	1.98	1.98
200	1.84	1.90	1.93	1.95	1.97	1.97	1.98
250	1.76	1.86	1.90	1.93	1.95	1.97	1.98
300	1.68	1.80	1.87	1.90	1.94	1.96	1.97
350	1.58	1.74	1.83	1.87	1.92	1.95	1.97
400	1.47	1.68	1.78	1.84	1.90	1.93	1.96
500	1.26	1.53	1.68	1.76	1.86	1.90	1.95
600	1.04	1.37	1.56	1.68	1.80	1.87	1.93
800	0.66	1.04	1.30	1.47	1.68	1.78	1.89
1000	0.38	0.74	1.04	1.26	1.53	1.68	1.84
1200	0.21	0.51	0.80	1.04	1.37	1.56	1.78
1500	0.08	0.26	0.51	0.74	1.12	1.37	1.68
2000	0.01	0.08	0.21	0.38	0.74	1.04	1.30
2500	0.00	0.02	0.08	0.18	0.46	0.74	1.26
3000	0.00	0.00	0.03	0.08	0.26	0.51	1.04

Table A.16. Molar Heat Capacity C_p° for Internal Rotor as a Function of Barrier (V), Temperature, Partition Function Q_f

$\frac{V}{RT}$	$(1/Q_f)$				
	0.0	0.2	0.4	0.6	0.8
0.0	1.0	1.0	1.0	1.0	1.0
0.5	1.1	1.1	1.1	1.0	1.0
1.0	1.2	1.2	1.2	1.1	1.1
1.5	1.5	1.4	1.3	1.2	1.1
2.0	1.7	1.7	1.5	1.4	1.2
2.5	1.9	1.9	1.7	1.5	1.3
3.0	2.1	2.0	1.8	1.6	1.3
4.0	2.3	2.2	2.0	1.7	1.4
6.0	2.3	2.2	1.9	1.5	1.2
8.0	2.2	2.1	1.7	1.3	0.9
10.0	2.1	2.0	1.5	1.0	0.7
15.0	2.1	1.8	1.2	0.7	0.4
20.0	2.0	1.7	1.0	0.5	0.2

From G. N. Lewis and M. Randall, revised by K. S. Pitzer and L. Brewer, *Thermodynamics* (2nd ed.), McGraw-Hill, New York, 1961

$Q_f = 3.6/\sigma\{I_r T/100\}^{1/2}$ with I_r in amu-Å and T in °K.

σ = symmetry of barrier (e.g., 3 for —CH$_3$, 2 for —ĊH$_2$)

Table A.17. Absolute Entropy of a Harmonic Oscillator as a Function of Frequency and Temperature[a]

Frequency (cm^{-1})	Temperature (°K)							
	300	400	500	600	800	1000	1200	1500
50	4.8	5.4	5.9	6.3	6.8	7.3	7.7	8.1
75	4.1	4.7	5.0	5.4	6.0	6.5	6.9	7.3
100	3.4	4.1	4.5	4.8	5.6	5.9	6.3	6.7
125	3.1	3.6	4.1	4.4	5.2	5.5	5.9	6.3
150	2.7	3.3	3.7	4.1	4.7	5.0	5.4	5.9
200	2.2	2.7	3.1	3.5	4.1	4.4	4.8	5.3
250	1.8	2.3	2.7	3.0	3.6	4.1	4.5	4.9
300	1.4	1.9	2.3	2.7	3.3	3.6	4.1	4.5
350	1.2	1.7	2.1	2.4	3.0	3.4	3.8	4.2
400	1.0	1.4	1.8	2.1	2.7	3.2	3.5	3.8
500	0.7	1.1	1.4	1.8	2.3	2.7	3.0	3.5
600	0.5	0.8	1.0	1.4	1.9	2.3	2.7	3.1
700	0.3	0.6	0.8	1.2	1.7	2.1	2.4	2.8
800	0.2	0.5	0.8	1.0	1.4	1.8	2.1	2.6
900	0.2	0.4	0.6	0.8	1.2	1.6	1.9	2.3
1000	0.1	0.3	0.5	0.7	1.1	1.4	1.8	2.1
1200	0.0	0.1	0.3	0.5	0.8	1.0	1.4	1.8
1500	0.0	0.1	0.2	0.3	0.5	0.8	1.1	1.4
2000	0.0	0.0	0.0	0.1	0.3	0.5	0.7	1.0
2500	0.0	0.0	0.0	0.0	0.1	0.3	0.5	0.7
3000	0.0	0.0	0.0	0.0	0.1	0.1	0.3	0.5
3500	0.0	0.0	0.0	0.0	0.0	0.1	0.2	0.3

[a] $x = 1.44\bar{\nu}/T$ determines $S°$, so that value of $S°$ must be the same for similar ratios of $\bar{\nu}/T$. Thus $S°$ of 500 cm^{-1} at 1000°K is the same as $S°$ of 250 cm^{-1} at 500°K or 1000 cm^{-1} at 2000°K. For $x \ll 1$, $S° = R + R \ln x = 1.99 + 4.575 \log x$.

Table A.18. **Approximate Moments of Inertia, Entropies, and Partition Functions Q_f at 300°K of Some Free Rotors[e]**

Rotor	(σ)[a]	Moment of Inertia[b]	Q_f (300°K)	$S^\circ_{f(\text{int})}$[c, d]		
				300°K	600°K	1000°K
—CH_3	(3)	3.0	3.7	5.8	6.5	7.0
—$\dot{C}H_2$	(2)	1.8	4.2	5.3	6.0	6.5
—$\dot{C}HD$	(1)	3.0	11.0	5.8	6.5	7.0
—C_2H_5	(1)	17.0	25.0	7.5	8.2	8.7
—i-Propyl	(1)	56.0	43.0	8.6	9.3	9.8
—t-butyl	(3)	100.0	20.0	9.2	9.9	10.4
—phenyl	(2)	88.0	29.0	8.9	9.6	10.1
—benzyl	(1)	~170.0	81.0	9.7	10.4	10.9
—OH	(1)	1.0	6.2	4.6	5.3	5.8

[a] The moment is taken along the —C axis, assuming tetrahedral angles for saturated bonds and 120° for angles about double bonded atoms. The value in parentheses is the symmetry number, σ.

[b] Units are given in amu-Å^2. It is assumed that the rotor is connected to an infinite mass.

[c] For units and equations, see Table A.16.

[d] $S^\circ_{f(\text{int})} = S^\circ_f + R \ln \sigma$.

[e] $S^\circ_f = R \ln Q_f + R\left(\dfrac{\partial \ln Q}{\partial \ln T}\right) = 4.6 + \dfrac{R \ln}{2}\left[\dfrac{I_r}{\sigma^2} \cdot \dfrac{T}{300}\right].$

Table A.19. Some Characteristics Torsion Barriers, V (kcal/mole) to Free Rotation about Single Bonds[a]

Bond	V	Bond	V
CH_3—CH_3	2.9	CH_3—OH	1.1
CH_3—C_2H_5	2.8	CH_3—OCH_3	2.7
CH_3—isopropyl	3.6	CH_3—NH_2	1.9
CH_3—t-butyl	4.7	CH_3—$NHCH_3$	3.3
CH_3—CH=CHCH$_3$ *cis*	0.75[b]	CH_3—$N(CH_3)_2$	4.4
trans	1.95	CH_3—SiH_3	1.7
CH_3—vinyl	2.0	CH_3—SiH_2CH_3	1.7
CH_3—CH_2F	3.3	CH_3—PH_2	2.0
CH_3—CF_3	3.5	CH_3—SH	1.3
CF_3—CF_3	4.4	CH_3—SCH_3	2.1
CH_3—CH_2Cl	3.7	CH_3—CHO	1.2
CH_2Cl—CF_3	5.8	CH_3—$COCH_3$	0.8
CCl_3—CCl_3	10.8	CH_3—allene	1.6
CH_3—CH_2Br	3.6	CH_3—(isobutene)	2.2
CH_3—CH_2I	3.2	CH_3—CO(OH)	0.5
CH_3—phenyl	0	CH_3—(epoxide ring)	2.6
CH_3—$CCCH_3$	0	$(CH_3)_2N$—$COCH_3$	20.0
CH_3—NO_2	0	CH_3—OCHO	1.2
CH_3O—NO	9.0	CH_3—ONO_2	2.3
CH_3O—NO_2	9.0	CH_3—O-vinyl	3.4
CH_3O—$CO(CH_3)$	13		

[a] For a more complete compilation, see J. Dale, *Tetrahedron* **22**, 3373 (1966). See also *Internal Rotation in Molecules*, W. J. Orville-Thomas (ed.), Wiley, New York, 1974.

[b] There seems to be a quite general "*cis* effect," such that F, Cl, CH_3, and CN, *cis* to a CH_3, lower the CH_3 barrier by ~1.4 kcal.

Table A.20. Decrease in Entropy of Free Rotor as Functions of Barrier Height (V), Temperature (T, °K) and Partition Function, Q_f^a

V	$1/Q_f$		
RT	0.0	0.2	0.4
0.0	0.0	0.0	0.0
1.0	0.1	0.1	0.1
2.0	0.4	0.4	0.4
3.0	0.8	0.8	0.7
4.0	1.1	1.1	1.0
5.0	1.4	1.4	1.2
6.0	1.7	1.6	1.4
8.0	2.0	2.0	1.7
10.0	2.3	2.2	1.9
15.0	2.7	2.6	2.2
20.0	3.1	2.9	2.4

[a] Values listed are $\Delta S = S_f^\circ - S_h^\circ$. From G. N. Lewis and M. Randall, *Thermodynamics*, McGraw-Hill, New York, 1961 (2nd ed.), see Tables A.16 and A.17 for definition and values.

Table A.21. Torsion Frequencies for Methyl-substituted Ethylenes (Pi Bonds)

Pi Bond	ω cm$^{-1\,a}$	ω cm^{-1} (Three electron Torsions)	$\mu_{(amu)}$
Ethylene (C_2H_4)	1000		
$\quad$ (E)$_t$ = CH$_2$$\cdot$CH$_2$		500	1.0
Propylene (C_3H_6)	800		
$\quad$ (P)$_t$ = CH$_2$$\cdot$CHCH$_3$		400	1.78
Isobutene (C_4H_8)	700		
$\quad$ (i-B)$_t$ = (CH$_3$)$_2$C$\cdot$CH$_2$		350	1.88
cis-2-Butene (C_4H_8)	400		
$(C\text{-}2\text{-}B)_t =$		200	8.0
trans-2-Butene	300		
$(t\text{-}2\text{-}B)_t =$		150	8.0
2-Methyl but-2-ene	250		
$\quad$ (2-MB-2)$_t$ = (CH$_3$)$_2$C$\cdot$CHCH$_3$		125	10.4
Tetramethyl ethylene	(210)		
$\quad$ (TME)$_t$ = (CH$_3$)$_2$C$\cdot$C(CH$_3$)$_2$		105	15.0

For (C-2-B)$_t$:

$$\left(\begin{array}{cc} CH_3 & CH_3 \\ & C\cdot C \\ H & H \end{array} \right)$$

For (t-2-B)$_t$:

$$\left(\begin{array}{cc} CH_3 & H \\ & C\cdot C \\ H & CH_3 \end{array} \right)$$

[a] It should be noted that, to a reasonable approximation (i.e., $\simeq \pm 50$ cm^{-1}), these torsional frequencies are consistent with the relation

$$\frac{\omega_1}{\omega_2} = \left(\frac{\mu_2}{\mu_1} \right)^{1/2}$$

where μ is given by the product of the sums of the oscillating masses on each end of the double bond, divided by the sum of the four masses. Thus for

$$\left(CH_3 - CH\cdot C \begin{array}{c} H \\ \\ Cl \end{array} \right)$$

$$\mu = \frac{(16)(36.5)}{(52.5)}.$$

[b] The three-electron torsion frequencies have been assigned a value of one half the corresponding double-bond torsion.

308

Table A.22. Bond-Dissociation Energies for Some Organic Molecules $R' - R''^{a}$

$R' \backslash R''$	(52.1) H	(19.8) F	(28.9) Cl	(26.7) Br	(25.5) I	(9.5) OH	(46±1) NH_2	(4±1) OCH_3	(34±1) CH_3	(26±1) C_2H_5	(18±1) $i\text{-}C_3H_7$	(8.0±1) $t\text{-}Bu$	(78.5±1) C_6H_5
(34±1)[b]CH_3	104	109	83.5	70	56	91.5	85	82	88	85	84	81	100
(26±1)C_2H_5	98	108	81.5	68	53.5	91.5	84	82	85	82	80	78	97
(21±1)$n\text{-}C_3H_7$	98	108	81.5	68	53.5	91.5	84[c]	82	85	82	80	78	97
(18±1)$i\text{-}C_3H_7$	94.5	105	81	68	53	92	84	82	84	80	77.5	74	95
(8.0±1)$t\text{-}Bu$	92	—	80	64	51	92	84	81	81	78	74	70	92
(78.5±1)C_6H_5	110.5	125	—	78	64	112	105	100	100	97	95	92	116
(45±1)$C_6H_5CH_2$	85	—	—	—	40	77	—	70	72	69	67.5	65	77
(40±1)allyl	87	—	82.5	—	43.5	80	—	(68)[c]	74.5	71.5	69.5	66	78
(−5)CH_3CO	87	118	—	—	51.6	108	101	97	81	78	80	—	95
(−4.5±1)CH_3CH_2O	104	—	—	—	—	43	—	38	81	81	82	82	99
(−1.5)$CH_3CH_2O_2$	90	—	—	—	—	(28)[c]	—	21	72	72	(71)	(69)	91
(67.5±2)$CH_2{=}CH$	108	—	91	—	60	—	—	95	99	96	94	92	112

[a] All values are in kcal/mole.
[b] Values in parentheses near radicals and atom are ΔH°_{f300}.
[c] Values in parentheses are estimates by the author.

309

Table A.23. Table of Ionic and Atomic Polarizabilities

Atom	Polarizability (Å^3)	Ion	Polarizability (Å^3)
H	0.4	Li^+	0.0
He	0.2	Na^+	0.2
		K^+	1.0
C	1.0	Rb^+	1.5
N	0.9	Cs^+	2.4
O	0.8		
F	0.4	F^-	0.85
Ne	0.4	Cl^-	3.0
		Br^-	4.2
Cl	2.2	I^-	6.2
Br	3.1		
I	4.8	$NH_4{}^+$	1.6
Ar	1.6	$OH_3{}^+$	1.2
P	1.3	OH^-	2.0
S	2.8	Mg^{++}	0.1
Kr	2.5	Ca^{++}	0.6
Xe	4.0	Sr^{++}	0.9
		Ba^{++}	1.6

Polarizabilities (α) (Å^3) are related to molar refractions [R] in cm^3 by the relation: $\alpha = 0.40[R]_\infty \times 10^{-24}\ cm^3$. Although both are tensor properties of molecules, their averages over all orientations of the molecule obey reasonably well ($\pm 10\%$) simple atom additivity. [For detailed discussion, see R. J. W. LeFevre, *Adv. Phys. Organ. Chem.*, **3**, 1 (1965)]. They obey bond additivity even better and polarizabilities parallel to a bond are usually about 30% larger than polarizabilities perpendicular to a bond.

INDEX